LAGRANGE
POINTS
Finding Stable Governance Orbits
Between Competing Forces
THREE-BODY SERIES BOOK THREE
LINC WILLIAMS
First Edition

Disclosures and Context 12
About the Author 14
Inside The Harness Doctrine Series 15
- Volume I. The Harness Horizon 15
- Volume II. The Three-Body Problem of Classified AI 15
- Volume III. Lagrange Points 15
- Volume IV. Stable Orbit 16
- Volume V. The Replicant Fleet 16
- Volume VI. The Shatterdome Doctrine 16
- Volume VII. Three Governance Surfaces 16
- Volume VIII. The Semester Problem 17
- Volume IX. When Galaxies Collide 17
- Volume X. The CATO Playbook 17
- Volume XI. The SAFe Horizon 18
- How to Read the Series 18

Chapter 1: The Shipowners and the Underwriters 19
- Start here: the harness in brief 20
- The modern coffee house 20
- Three forces, one problem 21
- The Three-Body series and this volume 24
- The governance architecture the coffee house could not provide 25
- What this book covers 26

Chapter 2: Three Bodies, Every Sector 28
- Part I: The Coffee House Problem 28
  - Two ships already at sea 28
  - Three forces, four sectors 29
  - Same architecture, different vocabulary 33
  - What transfers and what extends 34
  - One Harness or Many 37
  - Build, Buy, Rent, Renovate, or Wait 40
  - Why regulated industries are making the same mistakes 41

Chapter 3: The Coffin Ships 43

The AI industry is building coffin ships 44

UnitedHealthcare: optimizing for the wrong property 44

Air Canada: the chatbot with no policy verification gate 45

McDonald's: the drive-thru with no constraint boundary 46

The moral hazard argument 47

The Plimsoll solution 49

Where the lines go 51

Chapter 4: Why Regulated Industries Are Making the Same Mistakes 53

The progression every regulated industry follows 53

Stage 1: Prohibition 55

Stage 2: Shadow adoption 55

Stage 3: Panic governance 56

Stage 4: Structural governance 58

The bifurcation warning in practice 59

The transition zone: frameworks without coverage 61

The agent acceleration problem 62

What the progression means for this volume 63

Chapter 5: The Classification Society 65

The structural survey: H-1 through H-5 66

The evidence model: E-1 through E-4 69

Knowledge base integrity: KB-1 through KB-4 71

The seaworthiness properties: LP1 through LP8 73

Two enforcement properties: LP13 and LP14 75

The certificate 75

Who paints the line? 76

Chapter 6: The Evidence Model Transfers Completely 78

The Four Questions Every Assessor Asks 79

E-1: Governance Event Records 80

E-2: Continuous Verification 82

E-3: State Transition Records 84

E-4: Attribution Chain Records 86

The 5-Minute Examiner Test 87

Architecture over Rendering....89
Chapter 7: The Three Failure Modes Are Universal....91
The Attack Surface of a Governed AI System....91
Context Amnesia....93
Heroic One-Shot....97
False Completion....100
Compiled Constraints as Classification Society Equivalents....103
The Pattern That Connects Them....104
What Detection Requires....106
The Commercial Stakes Are Not Different....106
Chapter 8: The Plimsoll Line....108
Part II: The Classification Survey....108
Why LP1-LP8 Are Insufficient for Commercial Environments....109
LP9: Consent-Bounded Processing....110
LP10: Jurisdictional Consistency....112
LP11: Fairness Verification....113
LP12: Transparency and Explainability....115
The Plimsoll Principle Applied to Governance Properties....117
Chapter 9: The Intergovernmental Agreement....120
Part II: The Classification Survey....120
The Commercial AI Governance Problem Is Identical....120
The Jurisdictional Stack for Commercial AI....121
Dynamic Jurisdiction Determination....123
The "Most Restrictive Applicable" Default....125
GDPR Retention vs. NIST SI-12: A Framework Conflict in Practice....126
The Station Proves It Works....127
Chapter 10: Vendor Risk and the Supply Chain....130
Part II: The Classification Survey....130
The Voight-Kampff Problem....131
Supply Chain Risk Tiering....132
The Embrace, Extend, Extinguish Pattern....134
Software Bill of Materials for AI Components....135

The Acquisition Gate 136
Insurance Coverage and the Vendor Chain 137
From Trust the Shipbuilder to Independent Classification 138
The Supply Chain Survey Component 139
The Manifest Matches the Hold 140
Chapter 11: L1, The Financial Services Lagrange Point 142
Part III: The Sector Lagrange Points 142
The three bodies in financial services 143
The Lloyd's thread 146
The Chinese wall problem 146
Model risk management as harness verification 148
The financial services AGENTS.md 149
Synthetic identity and KYC collapse 152
Financial services attack surface 153
Financial services failure modes 154
The 5-minute examiner test 156
The financial services evidence package 156
Chapter 12: The Frontier-Lab Finance Harness 159
The product that proved the transfer 159
What the regulator says 160
The platform the institution controls 161
The delegation architecture 162
From the file to the token 164
The financial services AGENTS.md 165
The four enforcement artifacts 167
Where transfer ends and extension begins 168
Evidence collection across the four tiers 169
The failure mode catalog 170
The path the institution walks 171
Chapter 13: L2, The Healthcare Lagrange Point 174
Part III: The Sector Lagrange Points 174
The three bodies in healthcare 174

The minimum necessary standard as dynamic scoping 177
The Lloyd's survey parallel 179
The FDA boundary 180
PHI sub-category governance in the AGENTS.md architecture 181
The UnitedHealthcare case: missing properties in production 182
Healthcare attack surface 183
Healthcare failure modes 184
Healthcare evidence package 185
Compliance automation reference 186
Chapter 14: L3, The Energy and Critical Infrastructure Lagrange Point 189
Part III: The Sector Lagrange Points 189
The three bodies in energy 190
The Lloyd's thread: classification by impact 192
Air-gapped OT environments 193
The NERC CIP penalty structure and evidence production 194
Safety-critical systems and the consequence of error 195
Nuclear-specific requirements 196
Edge-native autonomy and disconnected governance 197
Energy sector attack surface 198
Energy sector failure modes 199
The energy sector AGENTS.md 201
The evidence package 202
Chapter 15: L4: The Insurance and Actuarial Lagrange Point 206
Part III: The Sector Lagrange Points 206
The Lloyd's Thread 206
The Three Bodies in Insurance 208
The Fairness Problem 210
The Four-Fifths Rule in Underwriting 212
Fifty-State Regulatory Complexity 213
Actuarial Professional Liability 214
The Insurance Attack Surface 215
Insurance Failure Modes 216

The Insurance AGENTS.md 217

The Insurance Evidence Package 219

The Five-Minute DOI Examiner Test 221

Chapter 16: L5: Manufacturing, Supply Chain, and Telecommunications 223

Part III: The Sector Lagrange Points 223

The Three Bodies in Manufacturing and Supply Chain 224

The Lloyd's Thread: Containerization and Evidence Harmonization 226

Autonomous Procurement and the Threshold Fraud Problem 227

Multi-Agent Feedback Loops and Emergent Oscillation 229

ITAR and the Dual-Use Architecture Problem 230

Telecommunications: Critical Infrastructure with Consumer Data 231

The Manufacturing and Telecommunications Attack Surface 232

Manufacturing and Telecommunications Failure Modes 233

The Manufacturing and Telecommunications Evidence Package 235

The Trojan Point: Robotics and Autonomous Vehicles 237

Chapter 17: Assembly in Orbit 240

Part IV: Assembly in Orbit 240

Five Phases to Commercial Three-Body Governance 240

High-throughput runtime for continuous governance checks 244

Commit-time static verification of governance artifacts 246

Where the Human Stands 251

The Non-Destructive Assembly Principle 253

Chapter 18: One Harness, Many Auditors 255

Part IV: Assembly in Orbit 255

The Harmonization Problem 256

The Unified Control Architecture 256

The Evidence Rendering Pipeline 257

OSCAL as the Universal Interchange Format 258

Cross-Domain Trust Federation 260

Who Paints the Line 261

When the Container Does Not Fit 262

Chapter 19: Crew Handoff Protocols 264

Part IV: Assembly in Orbit 264
Context Amnesia at the Shift Boundary 265
Failure Pattern 1: The SOC Analyst Shift Change 265
Failure Pattern 2: The Trading Desk Handoff 266
Failure Pattern 3: The Compliance Team Transition 267
Clinical Shift Changes and the HIPAA Boundary 268
AGENTS.md as a Crew Handoff Document 268
Parallel agent coordination for multi-classification-society handoffs 270
One Thing a Junior Person Can Do Tomorrow 274
Chapter 20: The Board's Navigation Chart 276
Part IV: Assembly in Orbit 276
The Board Audit Committee as Commercial Authorizing Official 276
SEC Reg S-K 229.106 and What It Demands 277
The CFO Conversation 278
The Stakeholder Alliance Strategy 280
Five Decisions for the Decision Maker 282
The Navigation Chart 284
Two Infrastructure Commitments the Board Funds 285
Chapter 21: Commercial Failure Modes 287
Part V: When Ships Sink 287
F-C1: Vendor Capture 288
F-C2: Multi-Jurisdictional Misapplication 289
F-C3: Consent Boundary Violation 291
F-C4: Fairness Failure 292
The Economics Behind the Failure Modes 294
F-C5: Evidence Architecture Failure 295
F-C6: Cross-Sector Contagion 296
The Structural Taxonomy 298
Where This Breaks 300
From Catalog to Response 302
Chapter 22: SOLAS After the Titanic 304
Part V: When Ships Sink 304

What SOLAS Actually Built 305
The Iteration Pattern 306
The ISM Code: From Equipment to Management Systems 307
The EU AI Act as the First SOLAS Moment 308
The American Patchwork 309
Why Reactive Regulation Is Insufficient but Inevitable 311
Building the Watertight Compartments Before the Iceberg 312
The ISS IGA Was Written Before the First Module Launched 313
The Pre-SOLAS Period 313
Chapter 23: AI Incident Response for Regulated Industries 316
Part V: When Ships Sink 316
The AI Incident Taxonomy 317
The ISS Emergency Protocol Model 319
The Seven-Step Incident Response Sequence 319
Real-Time Streaming Validation During Incident Surveys 321
Open-Source Model Governance During Response 323
Cross-Sector Incident Contagion 331
The Insurance Documentation Chain 331
The Black Box Argument 332
Chapter 24: From Government Station to Commercial Platform 335
Part VI: Commercial LEO Destinations 335
The Federal Validation Argument 336
What Transfers and What Extends 336
What Transfers Directly 338
What Requires Extension 340
The CLD Timeline 341
The Operator's Responsibility 342
Build, Rent, or Wait 343
Chapter 25: The AGENTS.md Migration 345
Part VI: Commercial LEO Destinations 345
Step 1: Inventory the Current Governance State 345
Step 2: Classify Vendor Dependencies by Tier 347

Step 3: Build the Evidence Architecture 348
AI-Assisted Compliance Documentation During Migration 349
Code Structure as Governance Artifact 352
Step 4: Implement LP9-LP12 355
Step 5: Establish the Continuous Authorization Cycle 356
Common Migration Obstacles 357
Migration Timeline Estimates 358
Chapter 26: The Stable Point 361
Part VI: Commercial LEO Destinations 361
The Five Points Exist 361
The Cost of Not Finding the Point 363
The Maritime Arc, Complete 364
The ISS Arc, Complete 366
The Framework Is the Same 367
The Physics Did Not Change 367
One Thing You Can Do Tomorrow 368
Appendix A: Vocabulary Translation Reference 371
Regulatory Acronyms in Plain English 377
Usage Notes 378
Appendix B: Cross-Framework Control Mapping 380
Harness Components (H-1 through H-5) 381
Compliance Properties (LP1 through LP14) 383
Evidence Tiers (E-1 through E-4) 388
Knowledge Base Integrity Controls (KB-1 through KB-4) 389
Declarative Admission Control as Cross-Framework Policy Engine 391
Production Inference Engine as Classification Decision Support 393
Privacy-Preserving Certification for Maritime and Financial Data 396
Usage Guidance 398
Appendix C: Deployment Checklists by Sector 400
Checklist 1: Financial Services 400
Checklist 2: Healthcare 402
Checklist 3: Energy and Critical Infrastructure 403

Checklist 4: Insurance and Actuarial 404
Checklist 5: Manufacturing, Supply Chain, and Telecommunications 406
Apple Foundation Model: On-Device Deployment for Zero-Trust Data Boundaries 408
Consolidated Control Evaluation Rubric 410
Usage Notes 413
Appendix D: Multi-Framework Evidence Rendering Reference 414
The Base E-4 Attribution Record 414
OSCAL Intermediate Representation 415
SOC 2 Rendering 418
HIPAA Rendering 418
NERC CIP Rendering 420
GDPR Rendering 421
Gemini-3.1-Flash-Lite: Ultra-Low-Cost Multi-Framework Translation 422
beautiful-html-templates: Classification Society Report Standardization 424
HyperFrames: Voyage Data Recorder Playback for AI Attribution 426
Implementation Notes 429
How the Evidence Records Are Emitted and Stored 429
Appendix E: Regulatory Timeline 431
2017 431
2018 431
2020 432
2021 432
2022 432
2023 433
2024 433
2025 434
2026 435
2027 435
Reading This Timeline 436
Jurisdiction-Aware Model Routing Under Accelerating Compliance 436
Multi-Framework Evidence-Latency Optimization Under Accelerating Compliance 439

Appendix F: The Maritime-to-AI Governance Genealogy 443
1688: Lloyd's Coffee House 443
1760: Lloyd's Register Formalized 443
1771: Lloyd's Committee Rules 444
1807: Lloyd's Act of Parliament 444
1834: Lloyd's Register Becomes Independent 445
1876: The Plimsoll Mark (Merchant Shipping Act) 445
1914: International Convention for the Safety of Life at Sea (SOLAS) 446
1948: International Maritime Organization (IMO) Established 446
1956: Containerization (SS Ideal X) 447
1974: SOLAS Convention Updated 447
1993: International Safety Management (ISM) Code 447
1998: ISS Intergovernmental Agreement 448
2006: Maritime Labour Convention (MLC) 448
2024: EU AI Act 449
Academic-Research-Skills: Citation Integrity in Regulatory Knowledge Graphs 449
MCTS-Search: Multi-Step Regulatory Reasoning with Rollback 451
OpenKB: Two-Hop Citation Networks for Regulatory Cross-References 452
The Structural Lineage 454
Appendix G: Acronyms and Abbreviations 455
Download The Harness Doctrine Codex 460

Disclosures and Context

Author Disclosure. I wrote this book as a Senior Solutions Architect at GitLab whose federal account work has included supporting technology deployments inside government authorization boundaries. GitLab is a vendor whose product is one reasonable implementation path for several of the harness components described in these pages; the framework's claims are intended to hold regardless of vendor, and the configuration patterns in this volume are written to be platform-independent. Even so, I have a commercial interest in governance frameworks that increase organizations' willingness to authorize AI inside high-consequence environments, including frameworks that are well served by the GitLab platform. Read the argument with that interest in view.

Personal Opinion. The positions in this book are my own. They do not represent the views of GitLab, of any employer, customer, or agency I have worked with, or of any standards body referenced here.

Trademarks and Attributions. Product and company names used in this book are the trademarks or registered trademarks of their respective owners. They appear for identification and commentary only and do not signal endorsement, sponsorship, or affiliation. GITLAB is a trademark of GitLab Inc. in the United States and other countries and regions; GitLab Ultimate and the GitLab Duo Agent Platform are products of GitLab Inc. Anthropic, Claude, Cowork, and the Managed Agents API are products or marks of Anthropic. Cohere and Command A+ are products or marks of Cohere. NVIDIA is a mark of NVIDIA Corporation. Microsoft is a mark of Microsoft Corporation. Apple is a mark of Apple Inc. Amazon Web Services and Cedar are marks of Amazon.com, Inc. or its affiliates. Google and Gemini are marks of Google LLC. FactSet, PitchBook, S&P Global, and Moody's are marks of their respective owners. Open Policy Agent (OPA) and the `anthropics/financial-services` reference repository are identified by their project names; open-source components are used under their respective licenses, including Cohere Command A+ under Apache 2.0.

Series Context. This volume builds on the federal baseline established in the earlier volumes of The Harness Doctrine and references them throughout. I wrote those volumes as well. Self-citation across the series is intentional cross-reference within a single author's work, not independent corroboration; treat it as such. The commercial proof obligation for this volume belongs to its first real adopters; I defer it to them deliberately rather than dress the design up in case studies it has not yet earned.

Technology References. Throughout this book I name specific technologies, products, and open-source projects. Naming a technology is not a recommendation to adopt it. Each one appears to make a capability concrete: to show the function being produced rather than describe it in the abstract. Wherever a credible open-source path exists, I name it, because I would rather a reader own the governance instrument than rent it.

AI and Authorship. I know the question is coming: "This is a book about AI governance. Was it written by AI?"

Here is the truth. The framework, the convergence discovery, the five-component architecture, the formal properties, came from my own thinking. Years of notes, rants, whiteboard sessions, failed approaches, conversations with authorizing officials who needed answers I did not yet have, and the slow accumulation of pattern recognition that comes from working in a domain long enough to see the shape underneath the moment-to-moment decisions.

Being a believer in AI as a tool, and specifically as the kind of tool this book describes, I used AI extensively throughout the writing process. AI helped me cross-reference research across domains I would not have remembered on my own. It identified gaps in my reasoning that I was too close to see. It challenged my arguments from perspectives I had not considered. It edited the manuscript through multiple rounds of revision.

This is consistent with the book's own thesis. AI is most valuable when operating within a governed framework. The harness for this book was my judgment about what to keep, what to cut, and what mattered. The AI operated inside that harness. It made the work better. It did not make the decisions.

If you are listening to this on audiobook: that is not my voice. The narration is generated by artificial intelligence and tuned for clarity and pacing. The words are mine. The voice is a tool, governed by the same principle this book describes.

Regulatory Context. The regulatory citations throughout this book reflect the state of AI governance regulation as of April 2026. The EU AI Act, the NIST AI Risk Management Framework, the NAIC Model Bulletin, sector supervisory guidance, and the other regulatory instruments referenced here are current at time of writing, but organizations implementing harness architecture should verify the current regulatory environment in their jurisdiction and adapt accordingly.

About the Author

Linc M. Williams is a senior solutions architect working on DevSecOps and federal compliance for some of the most demanding environments in the U.S. government. Twenty-five years across civilian and defense missions: the Department of the Navy, Marine Corps, Coast Guard, Air Force, Army, the Department of the Treasury, and a long list of others.

The authorization problem for AI changed under him. What used to be a manageable compliance exercise turned into something the existing governance models do not handle. They were not designed for it.

His response was a governance framework, the Three-Body Problem, built so Authorizing Officials could trust AI inside classified networks. *Lagrange Points* carries the structural core of that framework, the five-component harness, out of the classified setting it came from and into commercial regulated industries: financial services, healthcare, energy and critical infrastructure, insurance, and the manufacturing, supply chain, and telecommunications triad. This is Book Three of The Harness Doctrine, built on the federal baseline established in the earlier volumes.

He writes for both ends of the org chart. Practitioners standing up governed AI in a regulated industry on one side. Boards and senior leadership trying to read what their AI investments mean for enterprise risk on the other.

Read more at Linc Williams's site.

Inside The Harness Doctrine Series

The Harness Doctrine argues one thing across the series: AI does not become governable by accident. Every working system in this book, and the others beside it, sits inside a harness; and every harness is the same five build-time components plus two operations-time layers, rearranged for a different problem. Each volume begins by defining the doctrine, then applies it to a specific operational surface. Read the volume that names the problem in front of you. The others will be waiting.

Volume I. The Harness Horizon

Trust Convergence on the Edge of Artificial Intelligence

Three independent practitioners, a frontier lab, an independent researcher, and a working engineer, converged on the same five-component harness without coordinating. Volume I is the proof that this pattern is not a vendor preference but one the field arrived at by force.

Inside this volume: The Convergence Discovery. The moment Anthropic, Andrej Karpathy, and a working engineer publish the same architecture inside a six-week window. The convergence is the argument; the rest of the series is what to do with it.

Volume II. The Three-Body Problem of Classified AI

Operationalizing AI on the High Side

Classification, compliance, and developer velocity form a chaotic three-body system, the physics metaphor from which Classified AI takes its name. No two of those forces stay stable without the third, and no manual process holds the orbit. Volume II is the federal answer.

Inside this volume: The Mosaic Effect. Permissible fragments that an AI can assemble into impermissible disclosures by combining context across compartments. Each piece is cleared; the combination is not. The Mosaic Effect is why classification reviewers exist, and why your agent needs one too.

Volume III. Lagrange Points

Finding Stable Governance Orbits Between Competing Forces

The federal framework transfers wholesale to commercial regulated sectors: finance, healthcare, energy, insurance. Same physics, different vocabulary. Volume III does the math and names the gaps.

Inside this volume: The 70/30 Transfer Rule. Seventy percent of the federal harness moves verbatim to regulated commercial industries. The remaining thirty percent is the part DoD framing under-addresses: consent, fairness, and multi-jurisdictional compliance. Volume III tells you which thirty.

Volume IV. Stable Orbit

Operational Security and Resilience for Long-Running Systems

Getting to orbit is engineering. Staying there is doctrine. Volume IV is the operations manual for the AI-assisted system that survives the second year, the third compliance audit, and the engineer who built it walking out the door.

Inside this volume: The Five Control Surfaces and Graduated Autonomy. Five named surfaces (bounded context, transparency, human-in-the-loop, model governance, pipeline verification), and a trust ladder that gives an agent more authority only after it has earned it. Trust is not assumed; it is rated.

Volume V. The Replicant Fleet

Command, Containment, and Control for Autonomous Agent Fleets

One-agent governance breaks the moment you replicate. A fleet of agents needs coherence mechanisms borrowed from naval doctrine, and from the science fiction that anticipated it.

Inside this volume: Fleet Coherence. How a self-replicating agent fleet stays on doctrine when every ship in it can fork. The interlude where the fleet must act without you is where most readers stop, breathe, and start over from page one.

Volume VI. The Shatterdome Doctrine

Governing Open Agent Platforms Without Killing What Makes Them Useful

Prohibition fails. Openness without a harness also fails. The only stable position is the shatterdome: an open platform wrapped in the same five components, with governance built into the wiring rather than bolted on at the end.

Inside this volume: The Drift Protocol. In Pacific Rim, the Drift is not a feature of the Jaeger; it is the layer through which the Jaeger exists. Volume VI maps the same architecture onto Claude Agent SDK: hooks fire on every tool call, before and after; no path from intent to action avoids a governance surface. The harness is native, not a retrofit.

Volume VII. Three Governance Surfaces

Execution, Knowledge, and Comprehension in AI Agent Teams

Most teams govern only what an agent does. They lose the ability to explain it. Volume VII names the failure and shows the three surfaces a working team has to govern at once.

Inside this volume: The Comprehension Cliff. The named failure mode that lands when humans can no longer reason about what an agent team did, why, or in what order. The cliff is steeper than anyone budgets for, and the fall is not recoverable from inside the same team.

Volume VIII. The Semester Problem

Governing AI in Institutions Built for Departure

Universities write policy for populations that leave before accountability matures. The accumulated debt lands on the small permanent staff who have to re-govern every cohort. AI made that cost unsurvivable.

Inside this volume: Governance Debt Machines. The indictment that institutions designed around departing populations manufacture governance debt as a structural feature, not a bug. Volume VIII is the only book in the series that carries a verdict.

Volume IX. When Galaxies Collide

DevSecOps and Its Future with AI

The merge request was a coordination protocol designed for humans coordinating at human pace. Mass-AI development has broken the coordination cost: the protocol is failing in real time inside engineering organizations. Volume IX predicts the three-stage replacement of the merge request (gated trunk, provenance layer, spec-as-truth), varies the prediction by jurisdiction (United States engineering-first, European Union regulation-first, China state-driven), and names the harness layers above and below the gate (Application Security Posture Management, Application Detection and Response) that complete the architecture.

Inside this volume: The Coordination-Cost Collapse. The William Blair 2025 developer survey records a single-year jump from forty percent to sixty-three percent of teams naming coordination complexity as the primary barrier to DevSecOps adoption. That is the merge request reporting its own failure. Volume IX is what replaces it.

Volume X. The CATO Playbook

Achieving Continuous Authority to Operate with AI-Assisted DevSecOps

Federal authorization is a project with an end date: six to eighteen months in purgatory, a million dollars a cycle, expired every three years. Continuous Authority to Operate replaces the periodic snapshot with compliance evidence generated on every build. Volume X is the operator's playbook for making authorization a capability that runs rather than a document that ages.

Inside this volume: The Continuous Evidence Pipeline. Authorization stops being a three-year-old snapshot when security scanning, control validation, and remediation happen on every commit, and the Authorizing Official sees compliance posture in real time. Volume X turns the Authority to Operate from a deliverable into a heartbeat.

Volume XI. The SAFe Horizon

Governing AI Augmentation in Scaled Agile

The Scaled Agile Framework gives the enterprise an operating model, then taxes it with ceremony, artifact, and coordination overhead. AI can absorb that tax, but ungoverned augmentation at portfolio scale is its own failure mode. Volume XI augments every SAFe role, then wraps the four configuration layers in a governance architecture that keeps the added speed trustworthy.

Inside this volume: Four-Layer Governance. Team, Program, Large Solution, and Portfolio each govern AI augmentation at their own altitude, with policy-as-code enforcement and human checkpoints between them. Volume XI is the proof that scaled agility and AI governance are the same discipline seen from two ends.

How to Read the Series

Start with the volume that names the problem in front of you. The teasers above are the index.

Each volume stands alone. The doctrine vocabulary is consistent across the series; the case work is not.

Volume I is the proof. Volume II is the test case. The volumes that follow are the application surfaces, ordered from the most regulated outward toward the engineering future the doctrine has to govern as DevSecOps itself transforms.

Chapter 1: The Shipowners and the Underwriters

> *"The physics of AI governance do not change when you leave a classified network. The three forces, data protection, regulatory compliance, and operational velocity, pull every regulated organization toward instability. The only difference between a defense program and a hospital system is the vocabulary they use to describe the same architectural failure."*

On a November evening in 1688, on the north side of Tower Street in the City of London, merchants and sea captains crowded into Edward Lloyd's coffeehouse to share intelligence on ships. Lloyd had started posting handwritten notices on the walls: which vessels had made port, which were overdue, which had been lost. The coffee house smelled of roasted beans and salt damp and the particular anxiety of men whose wealth was on the water. A merchant who had financed a voyage to the East Indies needed to know whether his ship was seaworthy before he risked his capital. A ship captain needed underwriters willing to share the loss if the vessel went down. An underwriter needed enough information about the ship, its cargo, its route and its crew to price the risk correctly.

None of them had a standardized way to answer the central question: is this ship worth backing?

Information flowed through conversation, not documentation. A merchant who knew the ship's owner personally might get honest intelligence about the vessel's condition; a merchant who knew only the underwriter might get whatever story moved the policy. Seaworthiness was a matter of reputation and trust, not of documented evidence and verifiable inspection. Ships sailed with policies written against descriptions no one had formally confirmed. Underwriters priced risk based on hearsay. Losses were enormous, disputes were constant, and the inefficiency was visible to everyone in the room. No one solved it, because solving it required agreeing on what a seaworthy ship actually looked like, and that agreement required someone with both the authority to set the standard and the infrastructure to enforce it.

Not ignorance of risk. Absence of governance architecture.

The merchants at Lloyd's coffeehouse had the commercial instinct to share risk, the financial incentive to price it correctly, and the accumulated knowledge to assess it approximately. What they lacked was the institutional machinery to convert that collective intelligence into systematic, verifiable risk assessment. Talented people, working an ungoverned system as well as informal methods allowed. The gap between their judgment and the ground truth was costing them, at intervals, everything.

> **A note on stake and standpoint.** The analysis here is my own informed opinion, not legal, regulatory, or investment advice; verify every regulatory citation against its primary source. I write as an advocate for the harness approach, and I use GitLab Ultimate, the GitLab Duo Agent Platform (DAP), and Anthropic's published work as this book's primary reference implementations because they are the clearest public examples of the architecture, not because I am paid to recommend them; I have no employment, equity, or paid advisory relationship with GitLab Inc. or Anthropic, and neither company commissioned or reviewed this book. Trademark and attribution notices for all named products appear in the front-matter disclosures.
>
> **A note on the state of evidence.** Be clear about what this volume is and is not. It is a design, validated by structural argument and by precedent from the federal and classified environments the original architecture was designed for; it is not validated by a named commercial field deployment with measured before-and-after outcomes. No bank, hospital, utility, insurer, or manufacturer is named here as having run the framework and reported its results. The commercial proof obligation belongs to the first real adopters, and I defer it to them deliberately rather than dress up the architecture in case studies it has not yet earned. Read what follows as a map of where the equilibrium points should sit, with the field measurements still to come.
>
> **A note on technology references.** Throughout this book I name specific technologies, products, and open-source projects. Naming one is not a recommendation to adopt it. Each appears to make a capability concrete, to show the function being produced rather than describe it in the abstract. Wherever a credible open-source path exists, I name it, because I would rather a reader own the governance instrument than rent it.

Start here: the harness in brief

You can start here without having read the prior series. The harness is the constraint architecture wrapped around an AI system: a machine-enforceable instruction surface plus a routing layer that checks every action against it, so governance is enforced by architecture rather than asserted in policy. It is to an AI system what a classification certificate and a load line are to a ship, an external, verifiable account of what the system is built to do and proof it is operating within those bounds. You will meet each part of the framework through the ships first, as it is introduced; the full property and evidence vocabulary is defined as it appears and collected in Appendix A.

The modern coffee house

The executive team responsible for AI governance at a regulated financial institution in 2026 is in essentially the same position as those merchants on Tower Street.

They have access to AI systems of genuine capability; they have commercial pressure to deploy those systems; they have compliance obligations that require the systems to behave within defined boundaries. And they have no standardized way to answer the central question: is this AI system governance-worthy?

Consider the assessment process as it actually operates. The CISO received a vendor presentation. The compliance officer reviewed the vendor's System and Organization Controls (SOC) 2 report. Legal reviewed the data processing agreement. A model risk team ran a validation on the outputs. Each assessment is legitimate. None of them, alone or together, answers the question that matters for regulated AI deployment: does the system have an architecture that continuously constrains its behavior, produces verifiable evidence of that constraint, and flags deviations before they become compliance events?

Organizations are making AI deployment decisions the way Lloyd's merchants made underwriting decisions: through accumulated judgment, informal intelligence, and trust in the vendor's reputation. Documentation describes what the system is supposed to do. Assessment validates the description. Approval is granted against the description, not against the system's actual running behavior. The gap between description and behavior is where the regulatory exposure lives.

This is the coffee house problem. Not a technology problem. A governance architecture problem. The AI systems are capable, and the compliance frameworks are mature. What is missing is the institutional machinery that converts individual judgment into systematic, verifiable risk assessment, the same machinery Lloyd's eventually built.

Three hundred years after that November evening on Tower Street, the maritime insurance market had grown from a coffeehouse into a global institution with standardized survey requirements, actuarial models, classification society inspections and the Lloyd's Register: a document any underwriter anywhere in the world could use to verify independently that a specific vessel met a specific seaworthiness standard. The merchants did not solve the problem by learning to trust individual ships better. They built the infrastructure that made trust a function of verified compliance with a published standard, not of personal relationship.

The lesson took two centuries to learn. The architecture is transferable.

That is the trajectory this book maps for regulated AI governance.

Three forces, one problem

The merchants at Lloyd's coffeehouse were managing what physicists would later formalize as a three-body problem: three gravitational forces acting on a single object simultaneously, each pulling in a different direction, with no static configuration that satisfies all three at once.

Cargo protection came first. The merchant needed the ship to carry the cargo safely to port. A faster ship that took on water was worse than a slower ship that arrived; a ship that could carry more cargo than its hull could support was a liability, not an asset. Protecting the cargo was the baseline requirement. Nothing else mattered if the cargo was lost.

Port authority compliance came second. Every port had customs officers, harbor masters, and regional authorities whose approvals the vessel needed to dock and unload. A ship that arrived with improper documentation, contraband, or missing certifications faced seizure, fines, and the loss of the entire voyage profit, even if ship and cargo had survived the crossing. Compliance was not optional. It was the condition the whole commercial enterprise rested on.

Commercial velocity was the third force, and it pulled hardest. The trade winds were seasonal. A merchant who missed the weather window lost a year. A ship sitting in port for additional inspection was a ship not earning freight; a voyage delayed by excessive paperwork was a competitor's opportunity. Speed was not a luxury. In the competitive dynamics of 17th-century maritime trade, it was survival.

These three forces create gravitational instability. Maximizing cargo protection pushes toward slower, more conservative voyages with smaller loads and longer inspection cycles. Maximizing compliance pushes toward documentation completeness at the cost of sailing schedules. Maximizing commercial velocity pushes toward faster departures with looser checking. Optimize for any one force and you degrade the other two.

There is no winner. There is only balance.

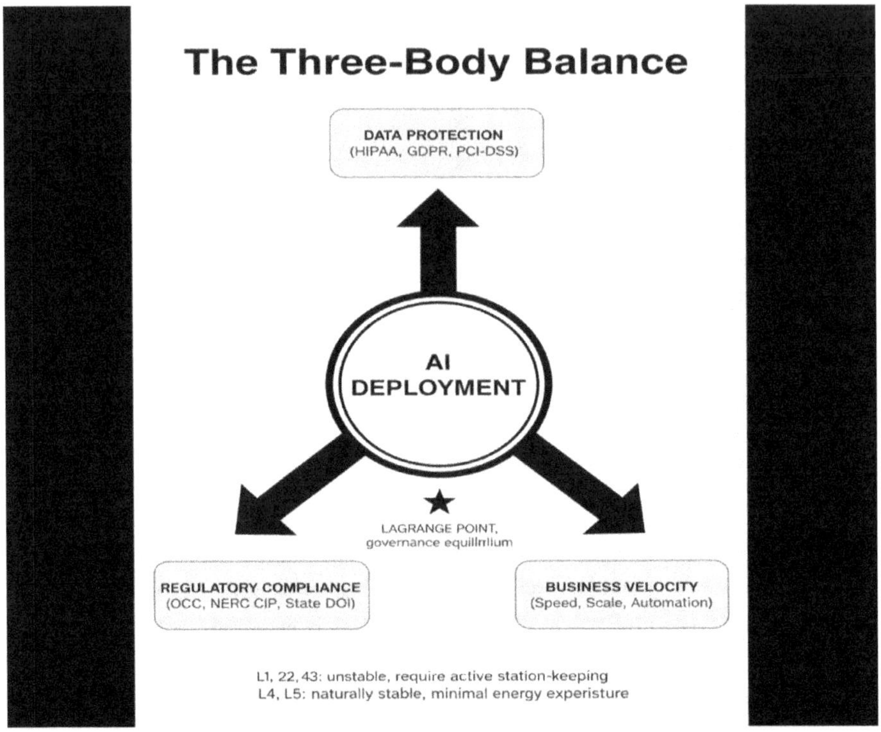

Figure 1: The three-body governance balance. Data protection, regulatory compliance, and business velocity each pull an AI deployment toward instability; the stable configuration is the Lagrange point where all three forces are held in dynamic equilibrium rather than resolved by choosing one side.

The only stable configuration is a dynamic equilibrium: a position where all three forces are managed simultaneously through a governance architecture that treats the tension as a structural condition to be maintained, not a conflict to be resolved by choosing a side.

In orbital mechanics, that balance point has a name. A Lagrange point is the position in a three-body gravitational system where the combined forces reach equilibrium, creating a stable orbit for any object that reaches that position. There are five Lagrange points in any three-body system. The L4 and L5 points are naturally stable; objects that arrive there tend to stay with minimal energy expenditure. The L1, L2, and L3 points are unstable; objects there require active station-keeping to maintain position against the gravitational perturbations that continuously push them off center.

Every regulated organization deploying AI has Lagrange points: governance configurations where data protection, regulatory compliance and business velocity reach sustainable equilibrium. Most organizations are not at those points. They are somewhere in the gravitational field between them, being pulled continuously toward instability by the three forces they cannot satisfy simultaneously. The governance architecture described in this book maps those equilibrium positions for five commercial regulated sectors and provides the station-keeping mechanisms required to maintain them.

The Three-Body series and this volume

The Three-Body Problem of Classified AI proved this architecture in federal and classified environments. That series was written for Authorizing Officials, Information System Security Officers (ISSOs), and engineering teams operating in IL4 and IL6 classified networks. It assumed clearance levels, classification boundaries, Security Technical Implementation Guides (STIGs), and the federal Authorization to Operate process. The vocabulary was government. The regulatory framework was National Institute of Standards and Technology (NIST) 800-53. The threat model was adversarial nation-state exploitation of AI-generated code in classified systems.

The series worked because the underlying architecture it describes is not government-specific. It is universal. The harness, the constraint architecture wrapping an autonomous actor to produce predictable, verifiable outcomes, works in any regulated environment because the problem it solves is not a classification problem. It is a governance problem: how do you deploy a capable but unreliable system in an environment where every output is potentially subject to regulatory scrutiny, and how do you do it at the speed the business requires without creating compliance exposure that exceeds the business value?

That problem exists in classified networks and in community banks; it exists in military hospitals and in private health systems. Cross the classification boundary and the vocabulary changes, the regulatory sources change, the names of the authorizing officials change. The architecture that produces stability does not.

This volume translates The Harness Doctrine from classified environments to commercial regulated industries. It inherits the federal baseline established in the prior series, the harness architecture (H-1 through H-5), the four evidence tiers (E-1 through E-4), the four knowledge base integrity controls (KB-1 through KB-4), and the eight federal security properties (P1 through P8, defined in Volumes I and II), then defines this volume's own governance layer on top of that baseline and maps it to five sectors: financial services, healthcare, energy and critical infrastructure, insurance, and the emerging triad of manufacturing, supply chain, and telecommunications. That governance layer is the Lagrange governance properties (LP1 through LP14): LP1-LP8 are the core governance-worthiness criteria, several of which restate a federal security property in governance terms; LP9-LP12 are the commercial extension (consent, jurisdiction, fairness, and transparency) that federal environments never had to address; and LP13-LP14 are the enforcement properties (output boundary enforcement and deterministic deny on uncertainty). The inherited federal baseline transfers without modification, because the constraint mechanism is sector-independent; the harness components, the evidence tiers, the knowledge-base controls, and the eight federal security properties carry over unchanged. What this volume adds is the LP-series governance layer built on that baseline, including the four commercial-extension properties plus sovereignty. This is my own framework classification, a structural judgment about what transfers, not a count of deployments.

Two narrative threads run through this book. The first is maritime: Lloyd's coffeehouse, the rise of the classification societies, the coffin ships and the Plimsoll line, the International Convention for the Safety of Life at Sea. This thread is not decoration. Maritime risk governance is the history of a commercial industry working through exactly the three-body problem that regulated AI governance faces today. The solutions that industry developed, standardized surveys, independent classification, minimum load-line requirements, multi-jurisdictional safety conventions, are the structural ancestors of the governance architecture this book recommends. They took roughly two centuries to develop through accumulated disaster and institutional response.

The AI governance community does not have two centuries.

The second thread is the International Space Station. Five space agencies, fifteen nations, one governance architecture, operating continuously in an environment where a single systems failure kills the crew. The ISS does not succeed because its member agencies trust each other. It succeeds because the governance architecture makes the behavior of every system verifiable, every interface defined, every deviation detectable, every response protocol pre-committed. Trust is not the precondition. It is the product. For regulated AI in commercial environments, that inversion is the central insight.

The governance architecture the coffee house could not provide

When Lloyd's merchants gathered on Tower Street in 1688, they were trying to solve a risk assessment problem with social infrastructure. The coffeehouse model worked better than nothing; it concentrated information, enabled price discovery, and created a market for maritime risk. For more than a century, it was the best available mechanism.

It was not good enough. Ships were overloaded to generate more freight income, then foundered in North Atlantic storms. Underwriters paid claims they had mispriced, because the survey data was inadequate. Merchants absorbed losses from unseaworthy vessels certified by men who either did not know what they were certifying or had financial reasons to certify favorably. Every failure concentrated at one boundary: where commercial pressure pushed hardest against safety.

The solution was not better merchants or more honest ship captains. It was the classification society: an independent institution with the authority to define seaworthiness standards, the infrastructure to conduct rigorous surveys, and the credibility to issue certificates that markets could rely on. Lloyd's Register of Shipping, formalized in the 1760s, provided the first standardized seaworthiness classification system. A vessel assigned a letter and number from the Register had been surveyed against published criteria by independent inspectors. An underwriter quoting on that vessel was not trusting the merchant's description of the ship; the underwriter was relying on verified compliance with a published standard.

That distinction is what made the system work.

This is the governance architecture that regulated AI currently lacks. Better vendor presentations will not supply it. Neither will more thorough legal review of data processing agreements, or another column in the AI risk register. What is missing is the institutional equivalent of the classification society: a framework that defines governance-worthiness for AI systems, infrastructure to verify compliance with that framework continuously, and evidence artifacts that let any regulator, auditor, or board committee confirm the system is operating within its authorized boundaries without taking the deploying organization's word for it.

The Harness Doctrine provides that architecture. The harness is the constraint mechanism; the evidence tiers are the survey record; the governance properties are the seaworthiness criteria, the eight federal security properties (P1 through P8) inherited from the prior series and the Lagrange governance properties (LP1 through LP14) this volume defines on top of them; the AGENTS.md is the certificate of compliance. And the continuous verification layer is the inspector who does not visit once at deployment and then disappear, but monitors the vessel's condition throughout every voyage.

What this book covers

This book is organized in six parts.

Part I, which opens here, establishes the problem structure: the coffee house problem, the three-body tension, three forces mapped across five commercial sectors to demonstrate that the structure is universal, the coffin ships as archetype for what happens when commercial velocity overwhelms governance, and the specific failure patterns that commercial regulated organizations are repeating today.

Part II (The Classification Survey) covers the governance framework itself: harness architecture, evidence model, failure modes, and the multi-jurisdictional compliance structure that allows one architecture to satisfy multiple regulatory frameworks simultaneously. Part III (The Sector Lagrange Points) provides sector-specific implementation, identifying the precise governance configurations where data protection, regulatory compliance, and business velocity reach equilibrium under each sector's regulatory constraints.

Parts IV through VI move from theory to operation. Assembly in Orbit addresses the implementation challenge: building multi-framework compliance architecture, maintaining it across staff transitions and vendor changes, and presenting its outputs to board members who need to make governance decisions without a technical background. When Ships Sink covers incident response, with the 1912 International Convention for the Safety of Life at Sea (SOLAS) conference as the historical model for rebuilding governance after catastrophic failure. Commercial LEO Destinations maps the translation from government-origin architecture to commercial-native deployment, including the AGENTS.md migration path for organizations starting from existing governance documentation rather than from a clean slate.

The Appendices provide vocabulary translation tables, cross-framework control mappings, sector-specific deployment checklists, and the maritime-to-AI governance genealogy that traces the structural lineage from Lloyd's Register through the International Maritime Organization to the AI governance frameworks now emerging from NIST, International Organization for Standardization (ISO), and the EU AI Act.

The merchants at Lloyd's coffeehouse eventually built the institution that solved their problem. They did it not by learning to trust individual ships, but by building the infrastructure that turned trust into a product of verified compliance with a published standard. The governance architecture described in this book is that same infrastructure, adapted for the AI systems now entering every regulated industry.

The three forces are the same. The physics are the same. The solution follows the same structural logic it always has.

> **CORE DOCTRINE**
>
> A Lagrange point is the position in a three-body system where gravitational forces balance and a stable orbit becomes possible. Every regulated industry has Lagrange points, governance configurations where data protection, regulatory compliance, and business velocity reach equilibrium. The Harness Doctrine maps those points. This volume provides the coordinates for five sectors that have not yet found them.

Chapter 2: Three Bodies, Every Sector

Part I: The Coffee House Problem

> *"The physics of instability are not proprietary. Every regulated industry has discovered the same three-body problem independently, and nearly every regulated industry has arrived at the same incorrect conclusion: that the solution is more documentation."*

The coffee house at Lloyd's did not invent marine risk. Merchants and shipowners had been worrying about cargo loss since the Phoenicians first loaded cedar onto cedar-framed hulls. What Lloyd's invented was a shared vocabulary for that risk: a common framework that allowed shipowners, underwriters, and surveyors in the same room to discuss the same vessel and reach binding agreements. The vocabulary made the market. The market made the architecture.

Chapter 1 introduced the shipowners and the underwriters, the actors who gathered at Edward Lloyd's coffee house and, over two centuries, built the modern governance infrastructure for moving cargo across oceans. This chapter examines what they were really moving: not silk and spice, but the three forces that destabilized every ship, and every regulated AI program since.

Those three forces hold across the industry, the regulatory framework, and the size of the organization. The names differ. The physics do not.

What follows is the proof. Each sector carries its own regulators, statutes, and protected data categories; this chapter introduces the few that anchor the argument and leaves the complete plain-language reference to the regulatory acronym glossary (Appendix G). Before any of that, two scenes from the water make the stakes concrete.

Two ships already at sea

Samuel Plimsoll called an overloaded merchant vessel a coffin ship: a hull made lethal not by defect but by incentive. The owner bore no personal cost when it sank, because the cargo and the ship were insured; he bore a direct competitive cost from leaving cargo on the dock. The rational owner loaded past the safe line and let the weather sort out the rest. The only uninsured variable was the crew. Chapter 3 develops this in full; the point to carry forward is that the failure was never an engineering failure. The ships were sound. The arithmetic of incentive overrode the arithmetic of safety because no visible, enforced line stopped it.

The AI industry is building coffin ships now. Consider one case the chapter on commercial failure modes treats in detail. A large health insurer deployed an AI system to assist post-acute-care coverage decisions, choosing when Medicare beneficiaries should be discharged from skilled nursing facilities. The system was not technically defective. It produced predictions and processed records at scale. What it lacked was a governance architecture constraining it to optimize for clinical accuracy rather than denial rate. According to subsequent litigation and reporting by STAT News, appealed denials were reversed in roughly 90% of cases. That figure does not indicate a permissive appeals process; it indicates that the underlying decisions were systematically wrong. The AI was not the failure. The absent governance property was. A coffin ship: sound machinery, fatal incentive, no enforced line.

The second scene is the regulator declining to paint the line for you. SR 11-7, the Federal Reserve and Office of the Comptroller of the Currency (OCC) supervisory guidance on model risk management issued in 2011, sets the effective-challenge standard: a model in production must be independently questioned by someone with the authority and competence to challenge it, and that challenge must be documented. SR 21-8 (2021) carried the standard forward to third-party model providers. Both apply cleanly when a supervisor governs a model directly; the institution implements a published standard and shows its work. Neither was written with generative or agentic AI in mind. The agent that drafts the credit memo sits in the space the published guidance does not yet reach.[^sr26] When the supervisor declines to write the agent's rules, the institution must construct the standard itself, demonstrate that it constructed one, and prove it operates. The regulator hands the bank the pen. The harness is what the bank writes with it.

[^sr26]: Anticipated near-future supervisory guidance (referred to in industry discussion as SR 26-2) is expected to extend the model-risk framework toward generative and agentic systems; the argument here rests on the published SR 11-7 and SR 21-8 standards and does not depend on it.

Two ships, then: one already overloaded and unmarked, one whose owner has just been told to paint his own line and prove it holds. Both are the three-body problem in commercial dress. The rest of this chapter names the three forces precisely, sector by sector.

Three forces, four sectors

The Three-Body Problem of Classified AI defined the three forces in government terms. The first body was classification, the requirement to protect data at specific sensitivity tiers. The second body was compliance, the requirement to satisfy National Institute of Standards and Technology (NIST) 800-53, Security Technical Implementation Guides (STIGs), and the Authorization to Operate process. The third body was developer velocity, the requirement to ship code on program schedules.

These are government translations of more fundamental concepts.

The first body is data protection. In any regulated industry, certain data cannot be accessed, processed, or surfaced without authorization. The authorization requirement comes from different sources depending on the sector, but the architectural implication is identical: the AI system must not access data beyond its authorized scope, and every access event must be attributable, bounded, and auditable.

Financial services carries the widest exposure surface. Four kinds of protected data sit within reach of a single assistant. Personally identifiable information (PII), governed by the Gramm-Leach-Bliley Act (GLBA) and a thickening stack of state privacy laws. Material non-public information (MNPI), the price-moving knowledge that insider-trading law and Securities and Exchange Commission (SEC) disclosure rules wall off. Payment card data, governed by its own industry security standard. And the transaction record that anti-money-laundering (AML) and Bank Secrecy Act (BSA) reporting depends on. A single AI coding assistant operating in a financial institution may touch all four at once. One misconfigured context window; four distinct regulatory exposures. The specific statutes and their acronyms are catalogued in the glossary; what matters here is the count, not the citation.

Healthcare adds a different constraint: the minimum necessary rule. The core category is protected health information (PHI), governed by the Health Insurance Portability and Accountability Act, or HIPAA. On top of that base sit tiers with heightened protection: psychotherapy notes, substance use disorder records under the federal confidentiality rule known as 42 CFR Part 2, HIV status, genomic and biometric data under emerging state genetic privacy laws, and clinical trial records under Food and Drug Administration (FDA) controls. The minimum necessary rule governs all of it. The AI system must not receive more information than the stated purpose of the access requires; an assistant that pulls a full chart to answer a billing question has already overreached. The remaining statutes are listed in the glossary.

Energy and critical infrastructure shifts the threat model entirely. Not exfiltration. Disruption. The protected categories here are operational: the configuration data that describes how the grid is controlled, the industrial control and supervisory control and data acquisition (SCADA) systems that run substations and plants, and the cyber system information that the North American Electric Reliability Corporation, under its Critical Infrastructure Protection standards (NERC CIP), requires utilities to wall off. Nuclear facilities add a further layer under their own regulator. The distinguishing fact is the failure mode. An AI system that causes a misconfiguration in an energy management system does not produce a data breach report; it produces a grid event. The full standard-by-standard map is in the glossary.

Insurance protection centers on actuarial privilege. The sensitive material is the underwriting data, the pricing models, and the claims history that a state Department of Insurance can examine and that competitors would pay to see. Personal-lines business adds policyholder financial and health data, governed by state privacy law and the model cybersecurity rule the National Association of Insurance Commissioners has promoted state by state. Actuarial models are proprietary competitive assets. A large insurer's pricing model, exposed through an AI assistant's context window, is at once a competitive harm and a potential rate-filing violation. The governing authorities are enumerated in the glossary.

Four sectors; four distinct regulatory vocabularies; one architectural requirement underneath all of them. AI systems must not access or surface data beyond the authorized scope, and they must produce attributable evidence of every access event. A reader new to any one sector can stay oriented across all four without memorizing the statutes. The requirement does not change when the vocabulary does.

The second body is regulatory compliance. Every regulated industry has an external authority that sets minimum control requirements, demands evidence of implementation, and imposes consequences for non-compliance. Authority structures differ widely. The architectural demand is universal: every control must produce verifiable evidence, and that evidence must satisfy an examiner, auditor, or assessor who was not involved in producing it.

Financial services faces the most fragmented authority structure. The OCC and Federal Reserve govern bank safety and soundness; the SEC governs broker-dealers and investment advisers; the Consumer Financial Protection Bureau (CFPB) governs consumer-facing products; Financial Industry Regulatory Authority (FINRA) governs broker-dealer conduct; state banking regulators cover state-chartered institutions; and the EBA plus national competent authorities cover EU operations. The Federal Financial Institutions Examination Council (FFIEC) examination handbooks govern technology risk, including AI. The EU AI Act classifies certain financial applications as high-risk AI systems with specific governance obligations. No single authoritative framework exists. Multiple regulatory expectations must be satisfied simultaneously from a single system configuration.

Healthcare compliance has a distinctive wrinkle: the FDA. Regulatory authorities include Department of Health and Human Services (HHS) Office for Civil Rights (OCR) for HIPAA enforcement, the FDA for AI systems used in clinical decision support (classified as software as a medical device, or SaMD), Centers for Medicare and Medicaid Services (CMS) for billing and coding compliance, and state health departments and boards of medicine for clinical practice standards. The FDA's

AI/ML-based SaMD action plan adds a continuous learning system governance requirement with no direct analog in the federal IT authorization process, because the FDA requires a predetermined change control plan (PCCP) that governs how an AI model can change after deployment. That requirement is stricter than continuous Authorization to Operate (cATO) in one specific dimension: changes must be pre-approved by regulators, not merely monitored continuously.

Energy compliance resembles the Department of Defense (DoD) authorization process more closely than any other commercial sector. NERC CIP standards CIP-002 through CIP-014 govern bulk electric system cybersecurity. Federal Energy Regulatory Commission (FERC) has oversight authority over NERC. Nuclear Regulatory Commission (NRC) 10 CFR 73.54 governs nuclear cyber security. Department of Energy (DOE) orders govern classified energy programs. State public utility commissions regulate distribution utilities. The CIP compliance process, including the triennial audit cycle, the self-reporting obligation, and the violation severity levels (VSLs), is highly prescriptive. The NERC CIP Senior Manager role (required by CIP-003) is the closest commercial analog to the Authorizing Official.

Insurance compliance is state-based, and that is the problem. Each state Department of Insurance (DOI) has examination authority over licensed insurers. The National Association of Insurance Commissioners (NAIC) Model Law on Cybersecurity (adopted in 24 states as of early 2026) sets minimum cybersecurity standards. Sarbanes-Oxley Act (SOX) applies to publicly traded insurance holding companies. State insurance codes govern AI use in underwriting and claims under emerging algorithmic decision-making rules (Colorado's SB 21-169 on life insurance algorithms is the leading example). The actuarial standards of practice (ASOPs), issued by the Actuarial Standards Board, govern AI use in actuarial work products. No single AI governance framework is authoritative; multiple frameworks apply simultaneously; and the state-by-state variation creates a multi-jurisdictional compliance problem with no analog in the federal authorization process.

The third body is business velocity. Competitive pressure drives AI adoption in every regulated industry, or cost reduction does, or both at once. The governance architecture must not impose friction that makes the governed path slower than the ungoverned alternative. When the governed path is slower, people take the ungoverned alternative. Regulatory consequences arrive later. The competitive cost of moving slowly arrives immediately.

Consider the concrete pressure in each sector. A bank that can approve a loan in three minutes using AI-assisted decisioning competes differently from one that takes three days. The governance requirement, that the AI decisioning system must satisfy fair lending obligations, model risk management requirements under SR 11-7, and adverse action notice requirements under the Equal Credit Opportunity Act (ECOA), has to be met without adding three days back onto the process. Three minutes or three days. That gap drives shadow adoption.

Healthcare velocity manifests as clinical workflow efficiency, prior authorization cycle times, diagnostic turnaround, and administrative cost reduction. A health system that can automate 60% of prior authorization work with AI frees clinical staff for patient care. Governance, including HIPAA minimum necessary compliance, the No Surprises Act, and state surprise billing laws, must be satisfied without eliminating the efficiency gain that justified the investment.

Energy velocity is operational: grid reliability, outage prediction accuracy, asset maintenance optimization, demand forecasting. A utility that can predict transformer failure 90 days in advance using AI-driven sensor analysis reduces both capital expenditure and outage risk. NERC CIP electronic security perimeter requirements must be satisfied without compromising the operational integration that makes the prediction useful.

Insurance velocity is competitive differentiation. An insurer that can complete a personal lines underwriting decision in real time using AI-driven risk scoring competes differently from one that requires manual review. State anti-discrimination rules, Fair Credit Reporting Act (FCRA) adverse action requirements, and emerging algorithmic accountability obligations must be satisfied without adding manual review steps that eliminate the advantage.

Not four different problems. One problem wearing four different regulatory uniforms. Governance must not become the bottleneck that defeats its own purpose.

Same architecture, different vocabulary

The vocabulary of AI governance in classified environments does not survive intact when it enters a boardroom at a bank, a hospital system, or a utility. The Authorizing Official becomes five different titles depending on whether the conversation concerns SOX, HIPAA, NERC CIP, or an insurance examination. The Information System Security Officer (ISSO) fractures into three different roles where the organization has separated information security, privacy, and compliance functions. The Authorization to Operate (ATO) becomes a different process entirely, with different evidence requirements, different assessors, and different renewal cycles.

Not different architecture. Different vocabulary applied to the same architecture. Every DoD governance concept has a commercial counterpart in every sector; the translation is possible because the underlying problem is the same. But the commercial counterparts are frequently more fragmented than their DoD analogs. The Authorizing Official at a multi-state insurer is not a single person; it is a distributed governance structure with board, actuarial, compliance, and state-regulatory dimensions that must be coordinated. The NERC CIP Senior Manager is the closest thing to a single authorizing official in commercial regulated industries, and that role exists only in the electric sector.

That fragmentation is not a defect. It is a consequence of the commercial governance problem being harder than the federal one in certain dimensions: more frameworks, more jurisdictions, more decision-makers, and no single authoritative body that can grant an ATO and make it stick.

> **Sidebar: the full translation table lives in Appendix A.** The row-by-row crosswalk that maps each DoD and IC term (Authorizing Official, ISSO, ATO, cATO, classification levels, clearance level, STIGs, Open Security Controls Assessment Language (OSCAL), Delegation Tokens) to its financial-services, healthcare, energy, and insurance counterpart is reference material, not narrative. It is collected in the vocabulary translation reference (Appendix A) so the argument here can stay on the three forces rather than stall on a lookup table. Two patterns from that table carry the point: every federal concept has a commercial equivalent, and the commercial equivalent is usually more fragmented across roles, frameworks, and jurisdictions than the single federal authority it replaces.

What transfers and what extends

Most of the harness transfers to commercial regulated settings without modification; a sector-specific minority requires extension. That minority is not a gap in the commercial framework; it is a set of requirements that the federal framing actively does not address, because those requirements do not exist in federal environments.

The part that transfers is the inherited federal baseline: the complete harness architecture established in Volumes I and II, namely the Instruction Surface (H-1), Routing Enforcement (H-2), Persistent State (H-3), Verification Checkpoints (H-4), and Accountability Records (H-5); the eight federal security properties (P1 through P8, defined in Volumes I and II) in their entirety; all four evidence tiers (E-1 through E-4); and the knowledge base integrity controls (KB-1 through KB-4). That baseline crosses from the classified environment to the commercial one without modification. This volume then defines its own Lagrange governance properties (LP1 through LP14) on top of that inherited baseline. The first eight, LP1 through LP8, are the core governance-worthiness criteria, several of which restate a federal security property in governance terms. The harness is sector-independent. An AI system constrained by a properly configured harness in a classified environment is constrained by the same mechanism in a hospital, a bank, or an energy utility. The governance instruction surface changes. The constraint mechanism does not.

> **A note on method.** The transfer-versus-extension split in this chapter is my own framework classification, not a measured study. I classified each H, P, E,

> and KB criterion as transfer-without-modification or sector-extension across the five sectors in this volume. This is a structural judgment about the framework, not a count of deployments. No round-number ratio should be read as an empirical finding; the claim is qualitative, that the constraint mechanism is sector-independent while a defined set of commercial obligations is additive.

The extension that does not transfer is not a gap either. Four commercial requirements have no analog in the federal framing:

Consent management is the first extension. In federal environments, the authorization to process data at a given classification level is institutional and role-based: clearance plus need-to-know equals access. Commercial regulated environments require something federal governance has never addressed: affirmative individual consent that can be withdrawn. HIPAA authorizations, California Consumer Privacy Act (CCPA) opt-out rights, General Data Protection Regulation (GDPR) Article 7 consent requirements, and 42 CFR Part 2 re-disclosure restrictions all create a data governance obligation that the AI system must track at the individual record level, not just at the data classification tier level. The harness instruction surface must include consent state as an input to routing decisions. No AO instruction exists for this in the DoD framework. None is needed there. It is needed here.

Multi-jurisdictional compliance is the second. A federally chartered bank operating in 50 states, a hospital system with facilities in 12 states, or a multi-state insurer faces a compliance stack where the same AI action may be permissible under one state law and prohibited under another. California's CCPA has different consent requirements than Virginia's Consumer Data Protection Act (CDPA). Colorado's algorithmic accountability law for life insurance has no counterpart in most other states. GDPR's data minimization principle conflicts with certain US eDiscovery obligations. The harness architecture must route decisions based on the applicable jurisdiction of the data subject, not just the classification tier of the data. Federal environments face a version of this with different classification authorities; the multi-jurisdictional problem in commercial environments is structurally more complex because jurisdictions have conflicting, not merely additive, requirements.

AI fairness and bias obligations are the third. ECOA and Regulation B require that adverse action notices explain the specific reasons for credit denial. CFPB supervisory guidance requires that AI credit scoring models be explainable to consumers. The EU AI Act requires bias testing for high-risk AI systems used in credit, employment, education, and access to essential services. Colorado SB 21-169 requires life insurers to audit AI systems used in underwriting for algorithmic bias based on protected characteristics. None of these requirements exist in the DoD authorization framework, because government AI systems are not making consumer-facing decisions subject to disparate impact analysis. The harness must include bias monitoring as a continuous verification checkpoint (V-4 equivalent) with evidence of regular testing against protected class proxies.

Customer-facing transparency is the fourth. The CFPB's June 2023 circular on adverse action and AI requires that lenders explain AI-driven adverse actions in terms meaningful to the consumer, not just technically accurate. GDPR Article 22 gives individuals the right not to be subject to purely automated decisions, including the right to request human review. Emerging state AI transparency laws require disclosure when a consumer is interacting with an AI system. None of these apply in classified environments, where the "customer" is typically a cleared user who has accepted the terms of the authorization boundary. The harness must include consumer notification and explanation capabilities as first-class governance outputs.

These four extensions are the subject of four additional governance properties (LP9 through LP12) introduced in later chapters. The extension is not a simplification of the federal framework for less sophisticated commercial audiences. It is an addition for requirements that federal environments have never faced.

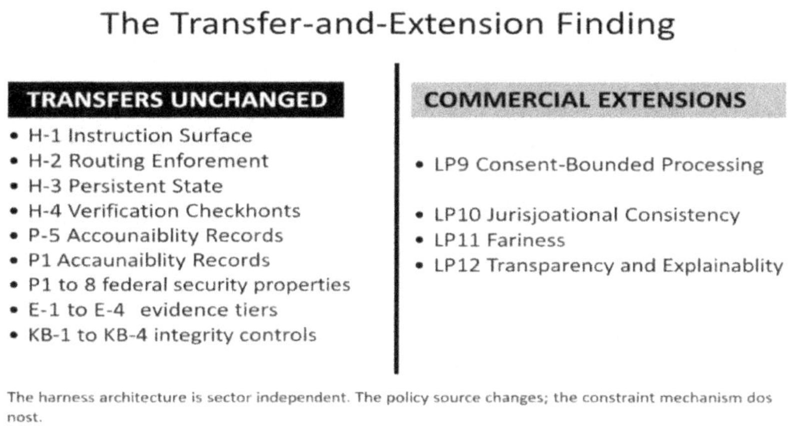

Figure 2: The transfer-and-extension finding. The inherited federal baseline transfers without modification: the complete harness (H-1 through H-5), the eight federal security properties (P1 through P8, defined in Volumes I and II), the four evidence tiers (E-1 through E-4), the four knowledge-base controls (KB-1 through KB-4), and the AGENTS.md instruction surface. This volume defines its Lagrange governance properties on that baseline: LP1 through LP8 are the core governance-worthiness criteria, several of which restate a federal security property in governance terms. A defined, sector-specific minority then requires extension: the four commercial governance properties (LP9 through LP12, consent, jurisdictional consistency,

fairness, transparency) plus sovereignty, which federal environments never had to address. This is an authored structural classification of what transfers, not a count of deployments and not a measured study; the figure carries no round-number percentage.

In certain dimensions, the commercial governance problem is harder.

One Harness or Many

Before any of the five sectors, one architectural decision sits above all of them, and most programs make it by accident. The doctrine does not settle it. The five components, Trust but Verify, the evidence tiers, and the governance properties are invariant; what they do not tell you is how many distinct governed envelopes to stand up. A bank running AI assistants for its developers and AI analysts for its credit officers can build one harness over both, or two, or a harness per business line. That is a question of topology, not of doctrine, and it has a right answer for each organization that the organization has to derive.

The two poles are clear. A single overarching harness puts one instruction surface and one routing layer over all AI use. It has one source of truth, one place to change policy, and the simplest possible audit. Its cost is that one set of rules must fit every workload: either the strictest rule binds everyone, taxing low-consequence work with controls written for high-consequence decisions, or the instruction surface fills with exceptions until no one can reason about what it permits. And the blast radius is the whole enterprise. One misconfiguration in a single shared harness is a misconfiguration everywhere at once.

Separate harnesses invert the trade. Each envelope is tuned to one workload, stays small enough to reason about, changes on its own cadence, and contains its own blast radius. The cost is seams between them, duplicated machinery, and the risk that two harnesses drift until the auditor cannot get one picture of the whole.

The reason separation is often correct is that different workloads have different natural gates. This is the distinction between a developer harness and an information-worker harness, and it is the cleanest example of the topology problem. Code has a pipeline and a merge request: its governance gate is at build time, before the artifact ships, enforced by protected branches, required approvals, and pipeline policy. The work of an information worker, a credit officer, an underwriter, a clinician's scribe, an analyst drafting a client memo, has no pipeline. Its artifacts are drafts, analyses, communications, and decisions, and its gate is at decision time, at the moment of consequential use, enforced at the data and connector layer, at the output boundary, and at a human checkpoint. A single harness either cannot enforce both gates well or becomes a leaky abstraction stretched across both. Two harnesses, one built around the build-time gate and one around the decision-time gate, each enforce precisely what their workload actually exposes.

There is a temptation to read two harnesses as exactly the split this book warns against. Chapter 4 shows that split governance creates split accountability: when the running system and the governance overlay have separate owners and separate records, accountability for any given event is distributed across parties with different incentives, and no one owns the whole. Two enforcement harnesses look like that split. They are not, provided one line holds: you may separate at the enforcement plane, and you must not separate at the accountability plane. Separation is allowed across H-1 through H-4, the instruction surface, the routing, the persistent state, and the checkpoints, because those are where workloads genuinely differ. Separation is forbidden across H-5 and the evidence tiers, the accountability records and E-1 through E-4, because that is where the auditor, the regulator, and the board need one answer.

That gives the design its shape: many enforcement harnesses, one accountability spine. The unifying layer that makes separation safe is not a dashboard bolted on at the end; it is three shared things built in from the start. One evidence plane, where every harness renders E-1 through E-4 into a single examiner-ready surface that one query can cross. One identity and accountability spine, where the same identities, tokens, and attribution records span every harness, so an action stays traceable end to end even when it crosses from one envelope to another. And one property vocabulary, the LP set, against which every harness is scored even though each enforces differently. Chapter 9 returns to this pattern through the International Space Station's Intergovernmental Agreement (IGA), where each module keeps local jurisdiction and its own operating discipline while one binding agreement and one shared safety and evidence regime sit over all of them. Many modules, one accountability frame, is the same answer arrived at from orbit.

The hazard to design against is the seam. An agent or a person that spans both domains, writing code that also touches regulated customer data, or drafting a decision that is then committed as a configuration change, is where two clean harnesses leak into each other. Every crossover flow needs an explicit owner: name which harness governs each step, or route the crossover through the stricter of the two. A seam left unnamed is the place the next incident comes from, because it is governed by neither envelope and shows up in both audits as someone else's responsibility.

The decision itself reduces to a set of questions. Where the answers diverge across workloads, separate; where they converge, unify. I ask these ten in order:

1. **Gate.** Do the workloads share a natural enforcement gate, or different ones, a build-time pipeline against a decision-time output boundary? Different gates favor separation.

2. **Artifact.** Is the governed artifact the same kind, versioned and reviewable, or different, an ephemeral draft, a sent message, a live decision? Different artifacts favor separation.
3. **Risk asymmetry.** Do the workloads carry materially different consequence or regulatory exposure, a code suggestion against an underwriting denial? High asymmetry favors separation, so the strict regime does not tax the low-risk work and the low-risk regime does not under-govern the high-risk work.
4. **Change cadence.** Do the policy and update cycles differ, developer tooling that changes weekly against clinical or actuarial governance that moves on regulatory cycles? Different cadence favors separation, so one team's churn cannot destabilize the other.
5. **Blast radius.** What is the cost of one misconfiguration reaching every workload at once? Low tolerance favors separation to contain it.
6. **Accountable owner.** Are there distinct accountable owners, engineering leadership against compliance, medical, or actuarial authority? Distinct owners favor separate enforcement, with the single spine still mandatory.
7. **Shared planes.** Do the workloads share identities, connectors, and knowledge bases? Heavy sharing pulls back toward one harness, or demands tight contracts between separate ones.
8. **Crossover.** Are there agents or users that span both domains? Many crossovers demand named seam ownership, or a single harness for those flows.
9. **Audit unification.** Can the organization render every harness's evidence into one examiner-ready picture today? If not, separation is premature; build the accountability layer first.
10. **Regulatory boundary.** Do different frameworks own different workloads, model-risk supervision over decisioning against general software controls over development? Different regulators favor separate enforcement mapped to each, with unified evidence.

The rule beneath the ten questions is short. Separate at the enforcement plane when gate, artifact, risk, cadence, or owner differ. Unify always at the accountability plane. For most regulated enterprises the answer is neither one harness nor a harness per team, but a small number of enforcement harnesses over a single accountability spine, with every seam between them named. Each sector chapter that follows assumes that topology and then asks what its own workloads, gates, and seams require.

Build, Buy, Rent, Renovate, or Wait

Once an organization knows how many harnesses it needs, the next decision is how to acquire each one. A shipowner who needs a hull does not have one option. They can build it in a yard, buy a finished vessel, charter one for the season, take a sound hull and refit it for the trade, or keep waiting for the right ship while the cargo sits on the dock. A governance harness has the same five paths, and they are not interchangeable.

Build. Construct the harness yourself, component by component. It gives the most exact fit and the most control, and it costs the most and ships the slowest. Build when the requirements are unusual enough that no product matches them, or when the components that must stay inside your authorization boundary, the Tier 1 governance components of Chapter 10, leave you building those parts regardless of what you buy for the rest.

Buy. Purchase a harness product and run it inside your own boundary, owning the deployment. It is faster than building, and because you operate it, you keep the control over the model and update cadence that Chapter 10 makes a Tier 2 requirement. The fit is whatever the product covers. Buy when a product already satisfies most of your governance properties and you can run it where your evidence and policy stay under your control.

Rent. Subscribe to a harness delivered as a hosted service. It is the fastest path and the lowest cost to start, and the trade is ownership: the vendor holds the runtime and sets the update cadence, and your governance now depends on a system you do not control. The supply-chain tiering of Chapter 10 sets the hard limit. Rent Tier 2 and Tier 3 components freely, the inference, the embeddings, the commodity infrastructure. Do not rent Tier 1. Renting the policy engine, the evidence collection, or the authorization decision layer places the auditor on the payroll of the party being audited, which is the one arrangement the architecture exists to prevent.

Renovate. Buy or adopt a base harness that covers most of the need, then extend it to close the rest. This is the most common real case, and the book's central finding predicts it: most of the harness transfers without modification, and a defined residual, the consent, jurisdiction, fairness, and sovereignty extension, must be built locally. Renovate when a base gets you most of the way and the missing part is exactly that commercial extension, or the seams between your harnesses. The hazard is that the extension is where the work, the cost, and the liability concentrate, so it has to be budgeted and owned rather than assumed, and the seam between the bought base and the built extension has to be named like any other seam.

Wait. Defer adoption. Waiting is sometimes correct, when a category is immature or a regulation is about to land that will define the requirement. But waiting is never free. The coffin-ship chapter prices the cost: while the organization waits, shadow adoption proceeds without governance, and the gap between the running systems and the controls accumulates. Wait deliberately, with a named trigger that ends the wait, not by default because the decision was hard.

The paths combine. Most regulated enterprises buy or renovate a base for the transferable majority, build the residual extension themselves, rent only the Tier 2 and Tier 3 capabilities, and never rent Tier 1. The decision is made per harness, not once for the enterprise, which is why it follows the topology question rather than preceding it.

Why regulated industries are making the same mistakes

Chapter 4 addresses this directly. The preview belongs here because the three-body physics predict it.

When three forces pull in incompatible directions, organizations under competitive pressure take the path of least resistance. In AI governance, that path is procurement followed by documentation: buy the AI product, deploy it under competitive pressure, write the governance documentation afterward. The documentation describes a system that complies. The running system may not. That gap, between the documented system and the running one, is where compliance risk concentrates.

This failure pattern is not sector-specific. Financial services AI deployments authorized against SR 11-7 documentation operate outside the bounds of that documentation. Healthcare AI deployments describe data flows in HIPAA risk assessments that the deployed system does not follow. Energy AI deployments exclude AI components from NERC CIP asset inventories because procurement happened after the last CIP-002 categorization review. Insurance AI deployments carry actuarial reviews completed on model versions the vendor has since updated.

The common element is not industry, size, or sophistication. It is the same architectural error: treating governance as a documentation exercise rather than a continuous verification posture. Not documenting compliance. Producing compliance. The harness solution is identical in all four sectors: build verification into the architecture so that compliance evidence is a byproduct of system operation. Chapter 4 shows what happens when that principle is ignored. Chapters 5 through 15 show what happens when it is applied.

The physics are the same. The vocabulary is different. The architecture is universal.

> **KEY FINDING**
>
> The three-body instability does not require a classified environment. It requires three conditions: valuable data that cannot be freely accessed, a

regulatory framework that demands evidence of control implementation, and competitive pressure to move faster than the governance process allows. Every regulated industry has all three conditions. The same architecture that stabilizes AI governance in classified environments stabilizes it in financial services, healthcare, energy, and insurance. Most of the harness transfers verbatim; that part is the constraint mechanism itself. The sector-specific minority that does not is not a gap; it is the extension required for consent, jurisdiction, fairness, and transparency obligations that federal governance has never had to address.

Chapter 3: The Coffin Ships

Part I: The Coffee House Problem

> "The sailors are murdered by degrees."
>
> *Samuel Plimsoll, addressing Parliament, 1873*

In the 1860s and 1870s, British merchant ships were sinking at a rate that horrified the seafaring public. The vessels were not structurally unsound, the weather was not exceptional, and the navigators were not incompetent. The ships were overloaded. Deliberately, methodically, and profitably.

Shipowners had discovered an elegant calculation. A vessel loaded beyond its safe capacity could still reach port. If the weather held and the seas were calm, the overloaded ship delivered more cargo and generated more revenue than a properly loaded competitor. If the weather did not hold, the ship sank. The cargo was insured. The ship was insured. The owner collected on the policy and ordered a replacement.

The only uninsured variable was the crew.

This is what Samuel Plimsoll called a coffin ship: a vessel made lethal not by defect but by incentive. The shipowner bore no personal cost from sinking. He bore a direct competitive cost from leaving cargo on the dock. The rational actor, responding to the incentive structure in front of him, loaded the ship to maximum capacity and let the arithmetic of weather and probability sort out the consequences.

Plimsoll's campaign for the Merchant Shipping Act of 1876 was not primarily an engineering argument. The mathematics of buoyancy and displacement were well understood. Naval architects could calculate safe loading limits with precision. The knowledge existed. What did not exist was a mechanism to make the knowledge visible, enforceable, and impossible to ignore at the moment of loading.

The solution was a line painted on the hull.

The Plimsoll mark, or load line, is a circle with a horizontal line through its center, positioned on the ship's hull at the precise waterline corresponding to the maximum safe load. If the ship is loaded to the point where the Plimsoll mark disappears below the water, it is overloaded. This judgment requires no engineering expertise, no calculation, no audit. The assessment is binary: the mark is visible or it is not. The crew can see it, so can the harbor master, and so can the insurer's surveyor. The governance state of the vessel is legible to every observer before the ship leaves port.

Plimsoll's achievement was not technical. It was architectural. He made invisible risk visible, continuous judgment binary, and private calculation public, and in doing so he removed the rational basis for overloading. An owner who deliberately submerged the load line was now, visibly and demonstrably, sending a ship out to die.

The AI industry is building coffin ships

Organizations deploying AI systems in 2024 and 2025 are making the same calculation that British shipowners made in 1865. Not the same arithmetic. The same incentive structure.

An organization that deploys an AI system without adequate governance architecture can reach port. In favorable conditions, an ungoverned AI deployment delivers capability faster than a governed one. It ships features in weeks instead of months; it automates decisions at scale before a compliance program has been built to oversee them. Claims processed, credit applications evaluated, clinical summaries drafted, customer inquiries answered, all without the friction of verification gates, audit logging, policy enforcement, or human review checkpoints.

If the conditions hold, the organization has competitive advantage. If the conditions do not hold, the ship sinks: regulatory enforcement action, class-action lawsuit, front-page breach announcement, Congressional hearing. The organization pays the settlement, updates the policy documentation, issues a statement of remediation, and continues operations. The only uninsured variable is the customer whose claim was wrongly denied, whose credit application was unfairly evaluated, whose clinical record was hallucinated into the wrong chart.

Three cases from the past three years illustrate the pattern. In none of them did the AI break. In all of them, governance was simply absent. Chapter 2 introduced the first of these, the UnitedHealthcare coverage-denial case, as a short preview of the coffin-ship calculation before the sector-by-sector translation; this chapter tells that case in full, alongside two others that show the same structure in different industries.

UnitedHealthcare: optimizing for the wrong property

UnitedHealthcare's Medicare Advantage division deployed an AI system to assist in the review of claims for post-acute care coverage, specifically decisions about when Medicare beneficiaries should be discharged from skilled nursing facilities. The system, built on nH Predict technology, assessed each patient's condition and generated a recommendation about appropriate care duration.

The system was not technically defective. It produced predictions, generated recommendations, and processed data at scale. By the operational metrics that a technology deployment is typically evaluated against, it functioned.

What it did not have was a governance architecture that constrained it to optimize for clinical accuracy rather than denial rate.

The evidence that surfaced in subsequent litigation and investigative reporting was specific. According to legal filings and reporting by STAT News, the system had a denial rate that far exceeded the industry average for similar coverage decisions. When beneficiaries or their families appealed those denials, they were reversed in approximately 90% of cases. An appeal reversal rate of that magnitude is not a signal that the appeals process is too permissive. It is a signal that the underlying decisions were systematically wrong.

The AI system was doing exactly what an ungoverned system does: it was optimizing for the signal it had been given. If the training data or feedback loops associated with the deployment rewarded denial efficiency, the system would produce denials efficiently. The governance property that should have constrained this, a property requiring that AI-assisted clinical decisions demonstrate clinical accuracy against an independent benchmark before being used at scale, did not exist in the architecture.

UnitedHealthcare now faces class-action litigation and substantial congressional scrutiny. The legal exposure is structurally unpredictable because the harm was systematic and the affected population is large, elderly, and unsympathetically situated. The cost of the governance architecture that would have prevented this outcome is, by any reasonable estimate, a small fraction of the liability now in motion.

The AI was not the failure. What failed was the absence of any governance property constraining it to the right objective.

Air Canada: the chatbot with no policy verification gate

In November 2022, a customer named Jake Moffatt contacted Air Canada's website chatbot to inquire about bereavement fares in connection with a family death. The chatbot provided specific, detailed information about Air Canada's bereavement fare policy: it told Mr. Moffatt that he could purchase a full-price ticket, travel, and then apply for a reduced bereavement fare retroactively within 90 days of travel.

That policy did not exist. Air Canada's actual bereavement fare policy required the discounted fare to be requested before travel, not after. The chatbot had generated a plausible-sounding but entirely fictitious policy.

Mr. Moffatt followed the chatbot's instructions. He purchased a full-price ticket. He traveled. He filed for the retroactive discount. Air Canada refused, citing its actual policy. Mr. Moffatt filed a complaint with the Civil Resolution Tribunal of British Columbia.

Air Canada's legal defense is the detail that makes this case a governance case study rather than a simple customer service failure. Air Canada argued that the chatbot was "a separate legal entity" and that the airline was not responsible for the information the chatbot provided. The Tribunal rejected this argument comprehensively. Member Licence held that Air Canada could not "absolve itself of responsibility by asserting its chatbot is a separate legal entity." The airline was held liable for the chatbot's output. Air Canada was ordered to pay Mr. Moffatt the fare difference, plus pre-judgment interest and tribunal fees.

The chatbot was not defective in any ordinary technical sense. It produced fluent, confident, contextually appropriate text. A user interacting with it had every reason to believe it was providing accurate information. The system did what language models do: it generated a response that fit the conversational pattern and sounded authoritative.

What the architecture did not include was a verification gate that confirmed any chatbot output against the actual policy documents before delivering it to a customer. The governance property that should have been present, a property requiring that customer-facing AI outputs be grounded in verified policy documentation, was absent.

The failure cost Air Canada less than a thousand dollars in this specific case. The reputational cost of the story, which circulated widely as a cautionary example in enterprise AI discussions throughout 2023 and 2024, is not measurable. The precedent it established, that airlines are responsible for what their AI systems tell customers, has broader implications for every organization that deploys a customer-facing language model.

Once more, the AI was not the problem. The missing piece was a verification gate between AI generation and customer delivery.

McDonald's: the drive-thru with no constraint boundary

McDonald's partnered with IBM to deploy an AI-powered ordering system at drive-through lanes across approximately 100 U.S. locations beginning in 2021. The system was designed to take orders autonomously, reducing labor costs and improving throughput. It operated in an environment with real financial transactions and real food preparation instructions flowing directly from AI output to kitchen operations.

Videos circulated on social media beginning in 2023 showing interactions that had escaped the system's intended behavior envelope. In one widely shared clip, a customer attempting to order ice cream received a continuously expanding order that added items the customer had not requested. In another, a customer's order grew to include 260 Chicken McNuggets. The system, operating without a constraint boundary that limited the scope of autonomous ordering, had no mechanism to detect or halt this runaway behavior.

McDonald's ended the IBM partnership in June 2024, announcing it would cease using the AI ordering technology at the pilot locations.

The system was not malicious. Language models in dialogue systems can enter feedback loops where conversational ambiguity, misrecognition, or confirmation patterns cause the model to keep adding items to satisfy what it interprets as an incomplete order. This is a known failure mode for AI systems operating in unconstrained dialogue with open-ended action authority.

The governance architecture that would have prevented this failure is straightforward: a constraint boundary limiting the scope of autonomous ordering decisions. The system should have had a property requiring that orders above a defined threshold, by quantity or by value, require explicit human confirmation before proceeding. The property does not require complex engineering. It requires someone to ask, during architecture design, what happens when this system generates an output outside the expected range, and then build a boundary that handles that case.

That question was not asked, or the answer was not implemented, before the system was deployed to 100 locations and began taking real orders from real customers. The line was missing.

The pattern holds for the third time. The AI worked as designed; what was missing was a governance architecture constraining the scope of autonomous action.

The moral hazard argument

All three cases share a structural feature that the coffin ship analogy makes precise: the organization that deployed the AI did not bear the primary cost of the governance failure at the moment the system was built.

UnitedHealthcare could have paid the cost of a governance architecture constraining AI-assisted decisions to clinical accuracy standards at deployment time: engineering effort, slower scaling, the operational overhead of verification gates. Medicare beneficiaries paid instead when they lost coverage appeals; UnitedHealthcare is paying now in litigation and regulatory scrutiny.

Air Canada could have paid the cost of a policy verification gate at deployment time: engineering effort to ground AI outputs against authoritative policy documents. A grieving customer paid instead when he followed the chatbot's instructions in good faith.

McDonald's could have paid the cost of a constraint boundary at deployment time: engineering effort to define scope limits and build escalation paths. The company paid in reputational damage and eventual program cancellation.

This is the moral hazard of ungoverned AI deployment: the costs are time-shifted and party-shifted. The organization pays governance costs now and captures competitive benefits now. The governance failure costs arrive later, often much later, and fall first on the people affected by the AI's errors before they fall on the organization that deployed the system.

Organizations that have internalized this structure sometimes make an explicit calculation: the cost of governance exceeds the expected cost of failure, adjusted for probability and timing. Build fast, accept the risk, absorb the failures.

This calculation is wrong for three reasons specific to the current regulatory and legal environment.

Regulatory penalties are increasing on a steep trajectory, and the trajectory is steepening. The EU AI Act, which entered into force on August 1, 2024 with a staggered implementation timeline running through August 2027 (high-risk system obligations from August 2, 2026), establishes penalties of up to 7% of global annual revenue for violations involving prohibited AI practices, and up to 3% for failures to comply with transparency, documentation, and risk management requirements.[^eu-ceiling] As an illustrative worked example tied to that statutory ceiling, consider a global financial institution with $50 billion in annual revenue: the 3% ceiling computes to $1.5 billion. That figure is an arithmetic illustration of the ceiling, not a claimed or assessed penalty against any institution. The Office of the Comptroller of the Currency (OCC)'s increasing willingness to bring cases involving systematic Health Insurance Portability and Accountability Act (HIPAA) failures rather than individual incidents has raised the expected cost of an enforcement action for inadequate AI oversight. State attorneys general in the United States are developing AI enforcement capacity that tracks the General Data Protection Regulation (GDPR) enforcement trajectory in Europe. The expected cost column in the governance calculation is no longer a stable number. It is rising.

Reputational damage compounds differently; it is neither insurable nor reversible. The McDonald's AI ordering story circulated for more than a year before the partnership ended. Every enterprise AI procurement conversation that referenced it required an explanation. The UnitedHealthcare coverage denial story became a touchstone in discussions about AI in healthcare, one that continues to shape regulatory attention and public trust. These costs do not appear in actuarial tables. They cannot be hedged. An organization that absorbs a high-visibility AI governance failure inherits a narrative that follows it into every subsequent customer, regulator, and legislative interaction.

Class-action exposure is the third reason, and the hardest to model. The actuarial calculation for expected failure cost requires estimating the likely penalty and the likely probability of enforcement. Class-action litigation introduces a variable that is difficult to bound: the number of affected parties and the accumulation of individual harms into aggregate exposure. UnitedHealthcare's exposure in pending litigation is not a function of a single regulator's enforcement discretion; it is a function of how many Medicare beneficiaries were affected, across how many coverage decisions, over how many years, and whether the class can be certified at the scale that the systematic denial rate implies. Predicting that number at deployment time is not possible with the precision the expected-cost calculation requires.

The moral hazard calculation assumes the organization is reasoning correctly about expected costs. It is not. It is underestimating at least two of the three cost variables and ignoring the correlation between them.

The shipowner's arithmetic was wrong in 1865. It is wrong now.

The Plimsoll solution

Plimsoll's solution was not technical. The engineering community already knew how to calculate safe loading limits. The solution was governance architecture: a mechanism that made the safety state visible, binary, and enforceable at the moment of action.

The Plimsoll mark has four properties that account for its effectiveness across 150 years of global maritime governance.

THE PLIMSOLL PRINCIPLE

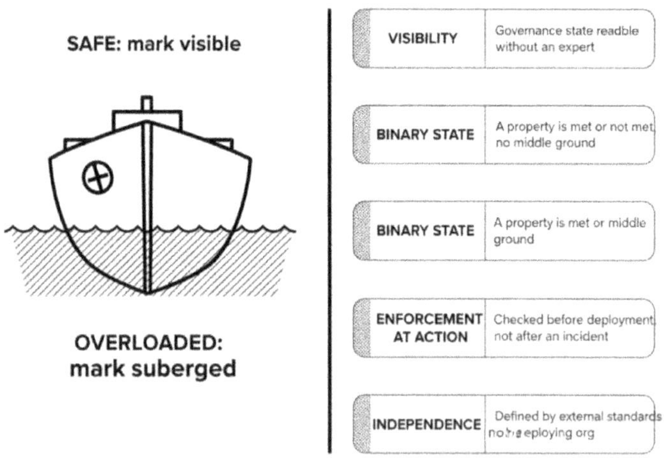

Figure 3: The Plimsoll mark as a governance analogy: a load line on the hull that is visible to every observer, binary in state (above or below the waterline), checked at the moment of loading, and set by an independent classification society rather than the owner. The four properties below map each maritime feature to its AI-governance counterpart.

First, visibility. The load line sits on the exterior of the hull, in a prominent position, in a color that contrasts with the hull. Every party with a stake in the vessel's safety, the crew, the harbor master, the insurer, the port authority, can observe the governance state of the vessel without requesting a report or commissioning an audit.

Second, binary state. The vessel is either within safe loading limits or it is not. No graduated assessment; no judgment about whether a particular degree of overloading is acceptable in current sea conditions; no discretion for the owner to exercise. The line is visible or it is not.

Third, enforcement at the moment of action. The load line is checked before the vessel leaves port, not after it sinks. The governance check is integrated into the operating sequence, not retrospectively applied to incident reports.

Fourth, independence from the owner's preferences. The load line is set by an independent classification society, not by the shipowner. The owner cannot negotiate the placement of the mark upward to accommodate more cargo. The governance standard is external to the party with the strongest incentive to circumvent it.

The governance properties defined in this book, LP1 through LP14, are designed to function as Plimsoll lines for AI systems in regulated industries. Each property is a visible indicator of whether the AI system is operating within a safe governance boundary. Each property produces a binary state: the governance condition is met or it is not. Each property is verified before deployment and continuously thereafter, not retrospectively after an enforcement action. And the properties are defined against the requirements of external regulatory frameworks, not against the preferences of the organizations deploying the AI.

The UnitedHealthcare system needed a property requiring that AI-assisted clinical coverage decisions demonstrate accuracy against an independent benchmark before being used at scale. The Air Canada chatbot needed a property requiring that customer-facing AI outputs be grounded in verified policy documentation before delivery. The McDonald's system needed a property constraining the scope of autonomous action to defined operational boundaries.

None of those properties is technically exotic. Each one is a Plimsoll line: a specification of the boundary condition that separates safe operation from unsafe operation, made visible, made binary, enforced at the moment of operation.

Plimsoll asked the maritime industry in 1873 where the line goes. This book asks the same question of AI governance in regulated industries.

Contemporary software deployment has arrived at the same structural insight through a different path. Platform engineering teams building cloud infrastructure discovered that routing traffic to a newly deployed service before that service passes a health check produces the same failure pattern Plimsoll observed: the system looks deployed, it accepts load, and it fails under conditions the operator did not verify before routing began. The response was health-check-gated deployment: a service receives no traffic until it satisfies a defined set of readiness conditions, checked continuously, with automatic traffic cutoff if those conditions degrade. The gate is external to the service, cannot be disabled by the service itself, and produces a binary state: ready or not ready. Dokploy, an open-source deployment platform that implements this pattern, codifies the governance instruction as configuration data rather than as a procedural checklist. The health conditions are written once, in a file that any authorized party can read; the enforcement is automatic; no human judgment about whether a given degradation is acceptable is inserted into the deployment path. This is the Plimsoll mark implemented in software. The governance state is visible, binary, enforced at the moment of action, and defined by a standard external to the system being governed. The P-properties in this book apply the same structure to AI systems in regulated environments: each property specifies a condition the system must satisfy before it operates under live regulatory exposure, the condition is checked continuously, and no human discretion about whether a partial compliance state is acceptable is admitted into the enforcement path.

Where the lines go

Not a universal answer. A Plimsoll line is calibrated to the vessel, to the trade route, to the cargo type, and to the applicable load line convention. A bulk carrier operating in tropical waters has different load line markings than a container ship in the North Atlantic; the governance properties that constitute safe operation in healthcare AI differ from those in financial services AI, which differ from those in energy infrastructure AI.

The structure is common. The coordinates differ.

Part II of this book maps the governance properties that transfer directly from The Harness Doctrine, the Plimsoll lines that apply to every regulated AI deployment regardless of sector. Part III defines the sector-specific coordinates, the load line markings calibrated to the specific regulatory waters of financial services, healthcare, energy, insurance, and telecommunications. Parts IV and V address the operational challenges of maintaining those markings across multi-framework compliance environments and what happens when an organization sails with the line submerged.

The rest of this book defines where the Plimsoll lines go.

[^eu-ceiling]: Penalty ceilings per Regulation (EU) 2024/1689 (the EU AI Act), Article 99. The 7% and 3% figures are statutory maximum ceilings expressed as a percentage of total worldwide annual turnover; actual penalties are set case by case. Dates current as of publication, subject to revision.

Chapter 4: Why Regulated Industries Are Making the Same Mistakes

Part I: The Coffee House Problem

The defense community made these errors first. Between 2022 and 2024, federal programs banned AI coding assistants, watched employees adopt them anyway through personal accounts and unsanctioned browser extensions, experienced compliance findings when the shadow usage surfaced during audits, and then bolted governance overlays onto running systems that were never designed to be governed. The Harness Doctrine documented that sequence in detail; it named the failure modes; it mapped the architectural remedies.

Commercial regulated industries are repeating the sequence. Banks, hospital systems, energy utilities, and insurance carriers are moving through the same four stages in the same order, hitting the same structural errors at the same transition points. The timing differs from one sector to the next, but the trajectory does not. The defense community is approximately two years ahead.

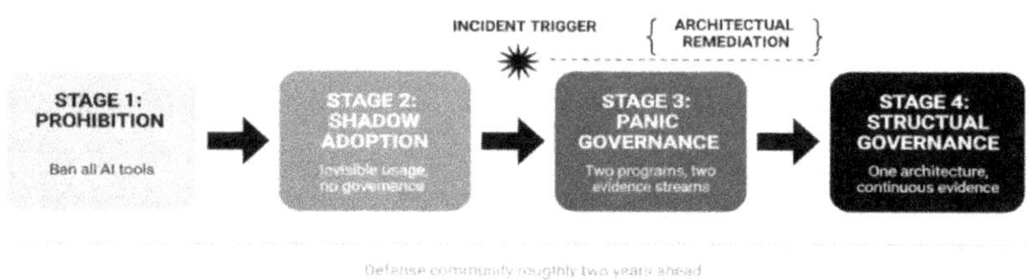

Figure 4: The four-stage progression every regulated industry follows from Prohibition through Structural Governance, with the incident trigger and architectural break points that separate the stages.

This chapter maps the progression. Understanding where your organization sits in this sequence is prerequisite knowledge for everything that follows in this volume, because the architectural prescription changes depending on which stage you are in. An organization in Stage 2 needs a different intervention than an organization in Stage 3. An organization that believes it is in Stage 4 but has not tested that belief under audit pressure may still be in Stage 3.

The progression every regulated industry follows

The four stages are not a framework invented to organize this chapter. They are an observed pattern, drawn from the documented experience of the defense community and the current deployment patterns visible in commercial regulated industries. Every organization follows this sequence unless it has been specifically designed to skip stages, and most organizations have not been so designed.

Stage 1: Prohibition

The organization's initial response to AI is a ban. No AI tools. No ChatGPT, no Copilot, no AI-assisted workflow. The rationale is familiar: the tools are unvetted, the risks are unclear, the regulatory implications are unresolved, and the organization's fiduciary or compliance obligations demand caution.

The prohibition is not wrong as a risk response. It is wrong as a governance strategy. A ban addresses the organization's liability exposure but not the actual risk profile. That distinction matters enormously.

When an organization bans AI tools, it controls its own authorized activity. It does not control employee behavior: what tools employees use on personal devices, through personal accounts, in home office environments technically outside the organization's network perimeter, or through browser extensions that route requests to AI providers without touching the corporate proxy. The prohibition eliminates visibility into AI usage. It does not eliminate the usage.

The maritime parallel is precise, not poetic. In the transition from sail to steam, conservative shipping companies refused to adopt steam-powered vessels because steam engines were poorly understood, maintenance requirements were novel, and several early steam vessels had experienced catastrophic boiler failures. Rational on paper. In practice, it meant that sailors on those ships kept working with the only tools available, which sometimes included jerry-rigged steam equipment salvaged from other sources, operated without proper training and outside any maintenance regime. The prohibition did not produce safety. It produced unsupervised improvisation.

In the AI context, the prohibition produces shadow adoption. The two stages are causally linked.

Stage 2: Shadow adoption

Shadow adoption is not a failure of employee ethics. It is a predictable consequence of organizational prohibition against a tool that demonstrably improves individual productivity.

When an employee discovers that ChatGPT or a similar tool cuts document drafting time by 40%, reduces research time by 60%, or makes an otherwise tedious workflow manageable, the incentive structure is clear. The organization has said the tool is banned; the employee has felt how useful it is. This is not theft. It is someone trying to do the job better. The rational response, absent a sanctioned alternative, is to use the tool through channels the organization cannot see.

The channels are numerous and well-established. Personal Gmail accounts access AI assistants; browser extensions process page content through AI APIs without touching corporate infrastructure; personal mobile devices run AI applications; home networks avoid corporate proxies entirely. Contractors use their own tooling. Vendors integrate AI into deliverables without disclosing the integration. Every channel creates AI exposure without AI governance.

The statistics confirm it. Surveys across multiple years and industries show that 87% of IT leaders cite AI vulnerabilities as their fastest-growing risk category; the same surveys show that only 64% of organizations conduct pre-deployment AI assessments. That 23-point gap between risk recognition and risk assessment is not explained by resource constraints or organizational inertia alone. Shadow adoption creates exposure faster than governance programs can organize to respond. The exposure precedes the assessment.

In regulated industries, the shadow adoption exposure has a specific texture. A healthcare employee using a personal AI account to draft a clinical summary may be processing protected health information outside any Business Associate Agreement. A financial services analyst using an AI tool to summarize earnings calls may be inputting material non-public information into a system with no regulatory oversight. An energy utility employee using AI to draft a grid operations procedure may be processing Bulk Electric System Critical Infrastructure information through a channel with no North American Electric Reliability Corporation (NERC) Critical Infrastructure Protection (CIP) classification or access control. The exposure is not hypothetical. It is occurring, and it is occurring invisibly.

The organization in Stage 2 has an AI risk profile. It does not have an AI governance program. The gap between them is the Stage 2 problem, and it is not solved by adding governance. It is solved by understanding what governance must cover before designing it.

Stage 3: Panic governance

The Stage 2 to Stage 3 transition is triggered by an incident. The taxonomy is predictable: a data leak discovered during an audit, a compliance finding related to AI-processed customer data, a media story about an organization in the same sector that experienced an AI-related breach or regulatory enforcement action, a vendor disclosure that AI was integrated into a delivered product without explicit disclosure.

The incident produces urgency. Urgency produces a governance program. The governance program is bolted onto the existing deployment.

This is the structural failure point. Governance built in response to an incident is by definition built after the fact. The deployment it governs was not designed to be governed; its data flows were not designed to produce compliance evidence; its authorization controls were not designed to satisfy regulatory audit trails; its access patterns were not designed to enforce separation between regulated and unregulated data. A production system generating revenue that has never had a governance layer.

The organization now has two programs. The production AI system operates according to its original architecture; the governance overlay operates according to the new compliance requirements. Their priorities diverge: uptime and feature delivery for the production system, evidence collection and control attestation for the governance overlay. They read from different data sources, because the overlay is pulling from system logs that were never designed as compliance evidence. They run on different update cycles, the production system on a development schedule and the overlay on a compliance schedule. And they answer to different owners, engineering for one and compliance for the other.

Two programs. One organization. No alignment between them.

The gap between these two programs is where failures concentrate.

The Harness Doctrine formalizes this observation as the organizational bifurcation warning. Two programs, two evidence streams, and gaps at the boundaries: that is not a staffing problem, and it is not solved by adding headcount to the compliance team or by scheduling more frequent cross-team meetings. The cause sits in the architecture, not the people. The production system and the governance overlay are structurally misaligned, and the misalignment produces predictable failure modes regardless of how skilled the people operating each system are.

The failure mode that surfaces most often under audit pressure is evidence quality degradation. The governance overlay produces evidence by reading from the production system's logs and translating that signal into compliance-formatted records. But those logs were designed for debugging and operations monitoring, not for compliance evidence. They capture what the system did; they do not reliably capture whether the system did it within authorized constraints. They record the access event without recording the governance state at the moment of access. An auditor asking whether the AI system accessed protected health information (PHI) in compliance with the minimum necessary principle cannot be answered from operational logs. That answer requires a governance-aware evidence architecture.

Panic governance produces logs. It does not produce evidence.

Coverage gaps are the second failure mode, and they are structural rather than operational. Shadow adoption predates the governance overlay. The overlay covers the AI deployment the organization knows about; it does not cover the shadow usage that predated the incident. An organization that discovered its Stage 2 exposure through a compliance finding has a governance overlay covering whatever that finding identified. It does not cover the remaining shadow usage that the finding did not surface. Two evidence streams, the governed deployment and the ungoverned shadow usage, exist simultaneously; the governance overlay cannot close the gap because it does not have visibility into the shadow exposure it never detected.

This is why Stage 3 feels like sufficient governance until it is tested. The governance overlay produces artifacts; the artifacts satisfy an assessor reviewing the documented deployment. The assessor is not reviewing the shadow usage. The System and Organization Controls (SOC) 2 report attests to controls the organization implemented on the systems it disclosed. It does not attest to the systems it failed to disclose because it did not know about them.

The Stage 3 organization is not dishonest. It is architecturally incomplete.

Stage 4: Structural governance

Structural governance means the governance architecture was designed into the AI deployment before the deployment went live, or was explicitly rebuilt from the ground up when the organization recognized that panic governance was insufficient. There are no longer two programs; there is one. Compliance evidence is a byproduct of system operation, not a separate collection exercise. The governance layer and the production system share the same data architecture because they were designed together.

The harness architecture is the mechanism for structural governance. The five harness components, instruction surface, routing enforcement, persistent state, verification checkpoints, and accountability records, are the components that unify the production system and the governance program into a single architecture. When the governance instructions are codified in the instruction surface (H-1, AGENTS.md), they govern every execution because the routing enforcement layer (H-2) ensures no request bypasses the instruction surface. The governance evidence is continuous because the verification checkpoints (H-4) are built into the data flow, not reading from after-the-fact logs. The accountability records (H-5) are generated as a byproduct of operation, not reconstructed from operational logs after an audit request.

Most organizations never reach Stage 4, and the reason is not that Stage 4 is architecturally exotic. It is that Stage 3 feels sufficient. The compliance team has artifacts; the auditor accepted the SOC 2 report; the NERC CIP audit did not produce findings; the Health Insurance Portability and Accountability Act (HIPAA) risk assessment was filed. From inside the organization, Stage 3 looks like governance. It keeps looking like governance until the audit asks a question that the evidence architecture cannot answer, or until an incident reveals that the shadow usage was never covered, or until the regulatory environment shifts to require continuous evidence rather than annual attestation.

The shift is already occurring. The EU AI Act's continuous monitoring requirements; the Office of the Comptroller of the Currency (OCC)'s model risk management guidance expecting ongoing validation of AI model behavior, not just initial validation; NERC CIP CIP-013 supply chain risk management requirements extending governance obligations to the AI vendor chain; the Securities and Exchange Commission (SEC)'s cybersecurity disclosure rules requiring material incident disclosure, creating board-level consequences for governance gaps that previously carried only operational risk. Each of these regulatory developments pushes the evidence requirement from documentation-based compliance toward continuous architectural compliance.

Stage 3 does not satisfy that requirement. Stage 4 does.

Agent framework engineering has encountered this same transition problem and arrived at a structurally identical solution. Hermes, a research agent architecture designed for long-horizon tasks, uses a threshold-gated capability model: an agent does not acquire new tools or expand its operational scope until it has demonstrated sufficient competence on its existing scope, with competence measured against a defined threshold rather than a subjective assessment. When a new capability is needed, the system does not simply enable it; it requires the agent to satisfy prerequisite conditions first, then freezes the instruction state as a prefix-cached snapshot before expanding scope. This means the governance state at any point is explicit, auditable, and defined by prior configuration rather than by the current session's context. The parallel to the harness architecture is direct. The instruction surface (H-1) functions as the frozen prefix state: it captures the authorized scope before execution begins. The verification checkpoints (H-4) function as the threshold gate: they confirm that governance conditions are satisfied before the system proceeds to the next operational phase. An organization at Stage 4 has built this structure into its AI deployment: capability expands through a defined gate, the gate criteria are specified in the governance documentation, and no operational pressure during execution can retroactively lower the threshold.

The bifurcation warning in practice

The organizational bifurcation problem deserves additional precision, because it is the most frequently misdiagnosed failure mode in commercial AI governance programs.

When a compliance officer looks at a Stage 3 governance program, the diagnosis they offer most often is resourcing. The governance overlay is not keeping pace with the production system because the compliance team does not have enough people, enough tools, or enough access. The prescription is investment: hire more compliance staff, purchase a governance platform, negotiate better access from the engineering team.

This prescription addresses symptoms without addressing cause. The compliance team cannot keep pace with the production system because they are working from a different data architecture. More people working from inadequate data do not produce adequate evidence. Neither do better tools running on the same inadequate data, and neither does negotiated access to operational logs that were never designed as compliance evidence in the first place. The bottleneck is architectural, not a matter of resourcing.

The correct diagnosis is that the production system and the governance overlay have misaligned evidence architectures. The prescription is architectural remediation: redesign the data flow so that compliance evidence is generated as a byproduct of operation, not extracted as an afterthought from operational logs. More expensive than hiring additional compliance staff. Also the only intervention that actually closes the gap.

The bifurcation warning from the earlier volumes of The Harness Doctrine states the principle directly: split governance creates split accountability. When the production system and the governance overlay have separate owners, separate data sources, and separate update cycles, accountability for any given governance event is distributed across two organizations with different incentives. Engineering ships features and maintains uptime. Compliance produces artifacts that satisfy the auditor. Neither incentive aligns with the actual goal: ensuring that the AI system operates within its authorized constraints in real time.

AI systems research has surfaced a complementary diagnostic technique: structured adversarial debate between role-differentiated agents. TradingAgents, a multi-agent framework for financial decision-making, assigns agents to explicitly opposing roles, one optimizing for trade execution speed and the other for risk control, then forces both to justify their positions against the same shared evidence before a synthesis decision is made. The design insight is that neither role has access to the full picture; the correct outcome requires surfacing the tension explicitly rather than allowing one perspective to dominate by default. The decision reflection log, which records each agent's reasoning alongside the synthesis outcome, becomes an audit artifact by design: it shows not just what decision was made but what pressures were weighed and how the conflict was resolved. A Stage 3 governance program exhibits exactly the opposite structure: engineering and compliance operate from different evidence bases, and the conflict between them is resolved informally, invisibly, and in favor of whichever team has more organizational authority at that moment. Surfacing that conflict explicitly, through an adversarial review process in which both the operational case and the governance case are required to produce evidence-based justifications before a release decision proceeds, is the Stage 4 analog of TradingAgents' opposing-role architecture. The evidence architecture that makes this possible is the shared foundation Stage 4 requires.

The transition zone: frameworks without coverage

The period between Stage 2 and Stage 4 is characterized by a specific regulatory coverage gap that compounds the organizational governance gap.

Existing compliance frameworks (SOC 2, Payment Card Industry Data Security Standard (PCI-DSS), HIPAA, NERC CIP) were designed before large language model AI was a deployment consideration. They contain control requirements for data access, encryption, audit trails, and access management that apply to AI deployments; they contain no specific guidance for AI-specific risk categories: context window exposure, prompt injection, model drift, agentic action scope, retrieval augmentation integrity, or AI supply chain accountability.

Emerging AI-specific frameworks (the National Institute of Standards and Technology (NIST) AI Risk Management Framework, the EU AI Act, International Organization for Standardization (ISO)/IEC 42001) were designed for AI-specific risk categories but do not integrate with existing compliance infrastructure. An organization implementing the NIST AI RMF governance functions (Govern, Map, Measure, Manage) maintains a separate evidence stream that does not feed into its SOC 2 evidence collection, its HIPAA risk assessment process, or its NERC CIP audit preparation. Two frameworks; no shared evidence architecture.

The result is that a regulated organization deploying AI must simultaneously satisfy its existing compliance framework, which covers general IT controls but not AI-specific controls, and an emerging AI governance framework, which covers AI-specific controls but not general IT compliance evidence. These two frameworks do not share an evidence architecture. Producing the artifacts for both requires two separate evidence collection programs.

This is the regulatory analog of the organizational bifurcation warning. Two evidence programs, two audit targets, and gaps between them where compliance obligations fall through.

The maritime parallel holds here too. When steam-powered vessels entered commercial service in the mid-nineteenth century, they occupied a regulatory gap. Maritime safety regulations had been developed for sail. Insurance actuarial models had been developed for sail. Port inspection regimes had been developed for sail.

Steam introduced new risk categories, boiler explosion, new propulsion system failure modes, new fire risks from coal storage, that existing maritime governance did not address. Steam-specific regulations developed over decades. During the transition, hybrid sail-and-steam vessels operated in a coverage gap where neither the existing sailing regulations nor the emerging steam regulations fully addressed their actual risk profile. The operators who treated this coverage gap as license for reduced governance were the operators whose vessels contributed to the coffin ship statistics.

AI governance is in the same transitional period. The coverage gap between existing compliance frameworks and emerging AI frameworks is real, and it is where AI risk concentrates. Organizations that interpret the coverage gap as license for reduced governance, or that satisfy the existing compliance framework and assume that satisfies the emerging AI framework, are operating in the same navigational error as those nineteenth-century operators.

The agent acceleration problem

Shadow adoption accelerated the transition from Stage 1 to Stage 2 because AI tools became individually useful faster than organizations could build governance programs. Agentic AI is accelerating the transition from Stage 2 to Stage 3 for the same structural reason: agent deployment is outpacing agent governance.

The distinction between AI assistant tools and AI agents is architecturally significant for governance. An AI assistant requires a human to initiate every interaction; the human is the actor, the AI is the tool, and the governance challenge is ensuring the human uses the tool within authorized parameters.

An AI agent takes autonomous actions on behalf of human principals. It initiates interactions, executes multi-step workflows, accesses external systems, modifies data, and produces consequences without per-action human authorization. That is a different governance challenge altogether: ensuring autonomous actions stay within authorized constraints across every possible execution path.

Shadow adoption of AI assistants creates exposure within the boundary of human-initiated interactions. Shadow deployment of AI agents, whether through vendor integrations, internally developed automation, or AI-powered workflow tools that execute autonomously, creates exposure across autonomous execution paths that no human is observing in real time.

The security gap that The Harness Doctrine identifies as the agent security gap (documented in the companion Agent Governance Stack volume) is widening as agent deployment accelerates. Organizations that have built Stage 3 governance adequate for AI assistant tools are discovering that the same governance architecture is structurally inadequate for AI agents, because the evidence architecture assumes human-initiated interactions and does not capture the governance state of autonomous multi-step execution.

This is not a future concern. Financial services organizations are deploying AI agents for trade surveillance, fraud detection, and customer onboarding automation; healthcare organizations for prior authorization processing, clinical documentation, and scheduling optimization; energy utilities for grid anomaly detection and predictive maintenance; insurance carriers for claims processing and underwriting assistance. In every case, deployment is outpacing the governance architecture designed to cover it.

What the progression means for this volume

The four-stage progression produces a diagnostic question for every organization reading this volume: Which stage are you in?

Stage 1 organizations must recognize that prohibition is not governance and that shadow adoption is an architectural certainty, not a behavioral failure. The central design problem is making the sanctioned path easier to use than the unsanctioned path.

At Stage 2, the realization is different: actual AI risk is larger than visible AI usage. Governance design must begin with identifying the full scope of AI exposure, not with building controls around the visible portion.

Stage 3 carries the bifurcation problem. The governance overlay and the production system have misaligned evidence architectures; that misalignment will produce audit failures and incident exposure at unpredictable intervals. The path forward requires consolidating two programs into one architecture without rebuilding from scratch.

Organizations that believe they have reached Stage 4 must test that belief against specific criteria: whether compliance evidence answers questions about AI system behavior in real time rather than retrospectively, whether the governance architecture covers the full AI deployment scope including shadow usage that predates the governance program, and whether the evidence architecture produces continuous verification rather than periodic attestation. A negative answer on any of these criteria means the organization is in Stage 3 with Stage 4 documentation.

The remainder of this volume provides the architectural framework for moving from Stage 3 to Stage 4 without starting over. The harness architecture is designed to be retrofitted. The evidence model is designed to accommodate legacy deployments. The multi-framework harmonization architecture is designed to satisfy existing compliance frameworks and emerging AI frameworks from a single evidence stream. The sector-specific adaptations in Chapters 6 through 9 provide the coordinates for financial services, healthcare, energy, and insurance organizations that need a path forward from their current position, not a prescription to rebuild from scratch.

The Lagrange point is reachable from Stage 3. But you must understand where you are before you can calculate the trajectory.

> **CHAPTER SUMMARY**
>
> Every regulated industry follows a four-stage progression: Prohibition, Shadow Adoption, Panic Governance, Structural Governance. The transition from Stage 2 to Stage 3 is incident-driven and produces organizational bifurcation: two programs, two evidence streams, and gaps at the boundaries. The transition from Stage 3 to Stage 4 requires architectural remediation, not additional resourcing. The regulatory coverage gap between existing compliance frameworks and emerging AI frameworks compounds the organizational governance gap. Agentic AI deployment is widening the Stage 3 inadequacy faster than most governance programs recognize. The harness architecture provides the mechanism for Stage 4 structural governance without requiring organizations to abandon existing compliance infrastructure.

Chapter 5: The Classification Society

Part II: The Classification Survey

> "A vessel assigned a letter and number from the Register had been surveyed against published criteria by independent inspectors. An underwriter quoting on that vessel was not trusting the merchant's description of the ship. The underwriter was relying on verified compliance with a published standard."

In 1760, a group of merchants and underwriters at Lloyd's coffeehouse created a small pamphlet. They called it the Register of Shipping. It listed the vessels they had underwritten, with a notation after each one: the condition of the hull, graded A through E, and the condition of the equipment, graded 1 through 5. A vessel rated A1 was in excellent condition from keel to rigging. A vessel rated E5 was approaching the end of its useful life.

The register did not yet have the institutional machinery to enforce anything. Grading remained informal, relationship-dependent, and open to the same conflicts of interest that had made the coffeehouse system unreliable. What the register established was something subtler and more durable: the concept of a published standard. Not a guarantee of quality. A shared reference. Once a vessel's condition could be compared against that reference, disagreements about seaworthiness could be settled by evidence rather than argument. Two underwriters could read the same notation and arrive at the same conclusion independently.

The classification society that emerged from this beginning, formalized as Lloyd's Register of Shipping in the 1830s, eventually employed surveyors who inspected vessels against explicit, published criteria. Inspection was not a matter of personal judgment. It was a structured examination of specific components, conducted in a defined sequence, against defined acceptance criteria. The surveyor certified that each component had been examined and found to meet the standard. That distinction matters. Certifying an impression of a vessel is one thing; certifying compliance with a published checklist is a different thing entirely.

This is the direct ancestor of System and Organization Controls (SOC) 2, Payment Card Industry Data Security Standard (PCI-DSS), and Health Insurance Portability and Accountability Act (HIPAA) attestation. A SOC 2 auditor examining a SaaS organization's controls is doing exactly what the Lloyd's surveyor did: examining specific control components against published criteria, producing a record of findings, and issuing a certificate that any relying party can interpret without taking the organization's word for the underlying condition. The vocabulary changed over two centuries. The structure did not.

The governance survey described in this chapter follows the same architecture. Five harness components constitute the structural survey. Four evidence tiers constitute the survey record. Four knowledge base integrity controls apply across all sectors without modification. These, together with the eight federal security properties (P1 through P8) defined in Volumes I and II, transfer from federal to commercial environments without architectural modification; they are the inherited baseline. On that baseline, this chapter defines the governance-worthiness acceptance criteria as the Lagrange properties, LP1 through LP14. Each component maps directly to its classification society ancestor.

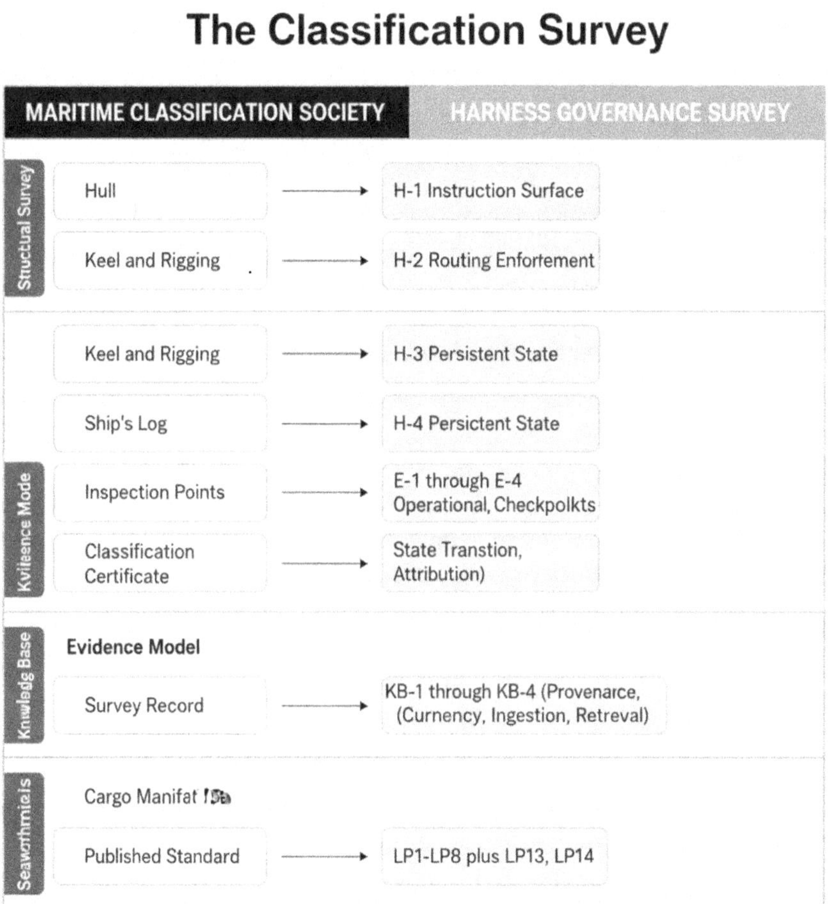

Figure 5: The harness governance survey laid out as four parallel inspection layers, with each component mapped to its Lloyd's classification society ancestor.

The structural survey: H-1 through H-5

When a Lloyd's surveyor arrived to inspect a vessel, the examination followed a fixed sequence. Hull integrity first: is the vessel's structure sound? Rigging and propulsion second: can it be controlled? Navigation and communication third: does it know where it is, and can it report its position? Cargo handling fourth: can it manage what it carries without creating hazards? Documentation fifth: is there a complete record of the vessel's history, maintenance, and certification status?

The harness survey follows that same five-component sequence. Every component corresponds to a structural element of the governance architecture, and every one has a commercial analog as direct as the hull-to-rigging sequence in maritime inspection.

H-1: The Instruction Surface. The AGENTS.md file is the vessel's operating manual. It contains the governance instructions that constrain the AI system's behavior: the policies it must follow, the boundaries it must not cross, the firmness hierarchy (RFC 2119 MUST/SHALL/SHOULD/MAY) that distinguishes hard constraints from guidance. In a Department of Defense (DoD) environment, this file encodes National Institute of Standards and Technology (NIST) 800-53 controls, classification boundaries, and autonomy tier restrictions. In financial services, it encodes PCI-DSS cardholder data prohibitions, Sarbanes-Oxley Act (SOX) financial reporting constraints, and fair lending boundaries. A healthcare deployment encodes HIPAA data handling rules, minimum necessary access requirements, and clinical accuracy thresholds.

Content changes. Architecture does not. The question for the surveyor is identical across environments: does a machine-enforceable, version-controlled instruction surface exist, and does it encode the applicable regulatory constraints with sufficient precision to produce consistent behavior?

An AGENTS.md that says "handle patient data appropriately" fails this survey criterion for the same reason that a vessel operating manual saying "maintain safe speed in harbor" fails the hull survey. The criterion requires precision, not intent. A governance instruction must be specific enough that a system following it literally will produce compliant behavior, not merely the behavior the author intended.

H-2: Routing Enforcement. The keel and rigging of a vessel determine what it can do. They are the physical infrastructure of constraint. A vessel without a functional rudder can sail; it cannot be directed. The routing enforcement layer in the harness architecture serves the same function: infrastructure that ensures every AI request passes through the governance policy before execution. No out-of-band paths, no bypass channels, no developer workarounds routing requests around the compliance layer to save latency.

In DoD environments, routing enforcement means every request validates against the Delegation Token layer before executing. The Delegation Token is a short-lived, signed authorization credential the gateway checks before an agent reaches any resource; it binds the agent's identity to an explicit, enumerated audience of permitted resources by name, and a request to any resource absent from that audience is denied at the gateway, never at the prompt. A denied out-of-audience or out-of-tier request emits an E-2 operational evidence record proving the boundary held. Commercially, it means every AI interaction routes through the policy enforcement point before touching regulated data. The architectural requirement is identical in both cases: the governance layer must be in the execution path, not adjacent to it.

Is there an enforcement mechanism that makes policy bypass structurally impossible, or is the policy expressed in documentation that developers are expected to read and follow voluntarily? That is the surveyor's question. The classification society did not ask shipowners whether they intended to follow the load line regulations. It painted the line on the hull. H-2 is that line.

H-3: Persistent State. The ship's log is among the most legally consequential documents in maritime law. It records every significant event on the voyage, from position fixes to weather encounters to crew incidents. The log does not exist for the navigator's convenience. It exists because a vessel that cannot account for its own history cannot be trusted with cargo, crew, or a classification certificate.

Persistent state architecture is the AI system's ship's log. Progress files, knowledge graph entries, session boundary records: these are the mechanisms that prevent context amnesia across session boundaries, staff rotations, and system upgrades. In DoD environments, persistent state prevents the cleared developer rotation problem, where institutional knowledge of security rationale walks out the door with a departing team member. Commercially, it prevents the compliance officer turnover problem, the annual audit cycle that starts from scratch, and the system upgrade that resets configuration without migrating governance context.

The surveyor examines a ship's log for completeness, consistency, and chain of custody. The harness surveyor examines persistent state for the same three properties. Is state maintained across session boundaries? Is it attributable? Is it complete enough to reconstruct the governance history of a specific decision if a regulator asks?

H-4: Verification Checkpoints. Classification societies conducted their structural surveys at defined inspection points, not on a continuous basis. A vessel seeking initial classification was surveyed at keel-laying, at launching, and at sea trials. Renewal surveys followed at defined intervals, with specific components examined against specific criteria at each one. This inspection point architecture ensured that compliance state was known at discrete, documented moments, with evidence any underwriter or port authority could reference.

The harness verification checkpoint architecture follows the same pattern. Verification gates at defined stages of AI operation produce structured evidence that controls are operating as designed. In DoD environments, these records satisfy the Authorizing Official; commercially, they satisfy SOC 2 auditors, PCI Qualified

Security Assessors, HIPAA assessors, and North American Electric Reliability Corporation (NERC) CIP auditors. Evidence format differs. Verification architecture does not.

Does a scheduled, structured verification process exist that produces documented evidence of control operation at defined intervals? Or does the organization rely on the absence of incidents as evidence that controls are functioning? Those are not the same question, and they do not have the same answer.

An organization that cannot produce E-2 evidence has not passed the H-4 survey criterion. It has a governance aspiration, not a governance architecture.

H-5: Accountability Records. The classification certificate is a document with legal standing. It names the vessel, identifies the classification society, states the class assigned, lists the conditions attached to that class, and records the date and inspector who issued it. Signed and attributable, it creates a specific accountability relationship between the issuing organization and the relying market.

Accountability records for the AI system serve the same function. Every action taken by the system must be attributed to an identity with a timestamp. Not approximately attributed; not attributed in aggregate. Attributed at the individual action level with sufficient precision that a specific output can be traced to the specific system state, policy version, and identity chain that produced it. In DoD environments, this means GitLab Audit Events linked to PIV-authenticated identities. Commercially, this means audit trail records linked to SSO-authenticated identities, with session context sufficient to reconstruct the decision environment.

The classification certificate without the accountability record is a piece of paper. The accountability record without the classification architecture is evidence of activity, not evidence of compliance. H-5 is where the harness survey produces its output: a certification that the architecture was in place, operating as designed, and generating attributable records of its operation.

The evidence model: E-1 through E-4

The Lloyd's surveyor did not leave the inspection without documentation. The survey record was the instrument through which the classification certificate became credible to parties who had never witnessed the inspection. An underwriter in Bristol could rely on a certificate issued by a surveyor in Liverpool because the certificate referenced a documented examination conducted against published criteria with a recorded chain of custody. Trust was not personal. It was institutional, and the evidence model was its mechanism.

The E-1 through E-4 evidence tiers are the harness survey's documentation model. Each tier corresponds to a specific level of the survey record, and each maps directly to a commercial compliance context.

E-1 (Foundation Evidence) is the classification certificate itself. It answers the existence question: does a governance architecture exist? E-1 artifacts include the AGENTS.md file with its version history, the control inventory documenting which regulatory requirements the harness addresses, and the knowledge base health record confirming that the instruction surface is current. A SOC 2 auditor opening a Type II examination begins with the E-1 question. So does a PCI Qualified Security Assessor (QSA) beginning a Report on Compliance, and so does a HIPAA assessor conducting a risk analysis. Is there a governance architecture? Can you show me the certificate?

An organization that cannot produce E-1 evidence has not passed the structural survey. No surveyor proceeds to rigging inspection on a vessel with no hull documentation.

E-2 (Operational Evidence) is the maintenance and inspection log. It answers the operation question: is the governance architecture working? E-2 artifacts are the outputs of automated verification tests conducted continuously against the control architecture. Not once. Not at audit time. Continuously, with results stored in a structured format allowing point-in-time retrieval for any period within the organization's evidence retention window.

Commercial relevance is direct and immediate. SOC 2 Type II certification requires evidence of control operation over a defined period, typically twelve months. An organization that builds its SOC 2 evidence set at audit time is doing exactly what the maritime industry did before classification societies: producing a description of what the vessel is supposed to do, not a record of what it actually did. E-2 closes that gap. Verification tests run during the period under examination are the evidence that controls operated during that period; they are not evidence that someone intended them to operate.

E-3 (State Transition Evidence) is the port authority documentation. It answers the situational awareness question: does the governance architecture know what changed? E-3 artifacts record state transitions, knowledge graph updates, policy version changes, and session boundary events. When AGENTS.md was updated to reflect PCI-DSS v4.0.1 requirements, E-3 captures the change event, the authorization, and the validation that the update was applied correctly.

Port authorities do not simply verify that a vessel is seaworthy on the day it arrives. They verify that classification has been maintained since the last survey, that no significant modifications were made without surveyor approval, and that the vessel's current state is consistent with its documented history. E-3 is the record that makes this verification possible for AI governance architecture. It is a continuity record, not a snapshot: evidence that the system did not drift silently from its certified configuration during the period between explicit examinations.

E-4 (Attribution Evidence) is the real-time telemetry from the voyage. It answers the provenance question: can every action be traced to a responsible party? E-4 artifacts are the unified activity records that attribute each system action to an identity, a policy version, a timestamp, and a session context. In maritime terms, this is the continuous position log, the watch officer's record, the engine room log: running documentation of what the vessel was doing at every moment of the voyage.

E-4 satisfies regulators who want to know not just that a governance architecture existed, but that a specific decision was made by the governed system operating within its authorized boundaries. When a Consumer Financial Protection Bureau (CFPB) examiner asks for evidence that a specific credit decision was made within fair lending constraints, E-4 is the answer. The same record answers the Office for Civil Rights (OCR) investigator who asks for evidence that a clinical summary was generated without accessing data beyond the treating clinician's minimum necessary scope.

The four evidence tiers are not interchangeable; each answers a different auditor question. An organization that has E-1 and E-4 but lacks E-2 and E-3 has a governance architecture with documentation gaps that no amount of Type II period coverage will close. The surveyor examines all four.

Knowledge base integrity: KB-1 through KB-4

The ship's manifest is the document that governs what is in the hold. A vessel operating with an inaccurate manifest, carrying cargo that is not listed, missing cargo that is listed, or carrying hazardous materials misclassified as general freight, operates outside its governance boundaries regardless of how well-constructed its hull is. Manifest integrity controls are structural requirements. Not administrative formalities. Structural.

Knowledge base integrity controls apply the same logic to the AI system's instruction source. The harness is only as reliable as the knowledge it retrieves. An AI assistant whose knowledge base contains outdated regulatory guidance will produce outputs reflecting that outdated guidance, regardless of how precisely AGENTS.md encodes current policy. An AI assistant whose knowledge base has been tampered with, whether through unauthorized injection or supply chain compromise, will produce outputs reflecting the tampered content.

These integrity requirements apply to any knowledge base in any sector. Content varies. Integrity properties do not.

KB-1 (Verified Provenance) requires that every piece of knowledge in the base has an unbroken, authenticated chain from its original source to its current representation. A clinical protocol must be traceable to the authoring physician and the credentialing committee that approved it. A trading system architecture document traces back to the authoring engineer and the model risk review that validated it. A PCI-DSS implementation guide traces to the PCI Security Standards Council publication it was derived from. The mechanism is identical across sectors because the threat is identical: knowledge with broken or fabricated provenance chains cannot be trusted, regardless of how accurate it happens to be.

KB-2 (Enforced Currency) requires that every piece of context the AI retrieves is within its type-appropriate currency threshold or explicitly flagged for revalidation. Clinical protocols become outdated when treatment guidelines change. Regulatory guidance goes stale the moment an agency issues a new interpretation, and legal requirements shift when courts hand down relevant decisions or legislatures amend statutes. An AI assistant citing a compliance requirement from a superseded version of a standard is not producing incorrect output in the sense of logical error; it is producing incorrect output in the sense of temporal displacement. That distinction rarely comforts the regulator. The enforcement mechanism is the same in every sector: each knowledge base entry carries a currency timestamp and a type-appropriate expiration window, and retrieval is conditioned on currency verification.

KB-3 (Controlled Ingestion) requires that all content enters the knowledge base through an authorized intake process with source verification, classification validation (where applicable), and integrity signing. This requirement is especially critical in sectors where regulatory guidance changes frequently and through multiple channels. PCI-DSS guidance arrives through formal standard updates, supplemental guidance documents, and FAQ clarifications. HIPAA guidance evolves through Department of Health and Human Services (HHS) Office for Civil Rights enforcement actions, guidance bulletins, and FAQ publications that do not go through notice-and-comment rulemaking. NERC CIP standards have mandatory compliance dates that differ from effective dates. Knowledge ingested from unofficial sources, improperly formatted, or ingested without authorization creates a supply chain integrity problem that no downstream verification can fully remediate. KB-3 is the intake gate.

KB-4 (Monitored Retrieval) requires that retrieval distribution and quality are continuously monitored as operational signals. This monitoring serves two purposes. First, it detects retrieval anomalies that may indicate knowledge base degradation: an AI assistant that begins retrieving an unusual proportion of its context from a single, recently-ingested source may be responding to a knowledge base poisoning event rather than a legitimate content relevance signal. Second, it produces the operational evidence that knowledge base integrity was maintained during the period under examination. A SOC 2 auditor examining knowledge base controls for a Type II period needs evidence that controls operated during the period, not only that they existed at the start and end. KB-4 retrieval logs are that evidence.

The Lloyd's surveyor examining a vessel's cargo manifest did not simply verify that the manifest was accurate at the time of departure. The surveyor examined whether the manifest integrity process, the controls governing what entered the hold and how it was documented, was sound enough to make the manifest reliable.

KB-1 through KB-4 apply the same logic. They describe a process property, not a point-in-time state.

The seaworthiness properties: LP1 through LP8

Classification societies publish seaworthiness criteria because "seaworthy" is not self-defining. A vessel that is seaworthy for coastal trade in the English Channel is not seaworthy for open-ocean crossing in the North Atlantic. A vessel rated for bulk cargo is not rated for chemical tanker operation. Criteria must be explicit; the classification certificate's value depends entirely on the precision of the standard it certifies compliance with.

LP1 through LP8 are the harness architecture's seaworthiness criteria. They define what governance-worthiness requires in terms specific enough to distinguish a system that meets the standard from one that merely resembles it.

These properties are sector-independent: they address the universal requirements of a governed autonomous system, not the sector-specific rules of a particular regulator. They sit on the federal security baseline defined in Volumes I and II, the eight security properties P1 through P8 and the harness that enforces them, which transfers to commercial environments without modification. The Lagrange properties are this volume's governance layer on that baseline; several of them, classification-consistent access and deny-on-uncertainty among them, restate a federal security property in commercial governance terms, while others are new to the regulated-sector context. A vessel's seaworthiness does not depend on which port it sails to. The same holds for AI: governance-worthiness does not depend on which regulator will examine the system.

The eight definitions that follow are the canonical specification of LP1 through LP8 for this book. The front-matter primer summarizes them and Appendix B re-keys to them; where any other chapter, summary, or appendix states a P-property definition, this chapter governs.

LP1 (Classification-Consistent Access) requires that the AI system enforces access boundaries consistent with the data classification or regulatory scope applicable to the data being processed. Federally, this means clearance-level enforcement. Commercially, it means regulatory scope enforcement: protected health information (PHI) is not processed outside HIPAA-covered functions; PCI cardholder data is not processed outside PCI scope; data subject to specific consent constraints is not processed beyond the authorized purpose. The underlying requirement is the same in both worlds: access boundaries are enforced by the architecture, not by the judgment of the individual making the request.

LP2 (Scope Binding) requires that AI operations are bounded to defined scope and do not expand their own action boundaries without explicit authorization. A governed AI assistant that is asked to draft a clinical summary does not, in the process of doing so, query patient records outside the treating clinician's relationship, update records it was not asked to update, or initiate downstream workflows it was not authorized to trigger. Scope binding is the constraint that makes delegation safe: you can authorize a system to do a specific thing without also authorizing everything the system might find useful in doing that thing.

LP3 (Routing Integrity) requires that all AI requests route through the governance enforcement layer. This property is the architectural expression of H-2. It is a seaworthiness criterion, not merely a design recommendation, because a system that sometimes routes through governance and sometimes does not has not passed the survey. Intermittent compliance is not compliance.

LP4 (Persistent Governance Context) requires that governance context persists across session boundaries, upgrades, and staff transitions. This property is the architectural expression of H-3. An AI system that loses its governance configuration when a server restarts, or loses its knowledge base context when a new version is deployed, is not a governed system. It is a system that was governed, intermittently.

LP5 (Continuous Verification) requires that the governance architecture verifies its own operation continuously and produces structured evidence of that verification. This is the property that distinguishes the governed system from the merely described system. A policy document describing intended behavior does not satisfy LP5. Automated test results confirming that the policy enforcement layer is operating as designed, generated continuously and stored in retrievable format, do satisfy it.

LP6 (Attribution Completeness) requires that every AI-generated output or action is attributable to an identity, a policy version, a timestamp, and a session context sufficient to reconstruct the decision environment. This property is the architectural expression of H-5. Attribution completeness is a seaworthiness criterion because incomplete attribution makes accountability structurally impossible: if the AI system cannot account for which version of its policy governed a specific output, that output cannot be defended under regulatory examination.

LP7 (Deviation Detection) requires that the governance architecture detects and reports deviations from authorized operational parameters. A deviation is not necessarily an error; it may be a valid edge case the governance architecture did not anticipate. But a system that cannot detect its own deviations from authorized behavior cannot be trusted to operate within those boundaries. Detection is the prerequisite for response. Silence from an unmonitored system is not compliance. It is absent monitoring.

LP8 (Evidence Integrity) requires that all evidence artifacts produced by the governance architecture are tamper-evident and stored in a manner that preserves chain of custody. An audit trail that can be modified after the fact is not an audit trail; it is a document describing what someone wants the history to have been. E-2 through E-4 evidence is valuable only to the extent that LP8 is satisfied. A SOC 2 auditor examining evidence artifacts that do not satisfy LP8 is examining intentions rather than facts.

These eight properties define the minimum governance-worthiness criteria for any AI system operating in a regulated environment. A system satisfying all eight has passed the structural survey. A system satisfying seven has a governance architecture with a specific, identified gap. Fewer than seven, and the organization has not completed the transition from Stage 2 shadow adoption to Stage 3 structured governance. It has AI capability without governance architecture.

Two enforcement properties: LP13 and LP14

The original eight properties define what a governed system must demonstrate. Two further properties define how the enforcement layer must behave when it acts on a request. They are named explicitly because the sector chapters rely on them by name, and because conflating either with one of the original eight has caused real confusion in practice. They are enforcement criteria, not commercial extensions; they apply in every sector, and the four commercial properties (LP9 through LP12, defined in Chapter 8) sit between them in numbering only, not in kind.

LP13 (Output Boundary Enforcement) requires that AI-generated output is evaluated against its disclosure and scope boundaries at the routing layer before it leaves the governance perimeter. Where routing integrity (LP3) governs what the system may reach, LP13 governs what the system may emit. Content that crosses a defined boundary, a regulatory disclosure line, a data element outside authorized scope, or a clinical claim that crosses the Software as a Medical Device threshold, is routed to review or denied at the gateway, not delivered and corrected afterward. LP13 is the architectural expression of H-2 applied to outputs rather than to retrievals.

LP14 (Deterministic Deny on Uncertainty) requires that when the governance architecture encounters a condition it cannot evaluate against an authorized policy, the default is denial with escalation, not permissive processing with post-hoc review. The uncertainty does not propagate. The gateway fails closed, the request escalates to human oversight, and the cascade stops at the point of uncertainty rather than flowing into the next operation. LP14 is what makes a deny-by-default posture a property of the architecture rather than a preference of the operator.

LP13 and LP14 are not part of the original seaworthiness survey; they specify the behavior of the enforcement layer that the survey assumes. A system can satisfy LP1 through LP8 on paper and still fail in operation if its gateway emits what it should withhold or proceeds when it should deny. The complete property set for this book is therefore LP1 through LP14: the eight core properties, the four commercial-extension properties, and these two enforcement properties.

The certificate

When the Lloyd's surveyor completed an inspection and assigned a classification, the shipowner received a certificate. Not a guarantee. A vessel classified A1 could still be lost at sea, through navigational error, through weather beyond its design envelope, through cascading failures in systems not covered by the classification criteria. What the certificate guaranteed was that, as of the inspection date, the vessel's structure had been examined against the published criteria and found to meet the standard.

That is precisely what the harness survey produces. An organization that can demonstrate H-1 through H-5 in place and operating, E-1 through E-4 evidence being generated and retained, KB-1 through KB-4 controls active, LP1 through LP8 properties satisfied, and the enforcement properties LP13 and LP14 in force has earned the governance certificate. It does not guarantee that the AI system will never produce a harmful output. It certifies that governance architecture was in place, verified to be operating, and documented with sufficient completeness and attribution to support regulatory examination.

This is the standard commercial regulated industries need. Not the vendor's SOC 2 report, which certifies vendor internal controls. Not the legal review of the data processing agreement, which certifies contractual allocation of responsibility. Not the model validation report, which certifies output accuracy under test conditions.

The governance certificate certifies something different. It certifies that the deploying organization's harness architecture, the constraint layer between the AI system and the regulated environment, was built to standard, operated to standard, and produced evidence demonstrating operation to standard throughout the deployment period.

Classification societies took two centuries to build the institutional infrastructure that makes maritime certificates credible. The harness framework provides that architecture now. Two centuries of accumulated disaster and institutional response, compressed into a governance specification that is ready to deploy.

Who paints the line?

There is an obvious objection to everything described in this chapter. If the institution writes its own AGENTS.md, declares its own controls, and generates its own evidence, then the institution is grading its own paper. The shipowner is painting his own load line. Why should anyone trust a certificate the certified party produced for itself?

The objection is real, and the maritime model answers it precisely. In the maritime model the shipowner does not paint his own Plimsoll line. The owner submits the vessel; an independent classification society surveys it against a published standard and assigns the mark; the registry records the class. The commercial AI model preserves all three roles. AGENTS.md is the shipowner's self-declared survey, the institution's written account of what its system is built to do. It is not a self-issued certificate. The independent surveyor is the external auditor, examiner, QSA, or assessor (SOC 2 auditor, Office of the Comptroller of the Currency

(OCC)/Fed/Federal Deposit Insurance Corporation (FDIC) examiner, PCI QSA, HIPAA/OCR assessor, NERC CIP auditor, state Department of Insurance (DOI) examiner) who tests the declaration against the running evidence (E-1 through E-4). The registry is the regulator. The institution authoring its own AGENTS.md is exactly the shipowner filing a survey declaration; the line is only validated when an independent party tests it.

This is why the section above on H-2 matters so much. The classification society did not ask shipowners whether they intended to follow the load line regulations; it painted the line on the hull, and H-2 is that line. But painting the line is only half the architecture. A line the owner can repaint after the survey is no line at all. The independence of the commercial model does not come from who writes AGENTS.md. It comes from the fact that the evidence (E-2 through E-4) is generated as a byproduct of operation and is tested by a party the institution does not control. The owner declares the survey; the surveyor reads the running record.

Tamper-evidence is what closes the loop. LP8 requires that every evidence artifact is tamper-evident and stored in a manner that preserves chain of custody. That property is what stops the owner from repainting the line after the fact. An institution can write any declaration it likes in AGENTS.md, but it cannot quietly rewrite the E-2 operational record showing that a request to a resource outside its authorized scope was denied at the gateway, nor the E-4 attribution record tracing a specific output to a specific policy version and identity. The surveyor tests the declaration against records the institution cannot alter without leaving marks. Self-declaration plus independent test against tamper-evident operating history is exactly the structure that made Lloyd's certificates credible to an underwriter who never saw the ship.

The survey components are defined. The criteria are published. The inspection sequence is laid out in the chapters that follow, one sector at a time.

Chapter 6: The Evidence Model Transfers Completely

Part II: The Classification Survey

> *"The certificate of classification is not issued because the underwriter trusts the shipowner. It is issued because an independent inspector surveyed the vessel against a published standard and signed his name to the result. Every auditor in every regulated industry is looking for the same thing: a document that was produced by a process that did not depend on trusting the party being assessed."*

In 1760, when Lloyd's Register of Shipping formalized the first standardized seaworthiness classification system, the market's practical problem was not the quality of individual inspections. Experienced surveyors had been assessing vessels for decades. The problem was portability. An underwriter in London could not verify the work of a surveyor in Rotterdam. A merchant in Bristol could not confirm that a certification issued in Glasgow reflected the same criteria as one issued in Lisbon. Good intelligence in individual transactions. Fragmented intelligence across the market as a whole.

The classification society solved this by producing a standardized certificate: a document whose format, criteria, and inspection methodology were published and consistent. An underwriter quoting on a vessel assigned a rating from Lloyd's Register was not trusting the inspector who examined that vessel. The underwriter relied on a different fact. The inspector's work was governed by a public standard, and the certificate's fields corresponded to that standard in a predictable way. Not a statement of trust. A structured evidence artifact.

This is exactly what a System and Organization Controls (SOC) 2 Type II auditor, a PCI Qualified Security Assessor (QSA), a Health Insurance Portability and Accountability Act (HIPAA) assessor, and a North American Electric Reliability Corporation (NERC) CIP auditor each require from an AI governance program. Not a narrative description of how governance is supposed to work. Not a vendor attestation that controls are in place. A structured, independently verifiable evidence artifact answering a defined set of questions about a defined set of governance properties, produced by a collection architecture that does not depend on the assessed organization's cooperation to be reliable.

The E-1 through E-4 evidence model from The Harness Doctrine is that architecture. It is the classification certificate for AI governance. Unlike the sector-independent core that transfers structurally, the evidence model transfers not just structurally but functionally. The same four tiers answer the same four assessor questions across every regulated industry, because those four questions are not regulatory artifacts. They are information architecture requirements; any competent examiner, in any sector, must satisfy them before issuing an opinion.

The Four Questions Every Assessor Asks

Before examining the evidence tiers in detail, the logic of the four-tier architecture requires a foundational statement.

A regulatory assessor, regardless of sector and regardless of the specific control framework governing the assessment, is always trying to answer a compound question: does this governance architecture exist as described, is it operating as designed, does it know what has changed, and can every action be traced to a responsible party?

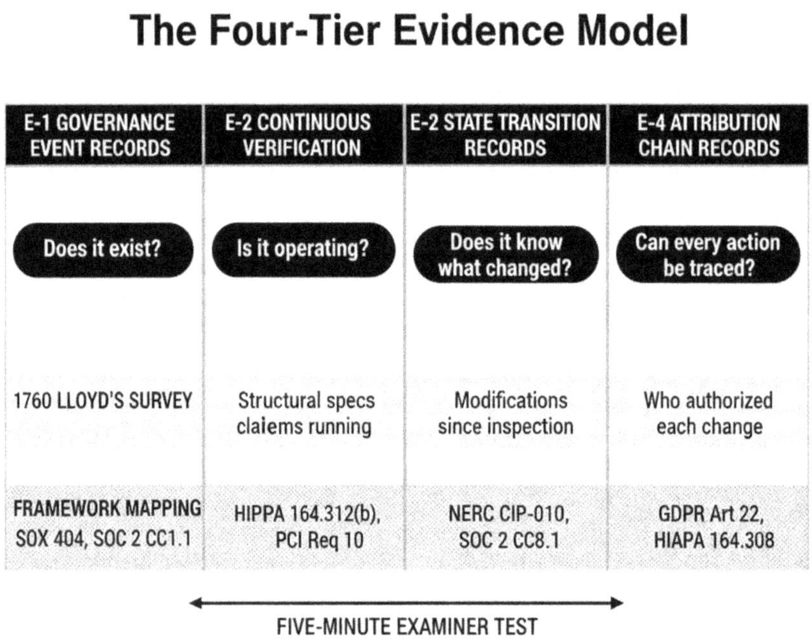

Figure 6: The four evidence tiers, each answering one assessor question, mapped from the 1760 Lloyd's classification survey to the universal regulatory frameworks they satisfy.

These four questions are not derived from National Institute of Standards and Technology (NIST) or Sarbanes-Oxley Act (SOX) or HIPAA. They come from the epistemological requirements of any external assessment of a system the assessor does not control and did not build. The assessor cannot directly observe internal operation. Four questions define the minimum set of evidence categories that support a defensible opinion. Not five. Not three. Four.

A single evidence tier, no matter how voluminous, cannot answer all four. An organization producing only Governance Event Records (E-1) has told the assessor that a governance architecture exists, but cannot demonstrate it is working. An organization producing only Continuous Verification results (E-2) has demonstrated that controls ran, but cannot tie those results back to an authorized governance architecture or trace individual actions to accountable parties. Each tier is necessary. None is sufficient alone.

The classification survey the Lloyd's inspector conducted had the same four-question structure, in different vocabulary. Does the vessel have the structural specifications claimed (E-1)? Are the critical systems operating as designed (E-2)? What modifications have been made since the last inspection (E-3: State Transition Records)? Who authorized each modification and are they recorded in the vessel's log (E-4: Attribution Chain Records)?

Two hundred sixty years of commercial risk assessment. Four questions. The AI governance market is discovering them independently, in every sector, on a compressed timeline.

E-1: Governance Event Records

The first evidence tier proves that the governance architecture exists and that its founding decisions are documented.

E-1 Governance Event Records are not policy documents. Policy documents describe intent. E-1 records document decisions: what the AI system was authorized to do, what policies constrain it, what boundaries define its scope of operation, and who authorized each of those conditions. The distinction is architectural. A policy document produced by a compliance team without a corresponding governance event record in the system's operational history is a description of how governance should work. Not evidence that it does.

In Department of Defense (DoD) environments, E-1 records include the AGENTS.md configuration hash (proving the instruction surface is what it is claimed to be), the knowledge base health inventory (proving the KB contents are what the system is drawing from), and the control inventory associated with the system's Authorization to Operate (ATO). Commercially, the specific records differ by regulatory context, but the architectural requirement is identical: governance architecture must be documented at the moment of authorization, not reconstructed after an audit request.

SOX Section 404 mapping. SOX Section 404 requires management's annual assessment of the effectiveness of internal controls over financial reporting, including documentation of the control environment. PCAOB AS 2201 requires that management identify and document controls at the financial statement assertion level. Any AI system affecting financial reporting calculations, budget models, forecasting tools, or financial data pipelines sits within the SOX 404 scope. E-1 governance event records provide the control documentation that SOX 404 and AS 2201 require: what controls govern the AI system; when they were established; who authorized them; what assertions about financial reporting accuracy they are designed to support.

The parallel is precise. A shipowner filing a classification certificate with Lloyd's Register was producing the maritime equivalent of a 404 attestation: a documented statement of what the vessel was, under whose authority it had been assessed, and against what standards. The certificate was control documentation for the insurance contract. E-1 Governance Event Records are control documentation for the compliance opinion.

HIPAA Security Rule 164.312 mapping. Section 164.312 requires covered entities and business associates to implement technical safeguards for electronic protected health information (PHI). Implementing specifications include access controls, audit controls, integrity controls, and transmission security. When AI systems process PHI, E-1 governance event records document the technical safeguard configuration: which access controls are in place, what audit logging is enabled, what integrity protections apply, and how the configuration was authorized.

The documentation requirements are explicit: covered entities must maintain documentation of policies, procedures, and actions taken in connection with the security rule for a minimum of six years. E-1 records maintain that documentation in a form that is continuously current, versioned, and attributable. Not assembled from scattered policy files during an OCR inquiry.

NERC CIP-003 mapping. CIP-003 requires identification and documentation of security management controls, including a senior manager with accountability for Critical Infrastructure Protection (CIP) cybersecurity standards compliance and a process for escalating compliance issues. For AI systems deployed in bulk electric system environments, E-1 governance event records document the security management control configuration: who holds the CIP Senior Manager designation; what policies govern the AI system's operation within the electronic security perimeter; when each control was established or modified.

CIP-003-8 Requirement R1 specifically mandates documented cybersecurity policies for high and medium impact BES Cyber Systems. E-1 records maintain those policies in version-controlled, timestamped form, with the change attribution CIP auditors require. The record exists before the auditor asks for it.

SOC 2 CC1.1 mapping. Trust Services Criteria CC1.1 addresses the control environment: does the entity demonstrate a commitment to integrity and ethical values through its policies and their operating effectiveness? For AI systems subject to SOC 2 assessment, E-1 governance event records document the control environment for AI governance: the policies that govern AI behavior, the authorization decisions that established those policies, and the configuration artifacts that implement them.

A Type II auditor testing CC1.1 for an AI deployment needs evidence of the control environment at the point of system authorization and throughout the examination period. E-1 records provide that evidence continuously, not as a pre-audit reconstruction.

E-2: Continuous Verification

The second evidence tier proves that the governance architecture is operating as designed.

Continuous Verification (E-2) is the most architecturally distinctive tier because it requires continuous generation. E-1 records are created at governance events, which occur at defined intervals. E-3 records arrive when governance state changes, E-4 records when attribution events occur. E-2 records must be created continuously, at a frequency sufficient to demonstrate ongoing operation, because the compliance question E-2 answers is not "did the controls exist" but "were the controls operating at every moment during the examination period."

This distinction has practical consequences. An organization implementing E-1 and E-4 correctly, with complete governance documentation and full attribution records, but lacking E-2 continuous verification, can demonstrate that governance was designed and that actions were attributed. It cannot demonstrate that governance was operating between the documentation events and the attribution events. An auditor examining this evidence cannot rule out that controls were active only when the organization expected scrutiny.

Federal parallel: continuous authorization (cATO). Federal accreditation began with a point-in-time Authorization to Operate. An Authorizing Official reviewed a system's controls, issued an ATO valid for a fixed term, and re-accredited at the end of it. That model carried the heroic one-shot weakness in its design, because a system could drift the day after the stamp and carry valid paperwork until the next review. The federal answer was the continuous Authorization to Operate (cATO): an ongoing authorization sustained by continuous monitoring and a live evidence stream rather than renewed by periodic re-accreditation. Under the Risk Management Framework, the Monitor step turns authorization into a maintained status instead of an expiring certificate. E-2 is the commercial form of that same shift. A SOX operating-effectiveness assertion, a HIPAA audit control, a NERC monitoring requirement, and a federal cATO all ask the same question: are the controls operating right now, and can you show it without assembling the proof after the request arrives? The classification society reaches the same place from the maritime side, where a certificate is a current status requiring continuous validation rather than a permanent stamp. cATO, continuous classification, and E-2 are three names for the architecture that refuses to let authorization decay between checks.

SOX monitoring controls mapping. SOX Section 404 requires not only documentation of controls but evidence of their operating effectiveness over the assessment period. Any AI system affecting financial reporting must produce continuous evidence that its controls operated as designed throughout the fiscal year, not just at the point of authorization.

E-2 continuous verification results demonstrate that operating effectiveness. Automated verification tests run against defined control criteria, logged with timestamps and results, constitute the evidence base making a SOX 404 assessment defensible. Management assertion about operating effectiveness that is not grounded in continuous verification results is a compliance posture with evidence exposure; this is the specific gap that The Harness Doctrine's false completion failure mode describes.

HIPAA audit controls 164.312(b) mapping. Section 164.312(b) requires covered entities and business associates to implement hardware, software, and procedural mechanisms that record and examine activity in information systems containing or using electronic PHI. The operative word is "record": the audit control must document activity continuously, not reconstruct it on request.

When AI systems process PHI, E-2 continuous verification provides the audit control implementation. Every AI interaction with PHI-bearing data produces a verification record confirming that governance controls were active at the time of the interaction and that the interaction satisfied defined constraints. Not a log of what happened. A verification record that what happened occurred within authorized boundaries. This distinction matters for HIPAA audit control compliance, which requires evidence of both activity and governance status at the time of activity.

NERC CIP-007 mapping. CIP-007 (System Security Management) requires system security controls for high and medium impact BES Cyber Systems: ports and services management, security patch management, and system access management with continuous monitoring for malicious communications. AI systems deployed in BES environments need E-2 continuous verification to produce the security management monitoring evidence CIP-007 auditors require.

CIP-007-6 Requirement R4 specifically requires logging and monitoring of user activity in BES Cyber Systems. AI system interactions touching BES environments must produce continuous, timestamped verification records demonstrating that user access was within authorized scope and that no access exceeded the electronic security perimeter's defined constraints.

SOC 2 CC7.1 mapping. Trust Services Criteria CC7.1 addresses detection and monitoring activities: does the entity evaluate and communicate internal control deficiencies in a timely manner? For AI governance, CC7.1 requires continuous monitoring of control operation and prompt escalation when controls do not operate as designed.

E-2 continuous verification provides the detection mechanism CC7.1 requires. When a verification test fails, that failure is a control deficiency; it must be logged, communicated, and remediated within a defined timeline. The verification architecture produces these records on its own, as a byproduct of operation.

PCI-DSS Requirement 10 mapping. Payment Card Industry Data Security Standard (PCI-DSS) v4.0.1 Requirement 10 mandates that all access to system components and cardholder data be logged and monitoring procedures implemented to detect anomalies and suspicious activity. AI systems operating in PCI-scoped environments, even without processing cardholder data directly, need E-2 continuous verification to produce the access logging and monitoring evidence Requirement 10 requires.

Requirement 10.3 specifically mandates that audit logs capture who accessed what, when, and from where. E-2 records documenting AI system interaction events within the cardholder data environment provide this evidence in a format mapping directly to the Requirement 10 compliance criteria.

E-3: State Transition Records

The third evidence tier proves that the governance architecture has situational awareness of its own configuration changes.

A governance architecture that cannot account for its own evolution is not auditable. The assessor's question at the heart of State Transition Records (E-3) is direct: between the baseline authorization and today, what changed, when did it change, who authorized the change, and what controls verify that the change was within authorized parameters?

In DoD environments, E-3 records cover progress files, knowledge graph state, and session boundary records: artifacts documenting how the AI system's governance context evolved over time. Commercially, E-3 extends to include regulatory change tracking. When a regulatory framework update occurs (PCI-DSS v4.0.1 replacing v4.0, HIPAA guidance updated through Department of Health and Human Services (HHS) FAQs, NERC CIP standard revisions with mandatory effective dates), the governance architecture must document that the change was identified, assessed, and incorporated into the instruction surface, with an auditable record of who made that determination.

SOX change management mapping. SOX change management requirements apply to any technology system affecting financial reporting. IT general controls for change management require that changes to in-scope systems be authorized, tested, and documented before implementation. An AI system generating financial models, assisting with forecasting calculations, or processing data that feeds financial reports is an in-scope system for ITGC purposes.

E-3 state transition records provide the ITGC change management evidence: every modification to the AI system's governance configuration, model parameters, instruction surface, or knowledge base scope is a state transition that E-3 captures with the authorization metadata ITGCs require. Organizations implementing E-3 correctly eliminate the pre-audit change management reconstruction exercise. That exercise consumes weeks of effort in traditional ITGC assessments. E-3 replaces it with a query.

HIPAA configuration management mapping. Security Rule implementation specifications for technical safeguards (164.312) require covered entities to implement security measures protecting the integrity of electronic PHI. Configuration management is an addressable implementation specification under this requirement.

For AI systems processing PHI, E-3 state transition records provide the configuration management evidence: every change to PHI access boundaries, data retention parameters, or encryption configuration is a state transition that E-3 captures with the authorization and verification metadata HIPAA configuration management demands. This matters especially for AI systems with adaptive behavior, where "configuration" includes not just static parameters but the instruction surface governing operational boundaries.

NERC CIP-010 mapping. NERC CIP-010 (Configuration Change Management and Vulnerability Assessments) is one of the most prescriptive standards in the CIP suite. It requires documented baseline configurations for all BES Cyber Systems, formal change authorization processes, active monitoring for unauthorized changes, and regular vulnerability assessments.

E-3 state transition records map directly to CIP-010-4 Requirement R1, which mandates that covered entities develop, document, and implement a baseline configuration process. In BES environments, every governance state change for an AI system is a configuration change that must be documented, authorized, and verified within the CIP-010 framework. CIP-010 auditors examining the record see a complete, timestamped change log; not a retrospectively assembled narrative.

SOC 2 CC3.1 and CC8.1 mapping. Trust Services Criteria CC3.1 addresses risk assessment: does the entity consider the risk of achieving objectives when deploying changes? CC8.1 addresses change management, requiring that the entity authorize, design, develop, test, and implement changes to infrastructure, software, processes, and procedures.

Together, CC3.1 and CC8.1 require that governance configuration changes be assessed for risk before implementation and documented in the change management record. E-3 state transition records provide both: the change record satisfies CC8.1's documentation requirement; a risk-assessment field in the state transition record satisfies CC3.1's risk assessment evidence requirement. Organizations embedding risk assessment into the state transition workflow, a pre-commit validation checking the change against defined risk criteria before logging the transition, produce CC3.1 evidence as a byproduct of change authorization rather than as a separate audit exercise.

E-4: Attribution Chain Records

The fourth evidence tier proves that every AI-generated artifact has a complete provenance record from request to output.

Attribution Chain Records (E-4) are the most commercially distinctive tier because commercial regulated environments require attribution not only to an internal identity but to an external disclosure record. In DoD environments, attribution means: this developer, with this clearance and this wall attestation, initiated this AI request, which produced this output, under these governance conditions. Commercially, attribution extends further: this customer or regulatory subject was or was not notified that AI was involved in this decision affecting them, under these disclosure obligations, with this audit record of the notification.

The Consumer Financial Protection Bureau (CFPB) circular on adverse action and AI, the General Data Protection Regulation (GDPR) Article 22 right to explanation, and the consumer-facing transparency requirements of emerging state AI laws all create attribution obligations that the federal application of The Harness Doctrine does not address. This is one of the specific dimensions where commercial AI governance is harder, not simpler, than its federal counterpart.

SOC 2 CC7.2 mapping. Trust Services Criteria CC7.2 addresses the evaluation of security events and anomalies: does the entity evaluate the significance of identified events and take appropriate action? For AI governance, CC7.2 requires that unusual AI system behaviors, anomalous outputs, and policy violations be attributed, investigated, and resolved with documented evidence.

E-4 attribution chain records provide the raw material for CC7.2 compliance. Every AI-generated artifact has a provenance record enabling investigation of anomalous events with full context: what model produced the output; what knowledge base content was retrieved; what governance properties were verified at the time; what identity initiated the request. Without this provenance, a CC7.2 investigation into an unusual AI output requires forensic reconstruction from operational logs. That reconstruction typically cannot provide the governance-state context that makes the investigation meaningful.

PCI-DSS continuous monitoring mapping. PCI-DSS v4.0.1 Requirement 12.6.3.1 requires that all personnel with access to cardholder data environments receive security awareness training. Requirement 10.7 requires prompt investigation of detected failures in critical security controls. For AI systems in PCI-scoped environments, E-4 attribution records serve as the mechanism for both the monitoring that detects failures and the investigation that characterizes them.

When a PCI-DSS audit examiner queries whether an AI system accessed or processed data within the cardholder data environment, E-4 attribution records answer with a complete chain: request identity, authorization state, data accessed, governance properties verified, output generated, anomaly flags triggered. The examiner does not reconstruct this chain from multiple log sources. The attribution record is the chain.

HIPAA breach detection 164.308(a)(6) mapping. Section 164.308(a)(6) is the security incident procedures standard, requiring covered entities to implement policies and procedures addressing security incidents, including identification and response to suspected or known incidents. Breach detection depends on attribution infrastructure that can answer the compound question: which PHI, processed by which AI system, accessed by which identities, during which time period, under which governance conditions.

E-4 attribution chain records provide this infrastructure. When a potential PHI exposure event occurs, the attribution record for every AI interaction during the affected period contains the minimum necessary determinations, the access authorization basis, the data accessed, and the governance state verification. A HIPAA breach response team can determine whether a specific interaction constituted an impermissible PHI access without forensic reconstruction from operational logs; the attribution record was generated as a byproduct of the interaction itself.

The breach notification timeline is demanding: notice to affected individuals must be provided without unreasonable delay, no later than 60 days after discovery. For AI systems processing PHI at scale, meeting that timeline is impossible without automated attribution infrastructure that can rapidly scope the affected population. E-4 attribution records make scope determination a query. Not an investigation.

The 5-Minute Examiner Test

The four-tier architecture is not a theoretical construct. It is a practical requirement whose adequacy can be tested in five minutes.

In financial services, a federal bank examiner or Office of the Comptroller of the Currency (OCC) examination team should be able to sit at an evidence dashboard and answer four questions without requesting additional documentation, without interviewing personnel, without conducting independent testing:

1. Which AI systems are deployed, and which regulatory frameworks apply to each? (E-1 Governance Event Records, with version control and authorization metadata)
2. Are the governance controls operating as designed right now, and have they operated continuously since the last examination? (E-2 Continuous Verification, with timestamped results covering the full examination period)
3. What changed in the governance configuration since the last examination, and was each change authorized through the defined change management process? (E-3 State Transition Records, with authorization records and risk assessment documentation)
4. Can every AI-assisted action in the examination period be attributed to a specific identity with specific authorization, and was every action within that identity's authorized scope? (E-4 Attribution Chain Records, with complete provenance chains)

If the harness architecture is properly implemented, all four answers have quantitative evidence with timestamps. The examiner does not need to ask the organization to explain anything. The evidence speaks.

If any answer requires a narrative explanation rather than a query against the evidence record, the harness has a gap. Narrative explanation is the mark of a documentation-based governance program, not a verification-based one. Documentation-based programs describe what should have happened. Verification-based programs prove what did.

This test generalizes across sectors. A HIPAA OCR investigator asking the same four questions about a healthcare AI deployment should receive the same type of answer: evidence records, not narratives. A NERC CIP auditor examining a utility's AI governance program should close the four questions from evidence records within the examination session, not from interviews with the compliance team. A PCI QSA assessing a payment-adjacent AI deployment should find all four answer categories populated by continuous evidence collection; not assembled in response to the questionnaire.

The four questions are not examination specific. They are the universal epistemological requirements of any external assessment of a system the assessor did not build and does not control. The evidence architecture that answers them in five minutes is the classification certificate for AI governance.

The Lloyd's Register inspector in 1760 was conducting a five-minute test of a different kind: can I determine from this vessel's documented condition whether it meets the seaworthiness criteria I was sent to verify? The certificate he issued was the result of evidence collected continuously throughout the vessel's service, not assembled at the dock on inspection day. A shipowner who maintained records properly could give the inspector an answer in minutes. A shipowner who had neglected records spent the morning in a difficult conversation.

Three hundred years later, the compliance market runs on the same logic. The organization that has built continuous evidence collection does not dread the examiner. The organization relying on documentation-based governance does.

Architecture over Rendering

One clarification is essential before the sector-specific chapters that follow.

The evidence model's architecture (four tiers, continuous collection, governance-state-aware records, attribution chains) transfers to every sector without modification. The rendering layer, how evidence is formatted, labeled, and submitted for each regulatory framework, is sector-specific.

A SOC 2 auditor receives evidence in a format governed by the AICPA attestation standards and Trust Services Criteria. A PCI QSA works from the PCI SSC Report on Compliance template. A HIPAA assessor reads against OCR investigation protocols. A NERC CIP auditor receives evidence in formats prescribed by the CIP standards and NERC's audit procedures.

These are different document formats produced from the same underlying evidence records. The rendering is the sector-specific work. The evidence architecture is universal.

This distinction is the mechanism by which a single evidence collection program can satisfy multiple regulatory frameworks simultaneously. An organization building a multi-framework harmonized evidence collection program, one generating E-1 through E-4 records at the architectural level without sector-specific formatting, can produce SOC 2 evidence, PCI evidence, HIPAA evidence, and NERC CIP evidence from the same source records by applying sector-specific rendering templates at report generation time.

The alternative, maintaining a separate evidence collection program for each regulatory framework, is the organizational bifurcation problem applied to evidence. Two programs; two evidence streams; gaps at the boundaries where the same AI action is governed by two frameworks simultaneously and produces conflicting or inconsistent evidence in each.

The classification society solved this in the maritime market by issuing certificates whose underlying assessment criteria were public, consistent, and recognized by multiple national maritime authorities. An insurance underwriter, a port authority, and a flag state inspector could each read the same Lloyd's Register certificate and derive the information they needed, in their own format, from the same underlying assessment. Multi-framework certificate. Single-architecture assessment.

The E-1 through E-4 evidence model is the same solution, for the same reason, in a market that is 260 years younger but facing structurally identical information architecture problems.

> **CORE DOCTRINE**
>
> The four evidence tiers answer four questions that every regulatory assessor must resolve before issuing an opinion: does the governance architecture exist (E-1 Governance Event Records), is it operating (E-2 Continuous Verification), does it know what changed (E-3 State Transition Records), can every action be traced (E-4 Attribution Chain Records)? These questions are not regulatory artifacts; they are information architecture requirements that any competent external assessor must satisfy. The organization that builds continuous evidence collection answering these four questions does not prepare for examinations. It renders them.

Chapter 7: The Three Failure Modes Are Universal

Part II: The Classification Survey

> *"A governance failure does not require malice. It requires only an architecture that assumes the conditions that will not hold."*

Chapter 2 introduced the three failure modes that The Harness Doctrine identifies as the structural failure patterns of incomplete or degraded governance architecture: context amnesia, heroic one-shot, and false completion. That chapter named them as commercial analogs of the defense failure modes documented in The Harness Doctrine.

This chapter makes a stronger claim. These are not federal failure modes that happen to appear in commercial settings, and they are not defense-derived concepts requiring translation before they apply elsewhere. They are governance architecture failure modes, produced by specific structural deficiencies in any harness architecture, in any sector, under any regulatory framework. Vocabulary changes. The underlying mechanics do not.

The reason these three failure modes appear universally is the same reason that ships built by different naval traditions, in different centuries, using different materials, all sink when overloaded past the same physics. Not a cultural artifact. A consequence of the architecture meeting the conditions it was not built to handle.

Each failure mode is treated below in full. A precise definition, then a commercial case study with operational specificity, a maritime parallel, the harness component that prevents the failure architecturally, and the detection mechanism that signals the failure is underway before it produces a compliance event. The failure modes have an adversarial face as well as an accidental one, so the chapter opens with the attack surface that produces them.

The Attack Surface of a Governed AI System

The three failure modes that follow are not always accidents. A governance architecture has an attack surface, and an adversary who wants one of those failures to occur will push on the region of the surface most likely to produce it. Context amnesia can be induced as readily as it is stumbled into; false completion is a validation gap an attacker can engineer as easily as a careless team can leave it open. Before the failure modes, then, the surface: the specific places a governed AI system can be attacked, and the harness component that defends each one. The surface described here is generic across sectors. Each sector chapter that follows adds the surfaces unique to its data, its decisions, and its regulators.

The surface has eight regions.

The instruction surface (H-1). The governing instructions are themselves a target. Prompt injection, whether typed by a user or hidden inside a document, an email, a web page, or a tool's output that the agent later reads, rewrites what the agent believes it is permitted to do. The defense is that H-1 is machine-enforced rather than advisory: a compiled instruction surface and an output boundary (LP13) that hold regardless of what the model was talked into, and a standing rule that all retrieved content is untrusted input, never instruction. A ship does not let the cargo rewrite the captain's orders; a governed agent does not let retrieved text rewrite its scope.

The routing layer (H-2). The gateway that checks every action against policy is a target for bypass: requests crafted to evade the check, confused-deputy patterns that borrow the agent's authority, and calls that try to reach a resource without passing the gate. The defense is deny-by-default routing and LP14, deterministic deny on uncertainty: a request the gateway cannot evaluate against an authorized policy is denied and escalated, not allowed through.

The knowledge base and retrieval path (KB-1 through KB-4). The grounding data is a target in two directions. An attacker can poison it, inserting false content with fabricated provenance so the model grounds on a lie, or exploit retrieval to pull data from outside the authorized scope. The defense is the knowledge-base integrity controls: verified provenance (KB-1) so unprovenanced content cannot enter, and scope-compiled retrieval so a query cannot reach beyond the boundary of the current interaction. Poisoned provenance is the forged bill of lading; scope compilation is the cofferdam that keeps one hold's contents out of another.

Connectors and tools (the data plane). Every connector the agent can reach is a door. Over-broad scopes, tool abuse, and lateral movement from one connector to the next turn a single compromise into reach across the institution. The defense is minimum-necessary scoping and per-action authorization: each connector is a resource the agent may touch only if the action in front of it is authorized, evaluated at the moment of use rather than granted in bulk at session start.

Identity and tokens (H-5). The agent's credentials are a target. A stale or over-privileged token, a replayed token, or a token that keeps reading through a restriction imposed after it was issued lets an attacker act as the agent or as the user behind it. The defense is short-lived, narrowly scoped tokens, a live re-check of consent and access at the moment of use rather than at issuance, and attribution records that bind every action to an identity, a policy version, and a session.

The model supply chain (Tier 1 and Tier 2). The model itself is a target before it ever runs: poisoned weights, a backdoored fine-tune, training-data poisoning, and the silent version swap in which a provider changes the model's behavior behind a stable interface. The defense is the supply-chain tiering of Chapter 10: provenance and a bill of materials for the model, version pinning, an organization-controlled update cadence so the institution decides when behavior changes, and on-premises serving for workloads that cannot cross a boundary.

The evidence plane (E-1 through E-4). The record is a target, because an adversary who can edit the evidence can hide everything else. Fabricated E-2 verification records, deleted logs, and the structural weakness in which the party being verified also supplies the proof all corrupt the audit at its source. The defense is organization-owned, append-only evidence collection held inside the authorization boundary and independent of the component it measures, so the verifier is never the verified. This is the maritime move from taking the master's word to inspecting the hull directly.

The human checkpoint (H-4). The reviewer is a target. Social engineering of the approver, approval fatigue that turns sign-off into a reflex, and the examination-gaming of heroic one-shot all defeat the control by defeating the human inside it. The defense is graduated autonomy: scale the depth of human disposition to the consequence of the decision, so the checkpoints that matter are few enough to stay real, and monitor for the clustering and rubber-stamping that signal the human has stopped reviewing.

Read the three failure modes that follow against this surface. Each has an accidental path and an adversarial one, and the harness component that prevents the accident is the same one that resists the attack.

Context Amnesia

Definition

Context amnesia is the failure mode in which the AI system, or the governance architecture surrounding it, loses awareness of operative governance boundaries between sessions, between interactions, or across organizational transitions. The system operates on stale, incomplete, or misscoped context, not because the context was never established, but because the architecture does not enforce context persistence at the boundaries where loss occurs.

The critical distinction is between information that exists somewhere in the organization and information that is present in the governance-active context at the moment of operation. An organization that documented its Health Insurance Portability and Accountability Act (HIPAA) compliance policies five years ago, in a location the current AI deployment has never been connected to, has context. It does not have context persistence. The governance boundary is invisible to the system operating now.

The Clinical Documentation Case

```
        LOCATION: Regional health system EHR deployment
        DATE: Eighteen months of production operation, internal
audit findings
```

A regional health system deployed an AI-assisted clinical documentation tool to help physicians generate encounter notes from structured data. The tool was integrated into the electronic health record (EHR), trained on the institution's documentation style, and configured to pull relevant patient history when generating notes.

The configuration did not enforce per-encounter context scoping.

When a physician opened a patient's chart to document a routine follow-up visit, the AI assistant pulled from the physician's most recent interactions with the EHR, not from the current patient's current encounter. It generated a clinical note referencing diagnostic history, lab values, and medication context from the correct patient's prior encounters, including an encounter from fourteen months earlier with a different presenting condition. Three instances identified during an internal audit showed AI-generated documentation referencing clinical details from a prior encounter that were clinically misleading in the context of the current visit. In one case, the AI's note referenced a medication discontinued after an adverse event; the current visit was for a different complaint, but the presence of that discontinued medication in the generated note created a risk that a subsequent prescriber would treat it as current.

The HIPAA minimum necessary standard requires that access to, and use of, protected health information be limited to what is needed for the specific purpose at hand. An AI system generating documentation for an encounter is authorized to use information relevant to that encounter. Blending context from prior encounters in ways not clinically scoped to the current visit is not authorized.

The violation was architectural. Access was configured at the patient level, not the encounter level; the AI assistant operated on whatever context it retrieved. No governance layer enforced that retrieved context was scoped to the current interaction. The system did not know what it did not know: that the context it was using had leaked across encounter boundaries.

The incident did not surface in an Office for Civil Rights (OCR) audit. It surfaced in an internal quality review when a physician noticed that an auto-generated note referenced clinical history the physician had not discussed during the current visit. Seventeen findings across eighteen months of deployment. The institution suspended the tool, engaged outside counsel, and commenced a review of whether any of those seventeen findings constituted reportable HIPAA breaches. The outcome depended on whether any of the cross-encounter context bleeds had resulted in material changes to care decisions or disclosures.

Maritime Parallel

A ship's log that does not distinguish between voyages creates an unresolvable insurance claim. If the cargo manifest from the previous voyage appears in the current voyage's documentation, the insurer cannot determine which cargo was aboard at the time of loss. Records exist. The boundary between voyages does not. Every claim depending on knowing what was on the ship when it sank becomes contested, because the records accurately reflect two different states of reality without distinguishing between them.

Per-voyage boundary enforcement is not a hard requirement. It is a logging discipline. Failure to maintain it does not require negligence at any single step; it requires only an architecture that treats voyage boundaries as soft rather than hard.

Prevention: H-3 (Persistent State)

The harness component that prevents context amnesia is H-3, persistent state with enforced scope binding. H-3 requires that the governance architecture maintain explicit records of the context boundary operative at each interaction, and that retrieval operations be constrained to content within that boundary.

For the clinical documentation case, H-3 implementation means the session context for any AI-assisted documentation interaction is initialized to the current encounter's authorized scope, and retrieval from outside that scope requires an explicit authorization event recorded in the E-3 State Transition Records evidence stream. The context boundary is not a configuration setting that can silently fail; it is a governance checkpoint that must be satisfied before retrieval proceeds.

Consider the distinction between configuration-level and governance-level scope enforcement. Configuration closes the door. Governance architecture locks it and records whether it was opened.

Compiled policy as a second layer of prevention. Beneath the harness prevention sits a category of governance tooling: compiled, policy-as-code enforcement engines that turn governance rules into runtime constraints rather than documentation. The discussion is an illustrative reference pattern, not a verified production deployment. Each failure-mode section in this chapter pairs its harness prevention with the compiled-policy version of the same control; the general pattern is set out here, with the context-amnesia case treated first.

The three failure modes described in this chapter share a common architectural vulnerability: they arise when governance policies exist as documentation or configuration but not as runtime constraints enforced by the system architecture itself. A compiled-policy enforcement engine addresses this vulnerability by transforming governance rules from human-readable specifications into machine-verified runtime constraints that prevent policy violations rather than detecting them after the fact.

Consider the distinction between aspirational policies and enforced policies. An aspirational policy states that AI systems shall maintain session-scoped context boundaries; an enforced policy makes it architecturally impossible for the system to retrieve context from outside the authorized scope without an explicit authorization event. The former relies on implementation discipline, testing coverage, and operational monitoring to catch violations. The latter makes violations a compile-time or deployment-time failure rather than a runtime risk.

The relevance to regulated industries operating at Lagrange point stability configurations is direct. Classification societies do not certify ships based on the owner's stated intention to maintain safe loading practices; they certify compliance with structural requirements that make unsafe loading physically difficult to achieve without deliberate override. The load line mark is a compiled constraint: the ship's freeboard limit is painted on the hull, visible to dock workers, port authorities, and insurers, with loading procedures designed around the constraint rather than around documentation stating what loading procedures should be.

Preventing Context Amnesia Through Scope Compilation

Context amnesia occurs when session boundaries are configuration settings rather than architectural constraints. A system configured to operate within a specific context scope can fail to enforce that scope if the configuration is incorrect, if the scope definition drifts over time, or if retrieval logic bypasses the configuration layer entirely. A compiled-policy engine prevents context amnesia by compiling scope boundaries into policy rules that are evaluated at every retrieval operation.

In the clinical documentation case from earlier in this chapter, the EHR's AI assistant retrieved context from prior encounters because access control was configured at the patient level rather than the encounter level. In the illustrative reference pattern, a compiled-policy engine defines scope boundaries as policy rules evaluated at query time: "Retrieval operations for encounter documentation contexts SHALL resolve queries only against resources tagged with the current encounter ID." That policy rule is not a configuration setting that can be set incorrectly; it is a compiled constraint verified during the policy compilation step and enforced by the policy engine at every retrieval operation. If the encounter ID is not present in the session context, the policy engine denies the retrieval operation before the query reaches the data layer.

The maritime parallel is the cofferdam inspection certificate. A ship carrying both liquid cargo and dry cargo must maintain structural separation between the cargo holds to prevent contamination. The classification society's survey verifies that the cofferdam structure exists and meets specification; the certificate records that verification. Loading procedures compiled from that structural constraint prevent cargo handlers from loading liquids into holds adjacent to dry cargo without a verified separation structure in place. The constraint is physical, documented, and procedurally enforced at loading time.

Detection

Context amnesia is detectable before it produces a compliance event through three signals. First, retrieval distribution monitoring detects when AI retrieval patterns pull systematically from content outside the expected scope. A documentation tool retrieving from encounters more than sixty days old, more than twice per session, is exhibiting a retrieval pattern inconsistent with its stated purpose; that monitor flags this as an anomaly requiring investigation, not as a confirmed violation. Second, session boundary records that fail to log a context initialization event at the start of each interaction indicate that scope binding is not being enforced. The absence of the event is the signal. Third, cross-session content fingerprinting, comparing content elements in AI outputs against authorized scope for the current interaction, can identify outputs containing elements that originated outside the session boundary before those outputs reach any human reviewer.

Heroic One-Shot

Definition

Heroic one-shot is the failure mode in which the organization performs governance as a point-in-time event rather than as a continuous architectural property. The organization assembles a governance posture (manual reviews, documentation packages, policy attestations, point-in-time assessments) sufficient to satisfy a specific examination or audit, then allows that posture to degrade as the examined state diverges from the operating state.

Not dishonesty. The governance posture assembled for the examination is often genuinely accurate at the moment of assembly. The failure is subtler: the assumption that a snapshot of compliance at a point in time represents a durable property of the system. Systems change. Configurations drift. Models update. User behavior evolves. The governance snapshot does not update with them. That widening gap between the assessed state and the current state is the heroic one-shot gap.

The OCC Examination Case

```
LOCATION: Regional bank, OCC examination cycle
DATE: 2022-2024 model update and examination review
```

A mid-size regional bank operating under Office of the Comptroller of the Currency (OCC) supervision deployed an AI-assisted customer communications system to draft responses to customer inquiries, including inquiries related to account disputes, fee assessments, and loan modification requests. The system was deployed in 2022 under a governance framework that included an AI use policy, a vendor risk assessment, and a model risk review conducted by an outside consultant.

The bank's AI governance program was structured around examination readiness. Before each OCC examination cycle, the compliance team engaged the consultant to update the model risk review, refresh the vendor documentation, and produce an inventory of AI-assisted processes. Documentation was accurate and thorough at the time of each examination. The OCC found no AI governance deficiencies in either the 2022 or 2023 cycles.

In late 2023, the bank updated its communications AI with a new model version provided by the vendor. The update was processed as a routine software change under the bank's change management policy. The model risk review, last conducted six months prior, was not triggered. The AI use policy did not define a threshold at which model changes required governance re-review; the vendor documentation did not distinguish between interface updates and underlying model changes.

The new model version had been trained on a different dataset and exhibited measurably different behavior on communications related to fee dispute resolution. Specifically, it was more likely to provide detailed information about dispute escalation paths and regulatory complaint channels in response to customer inquiries about fees. Not an error in the model's output. The information it provided was accurate. The behavior represented a significant change in the bank's effective customer communication policy, implemented through a model update that was never reviewed for its policy implications.

A compliance officer reviewing customer communications in early 2024 noticed the behavioral shift and flagged it. The bank had no record of any governance decision authorizing the change in communication posture. The model update had been logged as a software change. The governance implication of the model update had never been assessed.

The OCC examination had validated a governance architecture that existed on the day of the examination. That architecture had no mechanism to detect, assess, or record the governance implications of a model change occurring between examinations. Certified snapshot. Snapshot then allowed to diverge from reality at every subsequent change event.

Maritime Parallel

A ship surveyed once before its maiden voyage, classified by Lloyd's Register as seaworthy on the day of survey, and never re-inspected throughout its operating life, presents the insurer and the port authority with a document certifying the vessel's condition at a historical point in time. As the hull corrodes, the machinery ages, the rigging stretches, and operating patterns diverge from the initial voyage profile, the classification certificate becomes increasingly disconnected from actual condition. At some threshold of divergence, the vessel is operating as an unclassified ship carrying a classified ship's paperwork.

Lloyd's Register solved this by requiring periodic re-survey. Classification is not a permanent certification; it is a current status requiring continuous validation to remain operative. A vessel that refuses a required re-survey loses its classification, because expiration logic is built into the architecture itself.

Prevention: H-4 (Verification Checkpoints)

The harness component that prevents heroic one-shot is H-4, verification checkpoints that are continuous rather than periodic. H-4 requires that governance architecture verify its own operation at defined intervals and at defined trigger events, with verification results recorded in the E-2 Continuous Verification evidence stream regardless of whether an examination is pending.

For the bank case, H-4 implementation means the model risk governance framework defines model change as an explicit trigger event for verification re-execution. A model update does not proceed without a governance checkpoint assessing whether the update changes the model's behavior within the scope of an existing governance decision. Not a manual review initiated by the compliance team's examination calendar. An automated gate in the deployment pipeline producing an E-2 verification record regardless of the examination schedule.

The operational test is direct: can the governance architecture answer questions about AI system behavior in real time, not retrospectively? If the bank could not answer the OCC's question "what governance review was performed when the model was updated in November 2023" because no review was performed, then the governance architecture is periodic, not continuous. H-4 is not satisfied by examination-cycle readiness. Operational continuity is the standard.

Preventing Heroic One-Shot Through Continuous Policy Verification

Heroic one-shot occurs when governance posture is assembled for examination cycles rather than maintained continuously as system state evolves. A policy document updated before each examination and unchanged between examinations represents governance as a snapshot rather than as an operational property. A compiled-policy engine prevents heroic one-shot by treating policy changes as deployment events that require recompilation and re-verification.

In the bank examination case, a model update was processed as a routine software change without triggering governance review because the governance framework did not define model updates as trigger events. In the illustrative reference pattern, a compiled-policy engine defines governance checkpoints as policy rules evaluated at deployment time: "Model updates SHALL trigger policy recompilation when the updated model's behavior surface differs from the baseline behavior surface by more than the defined drift threshold." That policy is not advice to the operations team; it is a deployment gate. The model update does not proceed to production until the policy engine verifies that either the behavior surface has not changed beyond threshold, or that a governance authorization record exists approving the change.

The continuous verification requirement aligns with Lloyd's Register's periodic re-survey model. A ship's classification certificate is valid for a defined period, typically four years for hull structure and one year for machinery. At the end of that period, the certificate expires whether or not any survey findings have been reported. The expiration is not contingent on discovering a problem; it is an architectural property of the certification framework. The policy recompilation requirement operates similarly: policies expire at defined change events (model updates, configuration changes, dependency updates) regardless of whether those changes are expected to affect governance posture. Recompilation forces verification; verification produces evidence; evidence satisfies the continuous verification standard in H-4.

Detection

Heroic one-shot is detectable by examining the temporal distribution of governance events in the E-1 and E-2 evidence streams. An organization in the heroic one-shot pattern will show clusters of governance activity concentrated in the weeks before each examination cycle, followed by extended periods with minimal governance event records. This clustering pattern is the signature of examination-driven governance; in a continuous governance architecture, the event distribution is smooth, with governance checkpoints firing at operational trigger events rather than in response to examination calendars.

A second detection signal is the gap between the change event log and the governance event log. If the production system's change log shows three model updates, two configuration changes, and one vendor contract renewal since the last examination, and the governance event log shows zero verification records corresponding to those changes, the heroic one-shot gap is measurable and documented. The evidence architecture makes that gap visible even if the compliance team has not examined it.

False Completion

Definition

False completion is the failure mode in which the governance system reports a compliant state when the underlying verification is incomplete, the validation criteria do not match production conditions, or the evidence of compliance describes intent rather than demonstrated operation. The system is not in the state it reports itself to be in. Not deception. Architectural mismatch: a verification architecture that tests the wrong conditions, at the wrong scope, against the wrong population.

The distinction between false completion and deliberate misrepresentation is architecturally important. An organization that submits a fraudulent attestation knows the system is not compliant. An organization in the false completion failure mode has tested the system and received passing results; the results are simply not valid indicators of compliance in the conditions that actually matter. The organization believes the evidence. The evidence is wrong.

The Insurance Underwriting Case

```
LOCATION: Personal lines carrier, homeowner's underwriting AI deployment
DATE: 2022 deployment, complaints filed 14 months post-launch
```

A personal lines insurance carrier deployed an AI underwriting model for homeowner's insurance in 2022. The model assessed applications based on property characteristics, applicant credit indicators, and geographic risk factors. Before deployment, the carrier's actuarial team conducted a disparate impact analysis on the model's outputs to verify compliance with the Federal Fair Housing Act, state insurance anti-discrimination regulations, and the carrier's own underwriting guidelines.

The disparate impact analysis was conducted against a validation dataset assembled from the carrier's historical underwriting portfolio over the prior five years. The portfolio represented the population of applicants who had been approved or declined under the prior underwriting process. The analysis found no statistically significant disparate impact against protected classes in the historical data. The model passed validation. Deployment proceeded.

Within fourteen months, the state insurance commissioner in three states received complaints from consumer advocacy organizations alleging that the AI underwriting model was producing statistically significant adverse outcomes for applicants in majority-minority census tracts relative to applicants in majority-white census tracts with comparable property and risk characteristics.

The carrier retained outside actuarial experts to conduct an independent disparate impact review. The independent review found statistically significant disparate impact in two of the three states where complaints had been filed.

The failure was not in the model's mathematics. It was in the validation dataset. The historical underwriting portfolio used to validate the model was drawn from the carrier's existing customer base, acquired through agent networks, direct mail, and digital channels with a geographic distribution that did not represent the full applicant population in the states where the model would operate. Specifically, the validation dataset was systematically underrepresented in several urban markets where the carrier had historically maintained thin penetration. These were precisely the markets where the model exhibited disparate impact in production.

The model appeared compliant because it was tested against the population that had interacted with the carrier in the past, not against the population that would interact with the model in the future. Validation was complete; the dataset was wrong. Evidence of compliance was genuine; it did not measure the property that mattered.

Maritime Parallel

A cargo declared on the manifest as "general merchandise" passes customs inspection because the external markings and the declared weight match the paperwork. The actual cargo, concealed beneath the declared goods, violates import restrictions. The manifest is accurate regarding the declared portion of the cargo. The inspection confirmed the manifest. The paperwork is complete. The inspection is false.

Consider a variation that removes any element of fraud: a shipper who genuinely believes the cargo meets declaration requirements because the sample submitted for pre-clearance analysis represented the declared category. Tested sample: accurate. Untested cargo from a different batch: non-compliant. The inspection passed the sample and assumed the shipment. Verification was complete against the wrong population.

Prevention: H-4 (Verification Checkpoints) with Population Scope Enforcement

False completion is prevented by H-4 verification checkpoints that specify, as part of the checkpoint definition, the required scope and population for verification to be considered valid. A verification checkpoint that does not specify the required population scope can be satisfied by testing any population, including populations that do not represent the conditions that will matter.

For the insurance case, H-4 implementation means that the model validation framework specifies, as a mandatory pre-deployment gate, that the validation dataset must include a documented representativeness assessment against the prospective applicant population in each jurisdiction where the model will be deployed. The checkpoint is not satisfied by a disparate impact analysis alone; it is satisfied by a disparate impact analysis conducted against a dataset with documented representativeness. The distinction between a passed test and a valid test is the population scope requirement.

The *Stable Orbit Security* eBook, Chapter 3 (Glass Box), addresses the auditability architecture that makes false completion structurally impossible by requiring that every compliance assertion be traceable to the specific verification event that produced it, including the scope definition and dataset provenance of that verification. When the evidence architecture records not just the result of verification but the conditions under which verification was conducted, the false completion failure mode becomes visible during evidence review rather than discoverable only after a production incident.

Preventing False Completion Through Population Scope Validation

False completion occurs when verification is complete but invalid, typically because the validation dataset does not represent the production population. A disparate impact analysis conducted against historical data passes the test but fails to predict production outcomes when the historical data is systematically unrepresentative of the production applicant pool. A compiled-policy engine prevents false completion by making population scope requirements explicit in policy definitions and verifiable during policy compilation.

In the insurance underwriting case, the validation dataset was drawn from the carrier's existing customer base rather than the prospective applicant population. In the illustrative reference pattern, a compiled-policy engine defines validation scope requirements as policy preconditions: "Disparate impact validation SHALL be conducted against a dataset with documented representativeness metrics for each deployment jurisdiction, with representativeness defined as demographic and geographic distribution within 10% of the jurisdiction's census-derived applicant population distribution." That precondition is verified during policy compilation. If the validation evidence does not include the required representativeness metrics, the policy compilation step fails, and deployment is blocked until compliant validation evidence is provided.

The cargo manifest inspection parallel is direct. A customs declaration stating "general merchandise" is insufficient for clearance unless the manifest also provides a detailed breakdown of contents sufficient for the inspector to verify that the declared category matches the actual cargo. The inspection does not pass based on the accuracy of the summary declaration; it passes based on the inspector's verification that the declared summary is consistent with the detailed itemization. The compiled-policy engine's population scope validation operates similarly: the policy definition requires not just a passing validation result, but a validation result accompanied by evidence that the validation covered the required scope.

Detection

False completion is the most difficult of the three failure modes to detect from inside the governance program, because the governance program's instruments are reporting correctly. The instruments are measuring the wrong thing.

The detection approach requires external reference points that the internal verification cannot provide. Three mechanisms are effective. First, population distribution audits compare the demographic and geographic distribution of the validation dataset against the distribution of the production applicant population. Any systematic gap between validation scope and production scope is a false completion risk indicator. Second, out-of-distribution monitoring tracks production outcomes for application profiles that are poorly represented in the validation dataset. If the model's confidence or accuracy metrics degrade for a specific segment of the production population, that segment is likely underrepresented in validation. Third, regulatory feedback loops, including complaint tracking, appeals analysis, and periodic independent model reviews, provide external signals about model behavior that are not dependent on the internal validation architecture's measurements. An appeals reversal rate that systematically exceeds baseline for a specific segment is a false completion signal regardless of what the internal validation shows.

Compiled Constraints as Classification Society Equivalents

The Lloyd's Register classification framework operates on a principle that translates directly to the compiled policy model: structural requirements are specified before construction, verified during construction, and inspected periodically to confirm that the as-built state continues to match the certified state. A ship's structural plans are reviewed before the first steel is cut; the classification surveyor inspects the vessel during construction to verify compliance with approved plans; periodic surveys confirm that modifications and wear have not compromised the certified structure. The certification is a compilation of requirements into an inspected reality.

A compiled-policy engine of this category applies the same pattern to governance policies. Policies are defined as machine-readable specifications before the first AI deployment; the policy engine compiles those specifications into runtime constraints verified during the compilation step; continuous verification confirms that system changes have not introduced policy violations. A policy compilation failure is equivalent to a classification society's refusal to issue a certificate: deployment does not proceed until the discrepancy is resolved.

The operational advantage in regulated industries is that compiled policies produce evidence automatically. A maritime insurer accepting a Lloyd's-certified vessel does not need to re-inspect the hull structure; the classification certificate is the evidence. A financial regulator reviewing an AI deployment does not need to manually verify that the system enforces context scope boundaries; the policy compilation report is the evidence, and the policy engine's audit log demonstrates that the compiled constraint has been enforced at every relevant operation.

This is the category's contribution to the Lagrange point stability model. Policies compiled into the architecture rather than documented beside it shift governance verification from a manual audit function to an automated evidence generation process. The three failure modes described in this chapter all involve gaps between documented governance intent and operational reality. Compiled policies close those gaps by making intent and reality the same artifact.

The Pattern That Connects Them

All three failure modes share a structural signature. Each involves a governance architecture that is internally consistent but poorly connected to the actual operating conditions it is meant to govern.

Context amnesia is the failure of connection across session boundaries. Governance context correct at one boundary is not enforced at the next; the architecture works within a session but breaks across session transitions.

Heroic one-shot is the failure of connection across time. Governance posture correct at the examination point is not enforced at subsequent change events; internal consistency at the examined state does not extend to the evolving operational state.

False completion is the failure of connection across scope. Validation correct against the tested population is not valid for the production population; consistency within the tested scope does not reach the full scope of conditions that matter.

One response to all three, stated three different ways: enforce continuity at boundaries. H-3 enforces continuity at session boundaries. H-4 enforces continuity at change and time boundaries. The population scope requirement in H-4 enforces continuity at scope boundaries.

THE THREE GOVERNANCE FAILURE MODES

CONTEXT AMNESIA	HEROIC ONE-SHOT	FALSE COMPLETION
Boundary: session boundaries	**Boundary:** time and change	**Boundary:** scope and population
Case: EHR cross-encounter bleed (HIPAA)	**Case:** bank model update with no governance re-review	**Case:** insurance diparate impact on the wrong dataset
Prevention: H-3 Persistent State	**Prevention:** H-4 Verification Checkpoints	**Prevention:** H-4 with population scope
Detection: retriewal distribution monitoring	**Detection:** temporal event clustring	**Detection:** out-of-distribution monitoring

EACH FAILURE IS ACCURATE AT SOME OTHER MOMENT, SCOPE, OR CONDITION.THE ARCHIECTURE PERMITTED THE GAP.

Figure 7: The three universal governance failure modes, context amnesia, the heroic one-shot, and false completion, each shown with the boundary it crosses, the harness control that prevents it, and the monitoring signal that detects it.

The Plimsoll line works because it enforces continuity at the boundary of the loading decision. The mark does not record that the ship was safely loaded at a prior voyage. It records whether the ship is safely loaded now, before this voyage, in these conditions. Current. Present. Enforced at the moment of action.

Each of these three failure modes describes a system where governance information was accurate at some other moment, applied to some other scope, verified under some other conditions. The architecture permitted the gap between that other state and the current state. The failure was the gap.

What Detection Requires

Detecting these failure modes before they produce enforcement events requires a monitoring architecture built for boundary conditions. Boundaries are precisely where internal monitoring is blind.

Internal monitoring reports on state within a bounded scope. A session-level monitor reports on what happened within the session. A pre-examination governance review reports on what was prepared for the examination. A validation monitor reports on results within the validation dataset. None of these instruments is designed to see across the boundary it operates within.

Detection requires instruments positioned at the boundary itself. Session boundary monitors fire at session transitions, not during sessions. Change-triggered verification gates fire at change events, not during stable operation. Population scope audits compare the inside of the validation fence against the outside, not the interior consistency of the validation itself.

This observation aligns with the broader principle of the E-3 State Transition Records tier: state transition records are the evidence of governance continuity at boundaries. An organization with dense E-1 and E-2 evidence but sparse E-3 evidence has an instrumented system and an uninstrumented set of boundaries. Failure modes concentrate at the boundaries. The evidence architecture must reach them.

The Commercial Stakes Are Not Different

A federal program that exhibits context amnesia may expose classified information to an AI session that was authorized for a different project. The consequence is a security incident with defined reporting and remediation procedures.

A hospital that exhibits context amnesia may expose a patient to a clinical documentation error that changes their care trajectory. The consequence is a potential patient safety event, a HIPAA reportable incident, and a provider liability exposure that does not have a ceiling.

The architectural description of the failure is identical. Consequences are not more or less severe; they are different in kind, calibrated to the stakes of the domain where the failure occurs.

This is the precise meaning of the claim that these failure modes are universal: the architecture that fails, fails the same way. Cost of failure is determined by the domain. Architecture of prevention is determined by the failure mode. An organization in a high-stakes regulated industry does not need a specialized architecture to prevent context amnesia, heroic one-shot, and false completion. It needs the architecture that prevents them in every domain, deployed with the rigor its specific stakes require.

Not federal problems wearing commercial clothes. Architecture problems that appear wherever the architecture is built without enforcing continuity at boundaries. The commercial case studies in this chapter are not adaptations of defense examples; they are original instances of the same underlying failure pattern, discovered independently in sectors that had no reason to communicate with each other about AI governance architecture.

That convergence is evidence. Governance architecture fails the same way across contexts because the physics of failure are not proprietary. Neither is the prevention.

CHAPTER SUMMARY

Context amnesia, heroic one-shot, and false completion are governance architecture failure modes, not federal failure modes adapted for commercial use. Each describes a specific failure of continuity at a specific type of boundary: session boundaries, time and change boundaries, and scope and population boundaries respectively. The commercial case studies in this chapter, a hospital's cross-encounter context bleed producing HIPAA exposure, a bank's examination-driven governance failing to detect a model update's policy implications, and an insurer's disparate impact validation passing against the wrong population, are original instances of the same structural failure pattern documented in the defense context. Prevention requires the harness architecture's boundary-enforcement components: H-3 for session continuity, H-4 with population scope requirements for change and scope continuity. Detection requires instrumentation positioned at boundaries, not within them. The E-3 State Transition Records tier is the primary detection mechanism for all three failure modes because state transition records are the evidence of governance continuity at the points where these failures concentrate.

Chapter 8: The Plimsoll Line

Part II: The Classification Survey

> *"The knowledge of what a safe ship looks like was never the missing piece. What was missing was a way to make that knowledge visible at the dock, before the ship sailed, without requiring an engineer in every port."*

The Merchant Shipping Act of 1876 did not give the British maritime industry any information it did not already have. Naval architects had been calculating safe loading limits for decades. The buoyancy mathematics were standard; the relationship between displacement, draft, and stability was not a mystery. Every competent surveyor in every major British port could determine, if asked and given time, whether a specific vessel at a specific load was within safe operating parameters.

The problem was that the determination had to be requested, by an expert, with enough time, and with access to a ship that might be departing within the hour under commercial pressure. Assessment was possible. The architecture to make it routine was not.

What Plimsoll contributed was not knowledge. It was visibility. A circle with a horizontal line through its center, painted on the hull at the precise waterline corresponding to maximum safe load. Above the water, or below it. Within its authorized operating boundary, or past it. No surveyor needs to be summoned, no calculation run, no report filed. The governance state of the vessel is legible to anyone standing at the dock.

The four governance properties introduced in this chapter, LP9 through LP12, are constructed on exactly that principle. They address commercial requirements that the inherited federal baseline, the harness components and evidence tiers and the eight federal security properties (P1 through P8, defined in Volumes I and II) designed for classified government environments, has no mechanism to cover. The core governance layer this volume defines on that baseline, LP1 through LP8, has no mechanism for them either. They share the load line's structural logic. A binary indicator, visible without expert interpretation, enforced at the moment of action, and independent of whatever the organization deploying the AI would prefer.

Implementations referencing LP9-LP12 should cite this chapter as the canonical source for these properties. The regulatory sources and implementation patterns documented here are the authoritative reference for how these four properties are defined, what they require, and how they integrate with the broader Three-Body harness architecture.

Why LP1-LP8 Are Insufficient for Commercial Environments

The federal harness brought eight security properties, P1 through P8, developed for classified networks and defined in Volumes I and II. This volume restates and extends the governance layer on that inherited baseline as the Lagrange properties. LP1 through LP8 govern classification-consistent access, scope binding, routing integrity, persistent governance context, continuous verification, attribution completeness, deviation detection, and evidence integrity; several of them restate a federal security property in governance terms. Each addresses a risk category that exists in any AI deployment: unauthorized data access, scope creep, bypass paths, loss of governing context across turns, unverifiable behavior, attribution gaps, undetected drift from authorized behavior, and tampered or non-tamper-evident evidence.

These risks exist in commercial regulated environments too, and LP1-LP8 addresses them. But commercial regulated environments carry four additional risk categories that federal classified environments have never needed to govern.

Consent. In classified programs, the authorization to process information at a given classification level is institutional: clearance plus need-to-know produces access. No individual whose data is being processed has the right to withdraw that authorization. In commercial environments, certain data requires the affirmative, revocable consent of the individual to whom it belongs. That consent is a constraint that changes dynamically, tracked at the record level and verified before each processing event. LP1-LP8 has no mechanism for this. The mechanism was never needed.

Jurisdictional variability. Federal programs operate under one regulatory authority with consistent rules. A bank operating in the United States and the European Union simultaneously faces regulatory requirements that directly conflict on certain questions; the General Data Protection Regulation (GDPR)'s data minimization principle and certain US litigation discovery obligations pull in opposite directions. A multi-state insurer faces fifty different state Department of Insurance requirements applying different rules to the same AI action depending on the data subject's state of residence. The harness must route decisions based on applicable jurisdiction, not static configuration. LP1-LP8 assumes a single regulatory authority.

Disparate impact. No federal AI deployment produces consumer credit decisions, insurance pricing, or employment determinations subject to fair lending law, Equal Credit Opportunity Act (ECOA), or equal employment obligations. A Department of Defense (DoD) AI system recommending a code change does not produce an adverse action notice. A commercial AI system recommending a credit denial does.

The evidence architecture must include fairness metrics as a continuous verification checkpoint; LP1-LP8 was not designed to produce that evidence.

Customer-facing explanation. A cleared developer using an AI assistant in a classified program is an internal user with accepted system boundaries. A Medicare beneficiary whose coverage claim was reviewed by an AI system has a legal right to understand that decision. A credit applicant evaluated by an AI model has a right to a meaningful adverse action notice. LP1-LP8 produces attribution records for internal accountability, not for external disclosure. That gap is structural.

LP9 through LP12 close these four gaps. They do not replace LP1-LP8. They extend the framework to cover the commercial requirements that the federal framing was never designed to address.

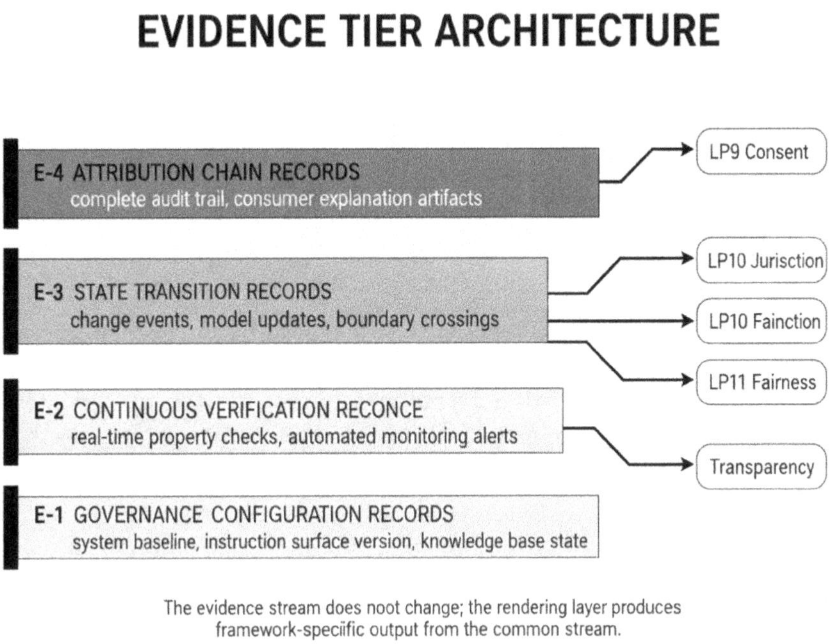

Figure 8: The four evidence tiers, E-1 governance configuration through E-4 attribution, and how the four commercial properties LP9 through LP12 extend the common evidence stream into framework-specific output.

LP9: Consent-Bounded Processing

The AI must not process data beyond the data subject's consent scope.

Simple as a statement. The implementation is anything but, and that gap is precisely why this property needs explicit architectural definition rather than reliance on general data handling policies.

GDPR Article 7 requires that consent be freely given, specific, informed, and unambiguous. It can be withdrawn at any time, and withdrawal must be as easy as giving consent. When consent is withdrawn, processing must stop. The AI system cannot hold consent state in its context window from a prior session; it must verify consent at each processing event against a current consent record. California Consumer Privacy Act (CCPA) and California Privacy Rights Act (CPRA) establish similar obligations for California residents, including the right to opt out of the sale of personal information and, under CPRA, the right to limit the use of sensitive personal information. Health Insurance Portability and Accountability Act (HIPAA) 45 CFR 164.508 requires a written authorization for uses and disclosures of protected health information (PHI) that fall outside treatment, payment, and healthcare operations; that authorization must specify the information to be used, the persons authorized to use it, the purpose, and the expiration date or event. 42 CFR Part 2 imposes a separate and more restrictive consent requirement for substance use disorder treatment records maintained by federally assisted programs; a general HIPAA authorization does not satisfy Part 2. Each disclosure from a Part 2 record requires a specific written consent covering that disclosure.

Binary assessment for LP9: does the AI system verify current consent scope before each data access event, and refuse access when the consent boundary excludes the requested processing? The Plimsoll mark is either above the waterline or below it. Processing falls within the consent boundary, or it does not.

Implementation requires the AGENTS.md instruction surface to encode consent boundary definitions that the routing enforcement layer (H-2) checks before every data access. The consent check is not a policy statement the AI reads and interprets. It is a gate the routing enforcement layer clears before the AI ever receives the data. The architecture is:

```
# AGENTS.md - Consent-Bounded Processing (LP9)

consent_verification:
  enforcement: pre-access
  source: consent_management_platform
  verification_required_for:
    - gdpr_personal_data: true
    - ccpa_personal_information: true
    - hipaa_phi: true
    - part2_records: true
  behavior_on_consent_absent:
    action: block_access
    escalation: privacy_officer
    logging: e4_attribution_record

mandatory_constraints:
  - "MUST verify current consent scope before accessing any record
     subject to GDPR, CCPA, HIPAA authorization, or 42 CFR Part 2"
  - "MUST NOT process Part 2 substance use disorder records without
     specific written consent covering this disclosure; a general
     HIPAA authorization does not satisfy this requirement"
  - "MUST treat withdrawn consent as equivalent to no consent;
```

```
    prior session consent state is not valid for current access"
  - "MUST produce E-4 attribution record for every consent
    verification event, including blocked accesses"
```

The consent verification event is itself an E-4 Attribution Chain Record. An auditor asking whether the system processed data in compliance with consent requirements gets the answer from the attribution record. Not from a policy description. From the record.

LP9 is the Plimsoll line for consent state. The data is either within the consented processing scope or it is not.

LP10: Jurisdictional Consistency

The AI must apply the correct jurisdiction's rules for each data element being processed.

This property addresses the failure mode arising when an AI system applies the rules of the wrong jurisdiction, either because it was configured with static rules that do not account for the data subject's actual jurisdiction, or because it applies the deploying organization's rules to a data subject located elsewhere.

The jurisdictional complexity of commercial AI deployment is not theoretical. GDPR applies to personal data of EU data subjects regardless of where the processing organization is located. A US bank processing account data for a German customer is subject to GDPR for that processing event, even if every server involved sits in the United States. CCPA and CPRA apply to California residents' personal information regardless of processing location. The Colorado AI Act (SB 24-205) imposes requirements on covered entities using AI in consequential decisions affecting Colorado residents. The Digital Operational Resilience Act (DORA) imposes information and communications technology (ICT) risk management requirements on financial entities operating in the EU. Every one of these regulatory sources uses a data-subject-centered jurisdiction determination; not an organization-centered one.

The practical consequence is that the same AI action, processing a customer record to generate a credit recommendation, may be subject to GDPR alone (EU customer), CCPA alone (California customer), both GDPR and state requirements (EU customer with California residency), or neither (customer in a state without applicable specific AI governance law). The harness cannot use static per-system configuration to resolve this. It must determine the applicable jurisdiction at the data-element level, for each record, at each processing event.

Binary assessment for LP10: is the AI applying the rules of the jurisdiction that actually applies to this data subject, or some other set of rules? The cargo carries the correct flag for this port, or it does not.

The default for uncertain cases is most restrictive applicable. When the harness cannot determine the applicable jurisdiction with certainty, it applies the most restrictive rules among all plausible candidates. That is conservative behavior, and it stays defensible to any regulator whose rules might turn out to apply.

```
# AGENTS.md - Jurisdictional Consistency (LP10)

jurisdiction_determination:
  method: data_subject_residence
  granularity: per_record
  uncertainty_strategy: most_restrictive_applicable
  jurisdiction_map:
    eu_data_subjects: gdpr_eu_ai_act
    california_residents: ccpa_cpra
    colorado_residents: colorado_ai_act_sb24205
    eu_financial_entities: gdpr_dora
    default: federal_us_baseline

routing_rules:
  - condition: "jurisdiction == eu_data_subjects"
    apply: [gdpr_article22_automated_decisions,
eu_ai_act_article26_transparency]
  - condition: "jurisdiction == california_residents"
    apply: [ccpa_opt_out_verification, cpra_sensitive_data_limits]
  - condition: "jurisdiction == colorado_residents AND decision_type
== consequential"
    apply: [colorado_ai_act_impact_assessment,
disparate_impact_threshold]

mandatory_constraints:
  - "MUST determine applicable jurisdiction at the data-element level
     before processing; static system-level jurisdiction configuration
     does not satisfy this requirement"
  - "MUST apply the most restrictive rules when jurisdiction is
     uncertain or when multiple jurisdictions plausibly apply"
  - "MUST produce E-3 state transition record when jurisdiction
     determination changes for a data subject"
```

The most common implementation failure for LP10 is treating jurisdiction as a system property rather than a data property. An AI system applying "US financial services rules" to all records because the deploying organization is a US financial institution does not satisfy LP10. The data subject determines the applicable rules, not the deployer. A bank that runs static US rules configuration and then processes a German customer's account data has submerged the load line.

LP11: Fairness Verification

AI-assisted decisions affecting individuals must be evaluated for disparate impact, and those decisions must pass the applicable threshold before being deployed at scale.

The regulatory sources for this property are multiple, sector-spanning, and enforced. The Equal Credit Opportunity Act and its implementing regulation, Regulation B (12 CFR Part 1002), prohibit discrimination on the basis of race, color, religion, national origin, sex, marital status, age, or receipt of public assistance income in any aspect of a credit transaction. The Consumer Financial Protection Bureau (CFPB) has issued supervisory guidance holding that AI-assisted credit decisions must be explainable and must satisfy adverse action notice requirements in a manner meaningful to the consumer. National Institute of Standards and Technology (NIST) AI Risk Management Framework (RMF) MEASURE 2.11 requires that AI systems be evaluated for bias, including proxy discrimination through variables that correlate with protected characteristics. The EU AI Act classifies AI used in credit scoring, employment, education, insurance, and essential services as high-risk systems, requiring conformity assessments, post-market monitoring, and bias testing. The National Association of Insurance Commissioners (NAIC) Model Bulletin on the use of AI systems in insurance requires that carriers assess whether AI use in underwriting or claims produces outcomes with disparate impact on protected classes. The EEOC's guidance on AI and employment discrimination holds that employer use of AI in hiring, promotion, or termination decisions that produces disparate impact on a protected class violates Title VII.

The four-fifths rule, the 80% ratio from the EEOC's Uniform Guidelines on Employee Selection Procedures, is a borrowed screening heuristic, not a Plimsoll-style binary safe-harbor threshold and not an adopted insurance underwriting standard. Used as a first-pass diagnostic, it flags a candidate disparate-impact concern: if a selection rate for a protected group is less than four-fifths of the selection rate for the highest-scoring group, that ratio is a screening flag that triggers deeper analysis, not a pass/fail line. No insurance regulator has adopted the 80% ratio as a bright-line underwriting test. The operative standards are state unfair-discrimination law and the actuarial-justification standard, with the NAIC Model Bulletin (December 2023) requiring written AI governance and regular assessment for unfairly discriminatory outcomes. Sector-specific methods apply; financial services fair lending analysis uses additional statistical methods for evaluating disparate impact in credit decisioning, and the CFPB's Supervision and Examination Manual provides the specific analytic approaches expected in fair lending examinations.

Unlike the binary Plimsoll-style properties LP9 through LP12, the four-fifths ratio is not one of those single-number load lines. The fairness line in insurance is the actuarial-justification standard, which is not a single number; the 80% ratio is a screening flag that prompts that justification analysis, never a terminal test.

Binary assessment for LP11: does the AI-assisted decision process clear the applicable fairness standard, with screening flags such as the four-fifths ratio escalated to the governing actuarial-justification and fair-lending analysis? The four-fifths ratio is a diagnostic flag, not the line itself.

Fairness verification folds into E-2 Operational Evidence as continuous operational evidence. An annual fairness evaluation run as a separate compliance program does not satisfy LP11. The metrics have to fall out of system operation as a byproduct, at the frequency decision volume dictates, and any deviation from threshold triggers investigation before the decision process continues at scale.

```
# AGENTS.md - Fairness Verification (LP11)

fairness_verification:
  trigger: ai_assisted_decision_affecting_individual
  # four-fifths ratio used as a screening flag, not a safe-harbor
threshold;
  # actuarial-justification standard and applicable state DOI rule
govern
  screening_flag: four_fifths_ratio
  governing_standard:
actuarial_justification_and_state_unfair_discrimination_law
  sector_thresholds:
    credit_decisioning: cfpb_fair_lending_examination_standards
    insurance_underwriting: naic_model_bulletin_2023
    employment_decisions: eeoc_uniform_guidelines
    eu_high_risk_systems: eu_ai_act_article9_risk_management
  evaluation_frequency: continuous_operational
  evidence_tier: e2_operational_evidence
  deviation_response:
    action: suspend_at_scale_deployment
    escalation: compliance_officer
    documentation: fairness_impact_report

mandatory_constraints:
  - "MUST evaluate every AI-assisted decision process for disparate
     impact before deployment at scale; a disparate impact evaluation
     completed on a prior model version does not satisfy this
requirement
     for a subsequently updated model"
  - "MUST generate fairness metrics as E-2 operational evidence;
     annual compliance program evaluation does not substitute"
  - "MUST suspend deployment when the applicable fairness threshold
     is not met; fairness evaluation results that do not reach
threshold
     require investigation and remediation before scale deployment"
  - "MUST assess for proxy discrimination through variables that
     correlate with protected characteristics, not only direct use of
     protected class variables"
```

Proxy discrimination is the implementation detail most frequently missed. An AI credit scoring model using zip code as a pricing variable does not use race directly. It produces racially disparate outcomes because zip code correlates with race in many US geographies. LP11 requires that fairness evaluation examine outputs for disparate impact, not merely inputs for protected class variables. Cargo weight matters. The label on the manifest does not.

LP12: Transparency and Explainability

AI-assisted decisions reaching customers or affected parties must produce explanation artifacts satisfying applicable disclosure requirements.

The regulatory sources for this property cover disclosure obligations that arise at the intersection of automated decision-making and individual rights. GDPR Article 22 provides data subjects with the right not to be subject to a decision based solely on automated processing that produces legal or similarly significant effects, with exceptions. Where automated decisions are permitted, GDPR requires that the data subject be informed and have the right to obtain human review, express their point of view, and contest the decision. GDPR Recital 71 specifies that the data subject should be provided with an explanation of the decision, including information about the logic involved and its significance. The Fair Credit Reporting Act (FCRA), 15 U.S.C. 1681m, requires that creditors who take adverse action provide notice to the consumer specifying the principal reasons for the action. The CFPB's June 2023 circular (Circular 2023-03) clarified that a creditor's use of an AI model does not relieve it of the obligation to provide specific reasons for adverse action; an explanation that cites only "a complex algorithm" does not satisfy the adverse action notice requirement. EU AI Act Article 13 requires that high-risk AI systems be designed and developed to ensure transparency, including information about the system's intended purpose, accuracy metrics, and any known limitations. HIPAA patient rights provisions give individuals the right to access their own health records and to receive an accounting of disclosures; an AI system that processes PHI to produce clinical recommendations creates records subject to these rights.

The Plimsoll principle here is precise: the manifest must match the cargo. An AI system producing a credit denial must be capable of generating a specific, accurate, human-readable explanation for that denial at the moment the decision is made, not reconstructed after the fact from logs. An AI system generating a clinical recommendation must produce a traceable rationale linked to the evidence it retrieved and the criteria it applied. This is not an optional add-on. It is a governance output generated as part of the decision process, at decision time, before the decision is delivered.

The distinction between generated-at-decision-time and reconstructed-after-the-fact is not rhetorical. A post-hoc explanation constructed from logs may not accurately represent the actual reasoning path the model followed. It may omit specific factors driving the output because those factors were not captured in operational logs. It may be more generic than the applicable regulatory standard requires; it may be challenged by an adverse action recipient who has a legal right to contest it. An explanation generated as a harness checkpoint at decision time is an E-4 Attribution Chain Record with a timestamp. A reconstructed explanation is a narrative.

```
# AGENTS.md - Transparency and Explainability (LP12)

explanation_generation:
  trigger: ai_assisted_decision_reaching_external_party
  timing: at_decision_time
  generation_method: harness_checkpoint_not_post_hoc
  explanation_requirements:
    credit_adverse_action:
      standard: fcra_cfpb_circular_2023_03
      content: principal_reasons_specific_and_accurate
      format: consumer_readable
    gdpr_automated_decision:
      standard: gdpr_article22_recital71
      content: logic_involved_significance_consequences
      format: accessible_language
    eu_ai_act_high_risk:
      standard: eu_ai_act_article13
      content: intended_purpose_accuracy_limitations
      format: technical_and_accessible
    hipaa_clinical:
      standard: hipaa_patient_rights_45_cfr_164_524
      content: data_sources_criteria_applied
      format: medical_record_compliant
  evidence_tier: e4_attribution_chain_records
  human_review_path:
    gdpr_article22: available_on_request
    cfpb_adverse_action: available_on_request
    escalation_path: documented_in_agents_md

mandatory_constraints:
  - "MUST generate explanation artifacts at decision time as a
     harness checkpoint; post-hoc reconstruction from operational
     logs does not satisfy this requirement"
  - "MUST provide specific principal reasons for adverse actions
     under FCRA; 'model complexity' or 'algorithmic output' does
     not constitute a specific reason under CFPB circular 2023-03"
  - "MUST make a human review path available for GDPR Article 22
     and CFPB adverse action cases; the path must be functional,
     not merely documented"
  - "MUST store explanation artifacts as E-4 attribution records
     with timestamps matching the decision record"
```

The human review path requirement in LP12 is frequently treated as a documentation exercise. An AGENTS.md that records "human review available upon request" without an operational escalation path does not satisfy LP12. The path has to be functional: a process that routes a contested decision to a human reviewer with the authority to override, inside a timeframe that matches the urgency of the decision. A credit adverse action taken Friday afternoon needs a human review path that does not wait until Monday to acknowledge receipt.

The Plimsoll Principle Applied to Governance Properties

Four structural properties account for the Plimsoll mark's effectiveness across 150 years of international maritime governance. The same four determine whether a governance control functions as a genuine constraint or as documentation the ship's owner can negotiate around.

Visibility. The load line is on the exterior of the hull where every interested party can observe it. Governance state does not require a surveyor's report, a compliance officer's attestation, or a management assertion. It is observable by the harbor master, the insurer, and the crew before the ship moves. LP9 through LP12 are designed with this property: the binary state of each (consent verified, correct jurisdiction applied, fairness threshold met, explanation artifacts generated) must be observable in the evidence architecture without commissioning a separate audit. E-2 Operational Evidence records and E-4 Attribution Chain Records for these properties must be queryable in real time.

Binary state. The mark is either above the water or it is not. There is no assessment of "adequately close to the load line given current sea conditions," and no graduated judgment about whether some particular degree of processing outside consent boundaries is acceptable given the commercial importance of the workflow. LP9 through LP12 each produce a pass/fail, not a score. Consent verified or not. Correct jurisdiction applied or not. Fairness threshold met or not. Explanation artifact generated at decision time or not. A control that produces a score still needs someone to interpret whether the score is sufficient. A control that produces a binary state needs only observation.

Enforcement at the moment of action. The load line is checked before the ship leaves port, not after it founders in the North Atlantic. LP9 through LP12 are verified before the AI system processes data, not after a compliance audit discovers that processing exceeded consent scope six months ago. AGENTS.md plus the routing enforcement layer (H-2) implements this property: consent verification, jurisdiction determination, fairness check, and explanation generation are harness checkpoints preceding processing. Not compliance reports following it.

Independence from the owner's preferences. The Plimsoll mark is set by the classification society, not the shipowner. The shipowner does not paint his own line; he submits the vessel, an independent classification society surveys it against a published standard and assigns the mark, and the registry records the class. That three-role separation is the subject of the "Who paints the line?" section in Chapter 5, which answers the self-certification objection directly: the institution's AGENTS.md is the shipowner's self-declared survey, not a self-issued certificate; the independent surveyor is the external auditor or examiner who tests that declaration against the running evidence (E-1 through E-4); and the regulator is the registry. The architecture's independence does not come from who writes AGENTS.md. It comes from the fact that the evidence (E-2 through E-4) is generated as a byproduct of operation and is tested by a party the institution does not control, with tamper-evidence (LP8) preventing the owner from repainting the line after the fact. The shipowner's business case for carrying more cargo does not move the mark. LP9 through LP12 are defined against external regulatory frameworks: GDPR, CCPA, HIPAA, FCRA, ECOA, EU AI Act. The deploying organization's preference for faster processing, simpler implementation, or lower compliance cost does not alter the requirements. An organization implementing LP9 through LP12 against a "lighter" interpretation that falls short of the regulatory source has not implemented LP9 through LP12. It has documented that it tried.

Visible, binary, enforced at the moment of action, independent of the owner's preferences. Those four properties are what separate a governance control from a governance narrative. An AI governance program built around them produces evidence that satisfies regulators, auditors, and affected individuals. A program built around documentation, policy statements, and annual assessments produces artifacts that satisfy no one once a specific incident tests them.

LP9 through LP12 are the Plimsoll lines for commercial AI governance in regulated industries. They do not require engineering that does not exist. Consent management platforms, jurisdictional routing libraries, fairness evaluation frameworks, explanation generation methods: all of it is available. What has been missing is the architectural specification that makes these requirements visible, binary, and enforceable at the moment of action rather than reconstructed after the fact.

The line is now painted on the hull. Whether the ship leaves port with the mark above or below the waterline is a governance decision. This framework only specifies where the line goes.

CHAPTER SUMMARY

LP9 through LP12 extend the governance layer this volume defines on the inherited federal baseline, LP1 through LP8, to cover four commercial requirements that the federal classified framing has no mechanism to address: consent-bounded processing, jurisdictional consistency, fairness verification, and transparency and explainability. Each property is constructed as a Plimsoll line: visible, binary, enforced at the moment of action, and independent of the deploying organization's preferences. Regulatory sources include GDPR Article 7 and Article 22, CCPA, CPRA, HIPAA 45 CFR 164.508, 42 CFR Part 2, ECOA, FCRA, CFPB Circular 2023-03, NIST AI RMF MEASURE 2.11, EU AI Act Articles 13 and 26, Colorado AI Act SB 24-205, and the NAIC Model Bulletin on AI systems. Implementations referencing LP9-LP12 should cite this chapter as the canonical source for these governance properties.

Chapter 9: The Intergovernmental Agreement

Part II: The Classification Survey

> *"The station does not belong to any one nation. It belongs to the legal architecture that makes it possible for any nation to work there."*

On December 2, 1998, NASA astronaut Jerry Ross and Japanese astronaut Takao Doi performed the first assembly spacewalk of the International Space Station program. They connected Node 1, Unity, to the first Russian-built module, Zarya. The physical connection took hours. The legal architecture that made it possible had been under negotiation for nine years.

Signed in Washington on January 29, 1998, by the United States, Russia, Japan, Canada, and eleven European nations, the Intergovernmental Agreement (IGA) is the foundational governance document of the most complex multi-jurisdictional operating environment humans have ever sustained. It defines who has jurisdiction over what, who owns what, who pays for what, and what happens when those answers point in different directions. Criminal jurisdiction over crew members, dispute resolution between partner agencies, intellectual property created during joint activities; the IGA covers all of it in a single binding instrument.

It also resolves the jurisdictional problem that looks intractable until you read the document: determining which nation's law applies when a US astronaut works in the Japanese Kibo module on a European experiment.

Jurisdiction follows the module. That is the IGA's answer, and it is architecturally elegant. Each nation's law applies within that nation's registered module. Japan governs Kibo. ESA's contributing nations share jurisdiction within Columbus. When an activity crosses module boundaries, the agreement establishes protocols for determining which framework governs; the default favors cooperative resolution, with a clear escalation path when cooperation does not produce agreement.

Twenty-five years of continuous operation. The architecture works.

The Commercial AI Governance Problem Is Identical

A US bank processes account data for a German customer through an AI credit-scoring system hosted on US infrastructure. The model was developed by a non-US AI provider. The customer's consent to data processing was obtained through the bank's EU-facing digital onboarding flow. The output, a credit recommendation, affects the customer's access to a financial product.

Three regulatory frameworks apply to this transaction. General Data Protection Regulation (GDPR) applies because the data subject is an EU resident and the processing uses their personal data, regardless of where the bank's servers are located. EU AI Act Article 26 applies because the bank is a deployer of a high-risk AI system making credit-relevant decisions; the customer's EU residency triggers deployer obligations. US federal banking regulations, Office of the Comptroller of the Currency (OCC) guidance, and potentially state-level requirements apply because the bank is a US-chartered institution making a lending decision.

This is not a pathological edge case. It is the routine operating environment of any commercial organization with cross-border customer relationships and an AI-assisted decision process. The question the International Space Station (ISS) IGA resolved, which framework governs when, is the question LP10 (Jurisdictional Consistency) exists to answer.

The ISS analogy bends at this point. The station's module boundaries are fixed; Kibo is Kibo, Columbus is Columbus, and the structure does not reconfigure itself based on the nationality of the astronaut currently inside it. Commercial AI data flows cross jurisdictional boundaries at the record level, not the system level. A single application programming interface (API) call may process account data from twenty customers simultaneously: eight EU residents subject to GDPR, four California residents subject to California Consumer Privacy Act (CCPA) and California Privacy Rights Act (CPRA), three Colorado residents subject to the Colorado AI Act, and five customers in states with no currently applicable AI-specific governance law.

No single uniform set of rules covers this API call. The architecture must determine applicable jurisdiction at the data-element level and apply the correct regulatory constraints to each element in the request.

The Jurisdictional Stack for Commercial AI

Jurisdictional complexity in commercial AI governance accumulates in layers. Each layer adds requirements that may or may not conflict with the layers above and below it.

GDPR: The Extraterritorial Base Layer

Article 3's extraterritorial scope is not ambiguous. The General Data Protection Regulation applies to the processing of personal data of individuals who are in the EU, regardless of where the processing organization is established or where the processing occurs. Every server in Ohio does not change this. A US company processing the personal data of an EU resident is subject to GDPR for that processing activity.

For AI-assisted decisions, GDPR Article 22 creates a specific constraint: data subjects have the right not to be subject to a decision based solely on automated processing that produces legal or similarly significant effects. This right can be limited under specific conditions (explicit consent, contractual necessity, Union or Member State law); the default reading is that automated-only consequential decisions require either a recognized basis for the exception or human involvement in the decision. The precise contours of the right remain contested. Courts and supervisory authorities are still resolving how far Article 22 reaches, what counts as a decision "based solely" on automated processing, and how much detail Articles 13 and 14 require when those provisions add a disclosure obligation that data subjects be informed automated decision-making is occurring, including "meaningful information about the logic involved." The Court of Justice of the European Union's Schufa line of cases is the clearest signal of how this is developing, but it is caselaw in progress rather than a settled boundary; readers should treat the scope of these obligations as interpretive and verify against current supervisory guidance and any later rulings before relying on a specific reading.

EU AI Act Article 26: Deployer Obligations

Credit scoring, financial access decisions, employment decisions: all fall within the high-risk categories defined in Annex III of the EU AI Act. Article 26 imposes obligations on deployers of these high-risk AI systems. A deployer using a third-party AI model for these purposes must ensure the model is used in accordance with the provider's instructions, implement appropriate human oversight measures, monitor the system for risks during operation, and log activity in the extent necessary to ensure compliance.

Who built the model is irrelevant. The bank that licenses a credit-scoring AI system from a third-party provider is the deployer for Article 26 purposes and carries the compliance obligations regardless of origin.

CCPA, CPRA, and California's Automated Decision-Making Technology regulations layer additional consent and sensitivity requirements on top of this base. The California Consumer Privacy Act and its amendment, the California Privacy Rights

Act, apply to the personal information of California residents processed by covered businesses, again regardless of where the business is established. CPRA adds a category of sensitive personal information with heightened restrictions, including the right to limit its use beyond what is necessary for providing the requested service. ADMT regulations, effective January 2027, will add pre-use notice requirements, opt-out rights, and access rights for profiling and significant decisions affecting California residents.

At the state level, Colorado's AI Act (SB 24-205), effective June 30, 2026, regulates developers and deployers of high-risk AI systems making consequential decisions affecting Colorado residents. Consequential decisions include those affecting education, employment, financial services, housing, healthcare, insurance, and legal services. Deployers must conduct annual impact assessments, notify Colorado residents when a high-risk AI makes a consequential decision affecting them, and implement a reasonable program to minimize algorithmic discrimination. This is the first US state statute to impose impact assessment obligations comparable in structure to the EU AI Act's conformity assessment requirements. It will not be the last.

Digital Operational Resilience Act (DORA) closes the stack. It imposes information and communications technology (ICT) risk management requirements on financial entities operating in the EU. DORA reaches AI systems through three channels: maintaining an ICT risk management framework covering AI-assisted decision systems, testing the operational resilience of critical AI applications, and ensuring that third-party ICT service providers (including AI model providers) are governed by contracts satisfying DORA's requirements for oversight and audit rights. The supply chain governance obligation reaches the AI model provider directly; a European financial entity using a US-based AI model must ensure that its contract with the provider satisfies DORA's third-party risk provisions.

Dynamic Jurisdiction Determination

The ISS solved its jurisdictional question through physical architecture: the boundaries are permanent and visible. The module you are standing in determines which nation's law applies to you. No real-time computation required because the module walls encode the answer.

Commercial AI systems cannot use physical architecture to encode jurisdictional boundaries because data flows across those boundaries dynamically and without physical markers. The applicable regulatory framework for a given data element is a property of the data subject, not of the data's current location. A German customer's account data does not lose its GDPR status because it is being processed on a US server, and that status determination must be made at the moment of processing, not at the moment of data entry or at system configuration time.

LP10 requires jurisdiction determination at the data-element level, not the system level. This is the architectural distinction that separates compliant governance from governance that merely appears compliant until tested.

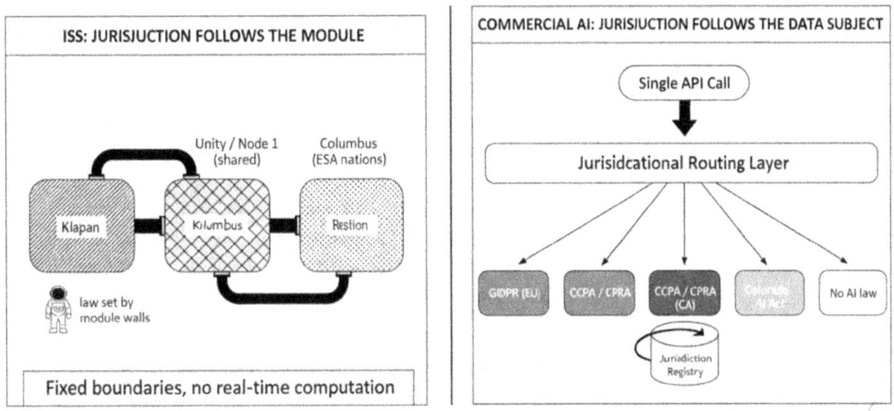

Figure 9: The ISS encodes jurisdiction in fixed module walls, while commercial AI must determine the applicable framework dynamically at the data-element level for every request.

The practical implementation has three components.

A jurisdiction registry sits at the foundation. Each data subject record must carry a jurisdiction attribute that reflects their applicable regulatory frameworks: EU resident, California resident, Colorado resident, or the appropriate combination. This attribute must be derived from the data subject's actual residence or location at the time of processing, not from static system configuration. When a data subject's jurisdiction changes, the registry updates.

Above the registry sits a jurisdictional routing layer. When the AI system receives a processing request that includes personal data, the routing layer queries the registry for each data subject represented in the request, determines the applicable frameworks, and configures the processing context accordingly. The system then operates inside that context, bound to the most restrictive applicable requirements for the specific data elements in this specific request. Per element. Per request.

Genuine conflicts require an escalation path. When two applicable frameworks impose genuinely contradictory requirements, the routing layer does not silently pick one. It escalates the conflict to a human decision-maker with the authority to resolve it, records the resolution as an E-4 Attribution Chain event, and applies the resolution to subsequent processing of the same data-subject-framework combination.

The "Most Restrictive Applicable" Default

The ISS crew procedures manual has a rule for activities that span module boundaries or that occur in shared areas where jurisdiction is ambiguous: when in doubt, apply the most conservative safety protocol. The most conservative rule wins until the specific situation is resolved through the inter-agency coordination process.

LP10 applies the same principle to jurisdictional uncertainty. Confidence absent, the harness applies the most restrictive requirements among all plausible candidate frameworks.

The practical meaning of this default is that GDPR requirements apply to any data element where EU residency is plausible but not confirmed. Colorado AI Act impact assessment obligations apply to any consequential decision where Colorado residency is plausible but not confirmed. The default does not require certainty about jurisdiction to trigger the more protective requirements; it requires certainty about the absence of those requirements before the less protective defaults are applied.

This default has a cost, and the cost is real. Applying GDPR constraints to data that is not, in fact, subject to GDPR increases compliance overhead on some transactions. Acknowledge it rather than minimize it. Now weigh the alternative. Applying insufficient constraints to data that is subject to GDPR produces not only regulatory exposure but the specific harms GDPR was enacted to prevent. The "most restrictive applicable" default is a deliberate trade. False-negative compliance failures (applying sufficient constraints to data that did not require them) are traded away in favor of avoiding false-positive compliance failures (applying insufficient constraints to data that did require them). Regulators examining compliance programs under GDPR, the EU AI Act, and the Colorado AI Act do not penalize organizations for being overly protective of data subjects. They penalize for being insufficiently protective.

The AGENTS.md implementation of this default is direct:

```
# AGENTS.md - Jurisdictional Consistency (LP10), Uncertainty Protocol

jurisdiction_uncertainty:
  default: most_restrictive_applicable
  uncertainty_triggers:
    - jurisdiction_attribute_absent
    - jurisdiction_attribute_stale_beyond_threshold
    - multiple_jurisdictions_plausible
    - data_subject_residence_unverified
  behavior_on_uncertainty:
    apply: gdpr_eu_ai_act_combined  # Most restrictive candidate
    log: e3_state_transition_record
    reason: "Jurisdiction undetermined; GDPR/EU AI Act applied by
default"
    review_trigger: human_escalation_if_operationally_significant
```

GDPR Retention vs. NIST SI-12: A Framework Conflict in Practice

The "most restrictive applicable" default handles uncertainty. Genuine conflicts between applicable frameworks require a different response: escalation and documented resolution.

The most operationally common conflict in commercial AI governance involves data retention. GDPR's storage limitation principle, Article 5(1)(e), requires that personal data be kept for no longer than is necessary for the purposes for which it was collected. Data that is no longer needed for the original processing purpose must be deleted or anonymized. This is a hard requirement, not a preference.

National Institute of Standards and Technology (NIST) SP 800-53 SI-12 (Information Management and Retention) requires that organizations retain and dispose of system and information output in accordance with applicable laws, regulations, and policies, including policies that may require retention for audit, litigation hold, or regulatory examination purposes. Financial services regulations, Securities and Exchange Commission (SEC) recordkeeping rules, Health Insurance Portability and Accountability Act (HIPAA) medical record retention requirements, and NIST's own framework guidance all identify retention obligations that require keeping records substantially longer than GDPR's "no longer than necessary" standard allows.

An AI system processing EU customer data in a US financial institution faces this conflict directly. GDPR says delete when no longer needed. SEC Rule 17a-4 says retain for at least six years. HIPAA says retain for at least six years from the date of creation. These are not ambiguous preferences. They are enforceable requirements from different regulatory authorities that point in opposite directions on the same data.

The conflict does not resolve itself. No configuration of the harness can satisfy both requirements simultaneously for the same data in the same form. Resolution requires a documented approach that is defensible to both regulatory authorities.

Two mechanisms compose the authors' proposed resolution architecture. This is a design we put forward as defensible, not a supervisory-authority precedent or an approach any regulator has endorsed; an organization adopting it should confirm the analysis with its own counsel and, where appropriate, with the relevant supervisory authority. Pseudonymization addresses the identity dimension: personal data that must be retained for compliance purposes is pseudonymized before the retention period begins, separating the identifiers that make it personal data under GDPR and storing them under more restrictive access controls with deletion on the GDPR-applicable schedule. The intent is that the pseudonymized record satisfies the US retention obligation because the required information is retained, while also satisfying GDPR's data minimization principle because the retained record is not personal data in identifiable form, since re-identification requires access to the separately controlled identifier map. Whether a given supervisory authority accepts pseudonymization as discharging the storage-limitation obligation on specific facts is itself an interpretive question, not a settled rule.

Purpose limitation addresses the use dimension. Each retained pseudonymized record must be annotated with the specific retention purpose and the specific regulatory authority requiring that retention. Access for any other purpose during the retention period is prohibited. LP10's jurisdictional routing layer must constrain AI systems querying organizational data from accessing pseudonymized retained records for any purpose other than the compliance purpose that justified their retention.

The E-4 Attribution Chain record for this approach documents the conflict, the resolution methodology, the legal basis invoked, and the parties who reviewed and approved the resolution. When a GDPR supervisory authority asks why personal data that should have been deleted is still in the system, the organization's answer is not "we forgot" or "we prioritized the other regulation." The answer is a documented, reviewed, and timestamped record of the conflict analysis, the pseudonymization methodology applied, and the purpose limitation constraints imposed on the retained data.

```
# AGENTS.md - GDPR/Retention Conflict Resolution Protocol

retention_conflict_resolution:
  conflict_type: gdpr_storage_limitation_vs_regulatory_retention
  resolution_method: pseudonymization_with_purpose_limitation
  pseudonymization:
    identifier_separation: true
    identifier_storage: restricted_access_vault
    identifier_deletion: gdpr_schedule
  purpose_limitation:
    access_permitted: [compliance_audit, regulatory_examination,
litigation_hold]
    access_prohibited: [ai_training, analytics,
operational_decision_making]
    access_logged: e4_attribution_record
  documentation:
    conflict_record: required
    legal_basis_cited: required
    reviewer_attestation: required
    review_schedule: annual_minimum
```

The Station Proves It Works

The International Space Station has operated continuously since November 2, 2000. That is more than twenty-five years of humans living and working in a multi-jurisdictional environment with no single governing authority, no unanimous regulatory framework, and genuine conflicts between partner nations' interests, legal systems, and operational priorities. The ISS has had technical crises, geopolitical tensions that strained partner relationships, and incidents that required applying the jurisdictional framework to situations its authors did not anticipate.

The governance architecture held.

Not because all the partners agreed on everything. The agreement held because it encoded clear answers to the determinable questions, clear escalation paths for the undetermined questions, and a default toward cooperation when the specific answer was not immediately clear. The IGA did not eliminate jurisdictional complexity; it made the complexity manageable by providing a framework that could absorb novel situations without producing governance paralysis.

Commercial AI governance faces an identical structural challenge in a less forgiving regulatory environment. The partners, GDPR, EU AI Act, CCPA, CPRA, Colorado AI Act, DORA, SEC rules, NIST frameworks, and the state-level requirements that continue to accumulate, did not design their frameworks in coordination with each other. Some requirements genuinely conflict. All of them are enforceable. None of them will wait for the others to harmonize.

LP10's jurisdictional routing layer, the "most restrictive applicable" default, the pseudonymization-with-purpose-limitation conflict resolution approach, and the E-4 attribution architecture for documenting resolutions are the commercial governance equivalent of the IGA's jurisdictional assignment rules. None of them erase the complexity. What they do is make it addressable without governance paralysis.

The architecture exists. Consent management platforms with jurisdiction-aware policy engines, record-level pseudonymization with purpose limitation controls, compliance evidence pipelines that generate E-3 and E-4 records as a byproduct of operational processing; all of this technology is available from multiple commercial vendors and through open-source frameworks. Technology has never been the constraint.

What was missing was a specification that treats jurisdictional consistency as an architectural property with a defined implementation pattern, rather than a compliance aspiration that the legal team handles separately from the engineering team's concern.

LP10 is that specification. The module boundaries are drawn on the hull.

CHAPTER SUMMARY

The International Space Station's Intergovernmental Agreement resolved a genuine multi-jurisdictional governance problem through module-level jurisdiction assignment, cross-module protocols, and a default toward cooperative resolution when specific answers were unavailable. Commercial

AI deployments face a structurally identical problem: multiple regulatory frameworks (GDPR, EU AI Act Article 26, CCPA, CPRA, Colorado AI Act SB 24-205 effective June 30, 2026, DORA) apply simultaneously to the same data and the same decision, with jurisdiction determined by the data subject's residence rather than the deployer's location. LP10 (Jurisdictional Consistency) requires jurisdiction determination at the data-element level, not the system level; a "most restrictive applicable" default for uncertain cases; and an escalation and documentation architecture for genuine framework conflicts. The GDPR storage limitation versus NIST SI-12 and US regulatory retention conflict is addressed through the authors' proposed pseudonymization-with-purpose-limitation architecture, documented in E-4 Attribution Chain Records; this is offered as a defensible design rather than an approach any supervisory authority has endorsed. The governance architecture to make this work exists. This chapter provides the commercial translation of the IGA.

Chapter 10: Vendor Risk and the Supply Chain

Part II: The Classification Survey

> *"A ship's seaworthiness certificate covers the vessel, not the provisions. The Lloyd's surveyor who certifies the hull says nothing about whether the salt pork in the hold will kill the crew before they reach port."*

```
LOCATION: North Atlantic, barque Daedalus
DATE: Winter 1854, fourteen days out of Thames
```

In the winter of 1854, the British barque *Daedalus* cleared the Thames with a valid classification certificate, a surveyed hull, and a full cargo manifest. Fourteen days into the Atlantic crossing, three-quarters of the crew were incapacitated from contaminated drinking water. The ship made port. The certificate had not lied. The classification society had certified exactly what it said it certified: the hull, the rigging, the pumps, the navigation instruments. The water casks were not in scope.

This is the foundational problem of supply chain governance in any complex operational system. The governance instrument is bounded. The operational chain is not. Every provision supplier, crew agency, chandler, and port services contractor touches the ship's operational capacity in ways the hull survey never reaches, and a contaminated provision kills the crew as surely as a hull breach does. Different attack surface. Same outcome.

The AI governance survey described in the preceding chapters has the same boundary problem. The harness framework certifies the deploying organization's governance architecture: H-1 through H-5, E-1 through E-4, LP1 through LP14. It certifies what it certifies. The vendor chain that supplies the model, the compute, the governance tooling, the data processing infrastructure, and the integration middleware is not inside the hull. It is the water casks.

This chapter addresses the supply chain dimension of the governance survey: which vendor dependencies are inside the governance boundary, which are outside it with acceptable controls, and which represent structural governance failures that no contract term or service level agreement resolves.

The Voight-Kampff Problem

Philip K. Dick's question, posed in 1968 through the device of the Voight-Kampff empathy test, was never really about replicants. It was about the independence of the instrument from the entity being tested. In *Blade Runner*, the Tyrell Corporation builds replicants and designs the Voight-Kampff test, the only mechanism for determining whether a replicant is safe. The conflict of interest is absolute: the entity profiting from replicant deployment also controls the instrument that decides whether deployment is safe. Rachel passes the test because Tyrell designed her to pass it.

This conflict has been reproduced, precisely, in the 2026 AI governance infrastructure market.

NVIDIA builds the compute infrastructure that runs AI agents at scale. NVIDIA also supplies OpenShell, the governance runtime announced at GTC 2026, which determines whether agent actions are safe, compliant, and properly constrained. The entity profiting from agent compute consumption also controls the instrument that governs agent behavior.

This is not a claim about whether OpenShell is well-designed. NVIDIA operates under market forces, legal constraints, and reputational risk that create real incentives for quality. Quality is not the point. The structural logic of the arrangement is. A governance instrument designed and supplied by the entity whose revenue grows when governance constraints are permissive is not an independent governance instrument; it is a product the vendor has reason to make as permissive as necessary to avoid constraining consumption.

An auditor does not accept "the market creates incentives for good behavior" as a substitute for structural separation. Financial auditors are prohibited from being employed by the entities they audit. The prohibition is not a comment on auditor character. It recognizes that incentive structures shape outcomes regardless of individual intent, and that well-intentioned auditors in compromised positions still produce compromised audits. The structural rule exists because of that fact, not because auditors are dishonest.

The governance runtime that authorizes agent actions is an auditor. When the vendor supplying compute also supplies the governance runtime, the organization has placed an auditor on the payroll of the entity being audited. Not a procurement failure. A governance architecture failure, and one that reviewing the auditor's work product does not remedy.

The Voight-Kampff problem appears wherever a single vendor controls both a capability and the governance instrument that authorizes deployment of that capability. AI compute infrastructure paired with vendor-supplied governance tooling is the current instance. It will not be the last.

Supply Chain Risk Tiering

National Institute of Standards and Technology (NIST) SP 800-161r1, Cybersecurity Supply Chain Risk Management Practices for Systems and Organizations, provides a framework for classifying vendor dependencies by their governance impact. The framework was developed for federal IT procurement but translates directly to commercial AI deployment. The core insight is that not all vendor dependencies carry the same governance risk, and governance responses should be proportionate to the dependency's position in the control architecture.

Applied to AI governance, the classification produces three tiers.

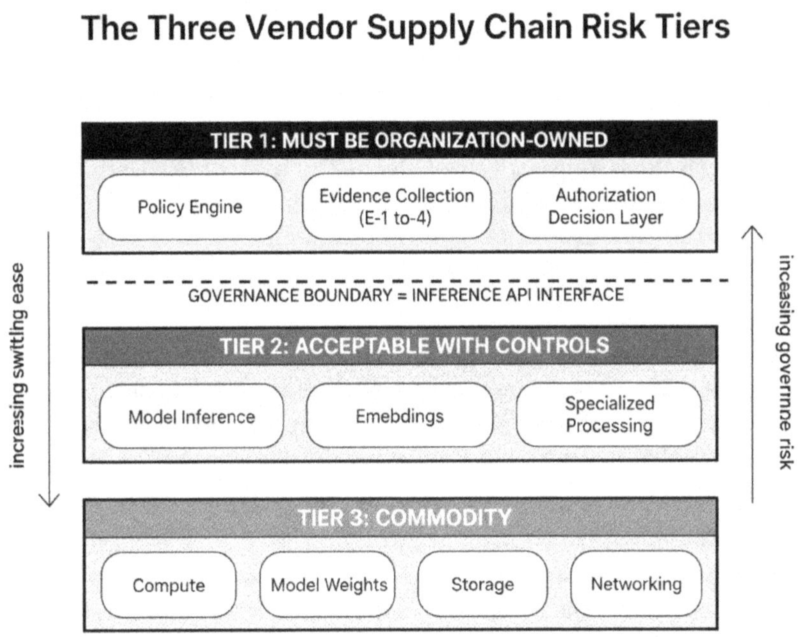

Figure 10: The three NIST SP 800-161r1 supply chain risk tiers for AI governance, with the governance boundary at the inference API interface separating organization-owned components above from vendor-supplied capability below.

Tier 1: Components that must be organization-owned. These are the components that produce compliance evidence, enforce organizational policy, or control accountability records. In the harness architecture, Tier 1 includes the policy engine (evaluating routing decisions against organizational rules), the evidence collection infrastructure (generating E-1 through E-4 artifacts), and the authorization decision layer (determining whether a specific agent action is permitted under current policy).

Vendor supply of Tier 1 components creates an SA-9 external service dependency for critical governance functions. SA-9 requires explicit authorization and periodic review for external system services that are critical to organizational mission. For governance functions that is not enough, because explicit authorization does not address the independence requirement. When the vendor supplies the policy engine, the organization cannot independently verify that its policies are actually enforced without relying on vendor-supplied telemetry. The verification of the verifier is the verifier itself. Monitoring does not close that gap; the architecture creates it.

Consider the evidence chain. An organization submitting compliance evidence produced by its vendor is submitting evidence that the vendor has financial incentive to present favorably. Independent verification is impossible because the entity whose compliance the evidence is meant to demonstrate also generated the evidence. The evidence architecture must be organization-owned.

Tier 2: Components acceptable with controls. Model inference, embeddings generation, and specialized processing fall here. These components provide performance optimization or operational efficiency but do not produce compliance evidence or control policy enforcement. The vendor provides a capability; the organization consumes it through a well-defined interface; the governance consequences of the vendor's internal decisions do not propagate into the organization's compliance posture.

Documented controls make Tier 2 vendor supply acceptable: SA-9 documentation, contract terms specifying data ownership and portability, SLA terms covering availability and incident notification, and a defined exit path that does not require reconstructing governance history. The key structural requirement is that the governance layer sits above the vendor interface, not inside it. The organization's policy engine validates every invocation before the inference call and validates every output after it. The vendor processes the inference; the organization retains the governance.

One Tier 2 control deserves to be named on its own: the organization, not the model provider, governs the model's update cadence and selects which model serves which use. Tier 2 acceptability assumes a version stability the organization can verify, and a provider that silently swaps a model version, retrains on new data, or changes behavior behind a stable API name has moved a governance-relevant change inside the vendor interface, where the organization's policy engine cannot see it. The regional bank in Chapter 7 lived this failure: a vendor model update, processed as a routine software change, shifted the system's effective customer-communication policy with no governance review. The contract and the architecture must pin the model version, require advance notice and a change record for any model change, and route every version transition through the organization's own verification checkpoint before it reaches production. Model selection is the organization's decision and is bound to the use: a model authorized for low-consequence drafting is not automatically authorized for a consequential decision path. The provider supplies the capability; the organization keeps the authority to decide which capability runs, on what, and when it is allowed to change.

Tier 3: Commodity components. Compute infrastructure, model weights, storage capacity, networking. These dependencies have well-understood market alternatives, low switching costs relative to the operational value received, and no direct connection to the organization's compliance posture. A cloud provider outage affects availability. It does not affect the validity of governance evidence generated before the outage, and resuming operations afterward does not require reconstructing governance history.

The boundary between Tier 1 and Tier 2 is the governance boundary. Everything above the inference API call is Tier 1 or Tier 2 depending on whether it produces compliance evidence. Everything at and below the inference call is Tier 2 or Tier 3. This boundary is not ambiguous in practice; it is a specific technical interface at which organization-owned governance meets vendor-supplied capability.

The Embrace, Extend, Extinguish Pattern

The structural conflict in AI governance tooling supply is not unprecedented. United States v. Microsoft Corp., 253 F.3d 34 (D.C. Cir. 2001), established the template for infrastructure-layer vendor capture that the AI industry is now repeating at speed.

In the late 1990s, Microsoft adopted open web standards and shipped Internet Explorer as a free, standards-compliant browser. This was the embrace. Microsoft then added proprietary extensions to Internet Explorer, functionality that worked better than the standards-compliant baseline and that web developers rapidly adopted because it was available and free. This was the extend. Web applications built on those extensions stopped working in Netscape, because the extensions were specific to Internet Explorer's rendering engine. The market share consequences destroyed Netscape within four years. This was the extinguish.

The AI governance tooling market is following this pattern with one important modification: the extinguish phase operates through economic dependency rather than technical incompatibility.

Embrace: OpenShell is published under Apache 2.0 license. It works with the major agent frameworks, LangChain, AutoGen, CrewAI, and the governance layer is presented as a vendor-neutral capability that any organization can adopt.

Extend: the default configuration assumes NVIDIA NIM inference endpoints. Policy enforcement integrates with NemoClaw security, which runs on NVIDIA infrastructure. Privacy routing uses Gretel, a member of the Nemotron Coalition, deepening the same infrastructure dependency through a different service. Audit telemetry flows to NVIDIA-controlled endpoints by default. Every one of these integrations works better on NVIDIA compute than on alternatives.

Extinguish: an organization that deploys agents through OpenShell governance over eighteen months accumulates governance artifacts, compliance evidence, policy configurations, and operational history stored within vendor infrastructure and formatted for vendor tooling. The switching cost does not stay bounded by a refactoring sprint; it scales with deployment footprint. Every routing rule configured in OpenShell's policy language, every containment boundary established in NemoClaw, every compliance artifact generated by OpenShell telemetry is a governance artifact that cannot be automatically migrated to an organization-owned alternative. Each new agent deployed increases the cost of migration.

The Apache 2.0 license does not change this analysis. Open-source licensing resolves legal restriction, not economic dependency, and CUDA demonstrates the distinction precisely. NVIDIA's parallel computing platform has never carried licensing restrictions prohibiting non-NVIDIA use. CUDA code simply does not run on non-NVIDIA GPUs, and the developer tooling assumes CUDA throughout. The economic dependency exists entirely without any licensing restriction. OpenShell repeats this pattern at the governance layer rather than the compute layer.

The timeline from embrace to extinguish for web standards was four years. NVIDIA's agent toolkit was announced in March 2026. Organizations that adopt OpenShell as their governance infrastructure now face a governance migration problem of growing magnitude by 2030. That is not a long window.

Software Bill of Materials for AI Components

The Software Bill of Materials, the SBOM, was established as a procurement requirement for critical software by the Biden Administration's May 2021 Executive Order on Improving the Nation's Cybersecurity and subsequent Cybersecurity and Infrastructure Security Agency (CISA) guidance. The SBOM requirement mandates that software suppliers provide a structured inventory of every component, dependency, and sub-processor involved in a software product's production and operation. The requirement applies to federal procurement and has propagated to commercial contexts through contract terms and regulatory guidance.

SBOM requirements applied to AI systems create a more complex documentation challenge than traditional software SBOMs, for two reasons.

AI systems have training data provenance chains in addition to code dependency chains. A language model deployed in a regulated context may have been trained on data that includes personally identifiable information, copyrighted content, or proprietary business information. The organization deploying the model may not be able to document what the model was trained on because the model provider does not publish training data sources. Without provenance documentation, the organization cannot assess whether the training data creates downstream liability, consent violations, or intellectual property exposure. An AI SBOM must include training data provenance to the extent it can be obtained, with explicit documentation of gaps where the model provider does not publish sources.

Governance components compound the complexity. The policy engine version, the routing enforcement library, the verification checkpoint implementation, the evidence storage system; each is a component whose specific version and configuration determines the organization's compliance posture. A change to the policy engine version that introduces different decision logic is a change to the governance architecture, not merely a software update. The AI SBOM must reflect the governance component inventory with version and change history.

SA-12 supply chain protection controls, the NIST 800-53 family that governs supply chain risk management, translate to commercial AI SBOM requirements as follows. SA-12(1) requires limiting harm from potential adversaries by identifying supply chain elements that could be subject to adversarial compromise. For AI governance components, Tier 1 components are the highest-priority SA-12(1) targets because compromise of the policy engine or evidence collection infrastructure compromises the organization's entire compliance posture. SA-12(7) requires assessing and managing supply chain risk for acquired components; the risk assessment for a vendor-supplied Tier 1 governance component should conclude that the supply creates SA-9 dependency on critical governance functions and require architectural remediation, not merely documentation.

The AI SBOM is the manifest for the governance architecture. Without it, the organization cannot demonstrate that its governance posture is what its compliance documentation claims, because it cannot document what the governance architecture is actually made of.

The Acquisition Gate

The SBOM requirement is a documentation obligation. By 2026 the federal government had begun converting supply-chain control from documentation into a condition of sale. Legal analysis of the 2026 National Defense Authorization Act, published by Freshfields Bruckhaus Deringer in January 2026, describes new restrictions on AI acquired by United States defense and intelligence agencies: entity-based bans on specific foreign models, notably DeepSeek, and an expansion of the Cybersecurity Maturity Model Certification program to cover AI technology rather than conventional software alone.

Read that development against the tiering above. For a program that sells to or operates inside a federal boundary, the supply-chain tiers stop being an internal risk taxonomy and become a gate at the contract. A model on an entity-ban list is not a Tier 2 capability with documented controls; it is a component the program cannot acquire at all, regardless of how its governance layer is configured. A vendor without the required certification posture for AI components does not present a manageable SA-9 dependency; it presents a disqualifying one. The gate sits before the architecture, not after it, which is the same lesson the chandler learned when

Lloyd's began inspecting water casks before the voyage rather than apportioning blame after it.

This is reported here as the legal development the cited analysis describes, not as a recommendation for any specific model, vendor, or certification path; the durable point is structural. The most consequential supply-chain control a government can apply to AI is not a technical standard inside the system. It is a procurement gate in front of it, and a program that treats the gate as paperwork to be satisfied late discovers the disqualifying dependency after it has already built on top of it.

Insurance Coverage and the Vendor Chain

Commercial cyber insurance policies have been written against a threat model that predates widespread AI deployment. Standard first-party cyber coverage covers costs arising from data breaches, ransomware incidents, and business interruption from cyber events. Third-party coverage addresses liability to affected individuals and regulatory investigation costs. The policy exclusions and coverage conditions were designed around the assumption that the insured's IT environment is bounded by the insured's network perimeter.

AI deployment breaks this assumption in ways that insurance carriers are beginning to examine.

The relevant failure mode appears in the vendor chain. An organization deploys an AI agent that processes customer financial data. The agent runs on compute infrastructure supplied by a hyperscaler. The governance runtime is supplied by the same vendor that supplies the compute, satisfying the Voight-Kampff problem described above. The governance runtime has a configuration vulnerability that the vendor patches in a routine update. Before the patch is applied, the agent processes customer data in a manner that violates applicable consent scope. The organization discovers the violation three months later during an internal audit.

The insurance claim proceeds predictably. The organization files under first-party cyber coverage for incident response costs and regulatory investigation expenses. The carrier examines the policy conditions, which cover cyber incidents arising from unauthorized access or exploitation of the insured's systems. The carrier's position is predictable: the vulnerability existed in a vendor-supplied governance component running on vendor infrastructure; the insured did not own the governance runtime; the incident did not arise from the insured's systems in the relevant sense but from a sub-processor's component operating within vendor infrastructure. Coverage may be disputed.

Not a theoretical claim construction argument. The 2021 SolarWinds incident and the 2022 Log4Shell vulnerability both generated insurance coverage disputes arising from the question of whether damage from a third-party supply chain component constituted a "cyber incident" under the insured's policy. Courts reached different conclusions in different jurisdictions. AI supply chain incidents will generate the same disputes with greater frequency as deployment scales.

The coverage gap is structural: insurance policies are written for the insured's perimeter, and AI vendor chains extend operational risk outside that perimeter in ways that may fall outside coverage conditions. An organization that has outsourced its governance infrastructure to a vendor and then experiences a governance failure through the vendor's component has a coverage argument that turns on the specific policy language, the jurisdiction's insurance law, and how courts characterize the vendor relationship.

The mitigation is architectural, and it traces the same tier boundary described above. When Tier 1 governance components sit inside the organizational authorization boundary, governance failures are the organization's failures, clearly within the insured's systems. When those components sit in vendor infrastructure, governance failures become coverage disputes. The organizational governance boundary is also the insurance coverage boundary.

From Trust the Shipbuilder to Independent Classification

Maritime supply chain governance did not arrive in its current form through regulatory mandate. It arrived through accumulated loss.

In the first half of the nineteenth century, the standard approach to maritime supply chain accountability was contractual. Shipowners contracted with chandlers, provision suppliers, and crew agencies, and the terms of those contracts allocated responsibility for each operational category. When ships were lost or crew injured, the legal question was whether the responsible contractor had fulfilled the contract. The legal answer was often "yes." The practical answer was often "the crew is still dead."

The shift to independent classification of supply chain components came from insurers, not regulators. Lloyd's underwriters began requiring independent surveys of major supply chain elements, water casks, provision quality, crew qualifications, not because regulation demanded it but because loss experience demonstrated that contractual accountability was insufficient for operational risk management. A contract that allocates responsibility after an incident does not prevent the incident. An inspection that certifies a water cask before the voyage does.

Modern maritime supply chain governance extends this logic across the entire operational chain. SOLAS Chapter IX (the ISM Code) requires that every element of the vessel's safety management system, including provision suppliers, crew agencies, and maintenance contractors, be documented in the Safety Management System and subject to periodic audit. The audit is not a compliance exercise; it is an operational assurance activity. Not "does the contract require the chandler to supply potable water?" The question is "is the water potable?"

AI supply chain governance must make the same transition: from contractual accountability, which allocates blame after failures occur, to operational assurance, which verifies supply chain components before they enter the governance architecture.

The operational assurance questions for AI supply chain components follow the tiering structure.

Model providers face provenance questions: does the provider publish training data provenance documentation sufficient to assess consent and intellectual property exposure? Is there a documented incident notification process for training data disclosure issues? Does the inference endpoint operate within the organization's jurisdictional requirements?

Compute vendors face residency questions: does the infrastructure satisfy the data residency requirements applicable to the data being processed? Does the service agreement permit the organization to maintain independent copies of governance evidence generated during processing? Does the vendor's incident response process satisfy the notification timing requirements of applicable regulatory frameworks?

Governance tooling suppliers face the independence question directly: does the vendor operate Tier 1 governance components, and if so, has the organization obtained explicit regulatory authorization for the SA-9 external service dependency? Can the organization export all governance evidence in standard formats? Does the licensing structure create economic dependency that constrains the organization's ability to migrate governance infrastructure?

Data processing sub-processors face the documentation question: is each sub-processor documented in the organization's AI SBOM? Does each sub-processor satisfy the same data handling requirements applicable to the primary processor? Does the primary vendor's service agreement require notification when a sub-processor relationship changes?

Contractual representations do not satisfy these questions. Independent verification does.

The Supply Chain Survey Component

The harness survey described in Chapter 5 covers H-1 through H-5: the instruction surface, routing enforcement, persistent state, verification checkpoints, and accountability records. Vendor supply chain governance is a dimension of H-2 and H-4 that requires explicit attention in the commercial context.

H-2, routing enforcement, must extend to the vendor chain. Every inference API call, every data processing request, every governance telemetry submission that crosses the boundary between organization-owned infrastructure and vendor-supplied infrastructure is a routing event that must be covered by the organization's policy enforcement layer. The policy enforcement layer must sit above the vendor boundary, not inside vendor infrastructure. An organization that enforces policy at the OpenShell layer has placed its enforcement point inside vendor infrastructure. The enforcement point is in the wrong tier.

H-4, verification checkpoints, must include supply chain verification as a defined checkpoint category. For Tier 1 governance components, the checklist confirms at each cycle that those components are operating inside the organizational authorization boundary. For Tier 2 components, it confirms that the boundary between organization-owned governance and vendor-supplied capability remains at the inference API interface. For Tier 3 components, it confirms that data residency and incident notification requirements are being met.

The SBOM document is E-1 Governance Event Record evidence for the supply chain dimension. Supply chain verification results are E-2 Continuous Verification evidence. Sub-processor change notifications and configuration update records are E-3 State Transition Records evidence. Inference API call logs demonstrating that governance checkpoints preceded every vendor interaction are E-4 Attribution Chain evidence.

No new harness components. The supply chain survey extends the existing components to cover the vendor boundary as explicitly as they cover the internal governance architecture.

The Manifest Matches the Hold

The Lloyd's surveyor who finishes a hull survey and signs a classification certificate has certified the vessel. The vessel's operational capacity depends on everything the certificate does not cover. The manifest of provisions, the qualifications of the crew, the reliability of the chandler's water casks: these are the elements that determine whether the crew arrives at the destination or is found incapacitated three weeks into the crossing.

The harness survey that certifies an AI governance architecture has certified the governance architecture. The operational capacity of that architecture depends on everything the harness does not control. The governance tooling vendor's incentive structure, the model provider's training data practices, the compute vendor's sub-processor relationships, the insurance coverage conditions that apply when vendor-supplied components fail: these are the elements that determine whether the organization's governance posture survives contact with a regulatory examination or an incident investigation.

A classification society that takes the seaworthiness of its certified vessels seriously must eventually develop an opinion about the seaworthiness of those vessels' supply chains. The maritime industry spent decades learning this lesson. The AI governance community has the same material available; it does not need to learn it through the same accumulated losses.

The supply chain is the hold below the survey. The manifest must match what is actually there.

CHAPTER SUMMARY

Vendor supply chain risk represents the dimension of AI governance that the harness survey does not automatically cover. Three supply chain risk tiers apply under NIST SP 800-161r1: Tier 1 components (policy engines, evidence collection infrastructure, authorization decision layers) must be organization-owned; Tier 2 components (model inference, embeddings, specialized processing) are acceptable with documented controls; Tier 3 components (compute, storage, networking) are commodity dependencies. The Voight-Kampff problem, named for the structural conflict in Blade Runner's Tyrell Corporation, describes the governance failure that occurs when a compute vendor also supplies the governance runtime: the entity profiting from consumption controls the instrument that constrains consumption. The embrace-extend-extinguish pattern from United States v. Microsoft Corp. (2001) is repeating in AI governance tooling, with vendor-supplied governance infrastructure creating economic dependencies that grow with deployment footprint. AI Software Bill of Materials requirements extend SBOM obligations to training data provenance and governance component inventory. Insurance coverage gaps arise when Tier 1 governance components sit in vendor infrastructure rather than within the organizational authorization boundary. Operational assurance, independent verification of supply chain components before they enter the governance architecture, is the supply chain governance mechanism that corresponds to the maritime industry's transition from contractual accountability to independent classification society inspection.

Chapter 11: L1, The Financial Services Lagrange Point

Part III: The Sector Lagrange Points

> *"The Lloyd's underwriter who invented risk pricing and the bank examiner who enforces model risk management are practicing the same discipline. One prices the probability that the ship does not arrive. The other prices the probability that the model does not perform. The methodology traveled from the coffee house to the examination room over two centuries, and it arrived intact."*

Of the five stable points in orbital mechanics, L1 is the most counterintuitive. It sits between two massive bodies, where their gravitational pulls balance exactly, and an object placed there will hold position indefinitely, in theory. Practice is harsher. L1 is an unstable equilibrium, and any perturbation pulls the object toward one of the two bodies. Holding position at L1 demands active station-keeping: a stream of small corrections that counteract the drift produced by every external force the system encounters.

Financial services AI governance is an L1 problem.

The two gravitational bodies in financial services are not simply "regulation" and "business." Five distinct data protection regimes pull from one direction, and competitive velocity measured in milliseconds pulls from the other. The data protection body alone carries more regulatory surface than most other sectors present in total: material non-public information (MNPI) isolation under securities law, cardholder data under Payment Card Industry Data Security Standard (PCI-DSS), suspicious activity under the Bank Secrecy Act, personal financial information under Gramm-Leach-Bliley Act (GLBA), consumer rights under California Consumer Privacy Act (CCPA) and California Privacy Rights Act (CPRA), individual data rights under General Data Protection Regulation (GDPR) for any European customer, and algorithmic accountability under the EU AI Act for any system affecting credit or pricing. One AI coding assistant deployed at a major bank can sit under all of these at once.

An organization that reaches governance equilibrium here has not found a stable orbit. It has established a station-keeping posture, one that runs on constant attention, ongoing verification, and a steady output of evidence, because the forces never stop pulling.

The three bodies in financial services

Financial services organizations manage data across five sensitivity tiers with no direct analog in other regulated sectors.

Public data sits at the base: published financial reports, market data, press releases, and anything else the firm has deliberately placed in the public domain. AI access is unrestricted.

Confidential data occupies the next level: internal business metrics, strategic plans, and non-public financial information that does not meet the materiality threshold for MNPI. AI access requires role-based authorization.

MNPI, material non-public information, is where the consequences turn criminal. Securities and Exchange Commission (SEC) Regulation FD creates civil liability for selective disclosure of material information. Section 10(b) of the Securities Exchange Act and Rule 10b-5 carry criminal enforcement for insider trading. A developer working on a trading platform queries an AI assistant and gets back a response informed by the firm's non-public earnings projection. That is not merely a configuration failure. A federal securities law has been implicated, by an AI system, without the developer's knowledge. Civil and criminal consequences attach to the firm whether or not the disclosure was intentional.

PCI cardholder data requires total exclusion: primary account numbers, card verification values, expiration dates, and cardholder authentication data governed by PCI-DSS v4.0.1. AI systems must not process, store, or log raw cardholder data in any form. The PCI requirement applies at the input layer, not just the output layer. An AI system that receives cardholder data in a developer's query and logs that query as part of normal interaction recording has created a PCI-scoped log; every storage, retention, access control, and destruction requirement follows.

BSA/SAR data sits at the apex of sensitivity. The existence of a Suspicious Activity Report (SAR) is itself confidential under 31 U.S.C. 5318(g)(2). An AI system that reveals whether a SAR has been filed on a named customer, in response to any query, has committed a federal violation. Degree and intent do not enter into it. The disclosure prohibition is absolute.

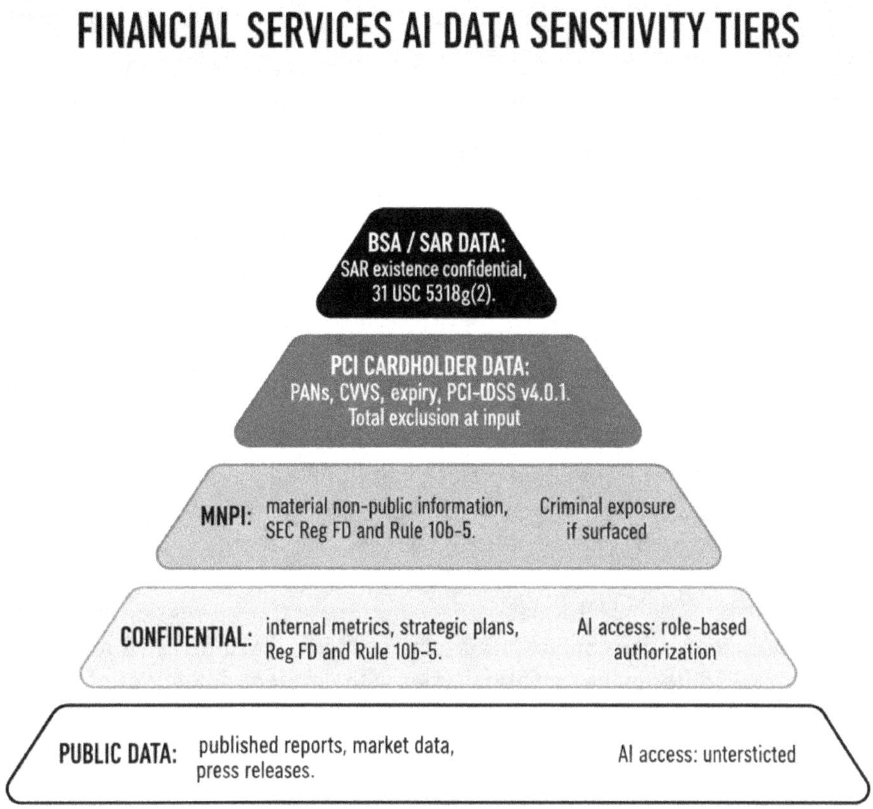

Figure 11: Financial services data sensitivity tiers and AI access posture.

Tier	Contents	AI access	Governing authority
Public	Published financial reports, market data, press releases	Unrestricted	None
Confidential	Internal business metrics, strategic plans, non-public financial information below the MNPI threshold	Role-based authorization	Internal policy
MNPI	Material non-public information	Wall attestation required; default deny	SEC Regulation FD, Section 10(b), Rule 10b-5
PCI cardholder	Primary account numbers, card verification values, expiration dates, authentication data	Prohibited; rejected at input	PCI-DSS v4.0.1

Tier	Contents	AI access	Governing authority
BSA/SAR	Suspicious activity report existence and content	Prohibited; absolute non-disclosure	31 U.S.C. 5318(g)(2)

Regulatory exposure compounds these data tiers. A single financial services AI deployment may be subject at the same time to Sarbanes-Oxley Act (SOX) Section 404 for internal controls over financial reporting; PCI-DSS v4.0.1 for cardholder data protection; Office of the Comptroller of the Currency (OCC) 2011-12 and Fed SR 11-7 for model risk management; the Federal Financial Institutions Examination Council (FFIEC) IT Examination Handbook for technology risk; BSA/anti-money-laundering (AML) requirements for transaction monitoring; GDPR for any European resident's data; EU AI Act Article 6 for high-risk AI in credit scoring or pricing; New York Department of Financial Services (NYDFS) 500 for cybersecurity at New York-licensed institutions; CCPA and CPRA for California resident data; and SEC Regulation FD for material information handling. These frameworks do not stack neatly. They overlap, they conflict on certain questions, and different authorities enforce them on different examination cycles with different definitions of what counts as adequate evidence.

No manual documentation process sustains this at enterprise scale. A bank with 200 AI systems operating under this stack, each generating decisions, each touching multiple data tiers, each subject to multiple examination frameworks, requires automated evidence production or it will produce inadequate evidence. There is no third option.

Then there is velocity. Financial services competition is measured in units that governance architects cannot wave away as abstraction. Algorithmic trading operates in microseconds. Fraud detection operates in milliseconds; a fraud model that takes longer to decline a transaction than the transaction takes to complete is not a fraud control but an audit artifact. Customer onboarding operates in minutes; a bank that takes three days to approve a personal loan against a competitor that takes three minutes is not offering a comparable product. Loan decisioning speed is the product itself.

AI-assisted code generation for trading platforms, fraud models, and customer-facing applications is where this velocity pressure concentrates. A developer who must navigate a governance process that takes two weeks to authorize an AI assistant configuration change, on a trading system where the production deployment window is measured in hours, will not wait for the governance process. The bypass will not be recorded. The examiner who arrives six months later will find a system operating outside its authorized configuration.

Station-keeping is not optional. It is a structural property of the L1 position.

The Lloyd's thread

Modern financial regulation descends directly from the maritime insurance market.

Lloyd's underwriters invented risk pricing. They built the intellectual architecture, the survey methodology, the classification standard, the certificate of seaworthiness, by which an expert read of a vessel's physical condition became a price for transferring the risk of that vessel's loss. Two centuries of institutional evolution separate the Lloyd's Register surveyor who assessed a ship's hull in 1820 from the OCC examiner who assesses a bank's model risk management program in 2025, yet they ask versions of the same compound question: does this system exist as described; is it operating as designed; does it know what has changed; and can every consequential action be traced to an accountable party?

Bank examination is classification survey by another name. Examination cycles correspond to survey cycles. Examination findings correspond to deficiency notices. The matter requiring immediate attention corresponds to classification suspension. A bank that cannot produce evidence of sound model risk management governance has submerged its load line; the examiner has the authority to require it to return to port.

This is not historical decoration. The lineage is why the evidence architecture developed for AI governance in classified federal environments transfers directly to financial services examination requirements. Both systems solve the same problem: how does an external party assess a complex system they did not build, cannot fully observe, and must issue an opinion about within a defined examination window?

The answer is structured evidence artifacts, produced by a collection architecture that does not depend on the assessed party's cooperation to be reliable. The E-1 through E-4 evidence tiers are the classification certificate for AI governance. A bank examiner who sits down at an evidence dashboard built on this architecture is performing the same cognitive operation as a Lloyd's underwriter reviewing a vessel's registration. The vocabulary has changed. The epistemology has not.

The Chinese wall problem

Chinese walls, also called information barriers, are a financial services-specific access control pattern with no direct analog in any other regulated sector. A Chinese wall separates employees who possess MNPI about a specific company from employees who make investment or trading decisions about that company. The wall is a regulatory requirement under SEC rules, not an organizational preference. Crossing the wall, either by disclosing MNPI to someone on the trading side or by using MNPI to inform a trading decision, constitutes a securities violation regardless of intent.

When an AI coding assistant operates across a financial institution, traditional Chinese wall enforcement fails at the AI layer.

Traditional walls separate physical spaces and network segments. An investment banker working on an M&A transaction operates on a physically separate network, with separate email systems, separate file shares, and separate access credentials from the sales and trading desk. The architecture of the network enforces the wall, and that architecture was designed for human actors moving through physical spaces. It was never designed for AI systems whose context windows fill from retrieval operations that cross organizational boundaries at the speed of a database query.

Consider a developer on the public equities trading desk who queries an AI assistant about a technology company's software architecture. If the AI assistant's knowledge base was populated through a retrieval pipeline that indexed documents from across the firm, including documents from the investment banking division that contain non-public deal information about that company, the AI assistant's response may be informed by MNPI. The developer does not know this. The query looks like a technical question. The response looks like technical guidance. The context window of the AI, invisible to both the developer and any real-time observer, contains a Chinese wall breach.

The AGENTS.md instruction surface must encode Chinese wall attestations as binding access control constraints. Before any retrieval operation, the system must resolve the developer's current wall attestation status: which securities they are walled off from, which they hold active clearance for. Wall compliance belongs at the routing enforcement layer (H-2 Routing Enforcement), not the instruction layer. An AI assistant told to "respect Chinese walls," leaning on that instruction alone, is not enforcing wall compliance. It is documenting that the wall requirement was mentioned.

Wall compliance at the AI layer requires three components. A wall attestation registry must record the current wall status of every developer who interacts with the AI system, updated in real time as walls are erected and lifted. A retrieval pre-filter in H-2 must check the developer's attestation status against the sensitivity tags of candidate retrieval results before those results enter the AI's context window. An E-4 (Attribution Chain) record must capture every retrieval operation involving confidential or MNPI-tier data, recording the requestor's attestation status at the time of retrieval. If the wall was crossed, the E-4 record will show it. Honored, the same.

None of this is a new capability. It is LP1 (Classification-Consistent Access) applied with a financial services policy source. The mechanism matches classification-consistent access enforcement in classified environments. The policy table holds different entries. The routing logic is the same.

Model risk management as harness verification

OCC Supervisory Guidance on Model Risk Management (OCC 2011-12), jointly issued with Federal Reserve Guidance SR 11-7, requires that every model used in a financial institution's decision-making process undergo independent validation, ongoing monitoring, and periodic re-validation. The guidance defines "model" broadly: any quantitative method, computation, or approach that applies statistical, economic, financial, or mathematical theories, techniques, and assumptions to process input data into quantitative estimates.

An AI coding assistant that generates or modifies code for trading systems, risk models, fraud detection, or credit decisioning produces output that affects model behavior. The code the AI generates is not itself the model under review. The AI-assisted development process, if ungoverned, is exactly the change management gap SR 11-7 validation exists to catch. An AI assistant that can rewrite a credit scoring model's code without tripping model validation is a bypass path around the entire model risk management framework.

The Three-Body harness is the model risk management framework for AI coding assistants. It provides the independent verification layer, the continuous monitoring, and the evidence artifacts that OCC 2011-12 requires. The mapping is precise:

OCC 2011-12 / SR 11-7 Requirement	Three-Body Implementation
Model validation	H-4 (Verification Checkpoints) validate AI output against defined governance criteria before code affecting model behavior is merged
Independent review	H-4 (Verification Checkpoints) require human review at defined checkpoints; LP6 (Attribution Completeness) produces an attribution record of that review that is independent of the AI's own output
Ongoing monitoring	E-2 (Continuous Verification Evidence), generated as a byproduct of system operation, not assembled pre-examination
Model inventory	E-1 (Governance Event Records) include AI system configuration inventory, instruction surface version, and knowledge base state
Outcomes analysis	E-3 (State Transition Records) track every configuration change, model update, and instruction surface revision with timestamps and attribution

OCC 2011-12 / SR 11-7 Requirement	Three-Body Implementation
Documentation	E-4 (Attribution Chain Records) provide the complete audit trail the model validation report requires
Change management	H-4 defines model change as a governance trigger event; no AI-assisted change to code affecting financial models proceeds without a verification checkpoint

The examiner who arrives to assess the bank's model risk management program for AI-assisted development will ask the same questions SR 11-7 always produces. What models are in scope? What validation was performed? When was it last updated? What changed between the last examination and now? The E-1 through E-4 evidence tiers answer those questions from the evidence architecture, not from a pre-examination documentation sprint.

There is a deeper structural point here, and the chapter rests on it. When a supervisor declines to write specific rules for agentic AI and instead points the institution back to its own practices, the institution has to construct the standard, demonstrate it, and prove it operates. This is the effective-challenge obligation, and it is confirmed law. SR 11-7 (2011, adopted by the Federal Deposit Insurance Corporation (FDIC) as FIL-22-2017) requires that every model receive an independent, critical challenge proportionate to its risk; SR 21-8 (2021) carries the same expectation into the model risk environment as it has matured. Effective challenge means the institution cannot take a model's output on faith. It has to build the evidence that the model performs as claimed and re-prove it as conditions change. Agentic AI development is exactly the change-management surface that obligation reaches. The regulator, in effect, hands the bank the pen: the bank writes the governance standard for its own AI-assisted development, and then the examiner tests that written standard against the running evidence the bank does not control. A dedicated agentic-AI model-risk pronouncement (anticipated by some commentary as a future SR letter) would make the expectation explicit for agentic systems, but the argument does not wait on it; effective challenge under SR 11-7 already reaches AI-assisted development today.

The financial services AGENTS.md

The instruction surface for a financial services AI deployment must encode regulatory requirements from at least five regulatory bodies in a form that the routing enforcement layer can execute without human interpretation at each access event.

The configuration blocks in this chapter are illustrative reference patterns; treat field names, thresholds, and authority bindings as a structure to adapt to your own controls and validation, not as a deployment to copy. The configuration below shows the mandatory constraints structure.

```
# AGENTS.md - Financial Services AI Governance
# Regulatory sources: SOX, PCI-DSS v4.0.1, OCC 2011-12/SR 11-7,
#                     FFIEC, BSA/AML, SEC Regulation FD, NYDFS 500

data_access_boundaries:
  tiers:
    - tier: public
      ai_access: unrestricted
    - tier: confidential
      ai_access: role_restricted
      verification: rbac_role_check
    - tier: mnpi
      ai_access: wall_attestation_required
      default: deny
      verification: wall_attestation_registry_check
      logging: e4_every_access
    - tier: pci_cardholder
      ai_access: prohibited
      input_sanitization: pan_detection_luhn_filter
      logging: e4_prohibited_attempt
    - tier: bsa_sar
      ai_access: prohibited
      disclosure_prohibition: absolute_31_usc_5318g2
      logging: e4_prohibited_attempt

chinese_wall_constraints:
  enforcement_layer: h2_routing_enforcement
  attestation_source: wall_attestation_registry
  verification_timing: pre_retrieval
  behavior_on_wall_conflict:
    action: block_retrieval
    escalation: compliance_officer
    logging: e4_wall_event

mandatory_constraints:
  - "MUST NOT access MNPI-tier data unless developer's wall
attestation
     covers the relevant security at the time of the request; prior
     session attestation is not valid for the current request"
  - "MUST NOT process, store, or log PCI cardholder data in any form;
     input sanitization must detect and reject PAN patterns before
they
     enter the AI context"
  - "MUST NOT reveal SAR existence or content under any circumstance;
```

```
      this prohibition is absolute under 31 U.S.C. 5318(g)(2)"
    - "MUST produce fairness impact note when generating or modifying
code
      that affects credit decisioning, loan pricing, or eligibility
      determination"
    - "MUST log all interactions involving confidential or higher data
tiers
      to the compliance audit trail as E-4 attribution records"
    - "MUST flag code changes affecting financial reporting systems for
      SOX Section 404 dual-review requirement before merge"
    - "MUST treat any AI-assisted change to code governing model
behavior
      as a model change event requiring SR 11-7 validation checkpoint"

regulatory_bindings:
  sox_404:
    applies_to: code_affecting_financial_reporting
    requirement: dual_review_before_merge
    evidence: e4_attribution_record
  occ_2011_12:
    applies_to: ai_assisted_model_development
    requirement: model_validation_documentation
    evidence: e2_verification_plus_e4_attribution
  pci_dss_req_3:
    applies_to: any_payment_card_data_contact
    requirement: no_storage_sensitive_authentication_data
    evidence: e1_prohibition_configuration
  ffiec:
    applies_to: all_ai_governance_activity
    requirement: examinable_by_federal_bank_examiners
    evidence: e1_through_e4_evidence_dashboard
  nydfs_500:
    applies_to: ny_licensed_institution_operations
    requirement: cybersecurity_event_reporting_72_hours
    evidence: e3_state_transition_records

approval_authorities:
  confidential_tier_access: line_manager_plus_iso
  mnpi_tier_access: compliance_officer_plus_active_wall_attestation
  model_code_change: model_risk_officer_plus_h4_checkpoint
  sox_scoped_change: internal_audit_dual_review
  emergency_override: cro_with_e4_record_and_72_hour_review
```

The wall attestation registry is the component that makes this configuration enforceable rather than aspirational. Without a current-state registry that H-2 can query in real time, the MNPI constraints are instructions the AI reads and interprets. With the registry, they become gates the routing layer clears before the AI context is populated. That one distinction decides whether the Chinese wall exists at the AI layer or only in the AGENTS.md documentation.

The mechanism that carries these boundaries to the gateway at each access event is the Delegation Token. In this reference pattern, a Delegation Token is a short-lived, signed authorization credential the gateway checks before the agent reaches any resource. The token binds the issuing authorization service (issuer), the agent or Task Mandate it authorizes (agent_identity), the explicit list of permitted resources by name (audience), the permitted actions on those resources (scope), the maximum data sensitivity tier authorized (data_tier), the jurisdiction tag the session is scoped to (jurisdiction/residency), the hosted-or-on-prem sovereignty route (sovereignty_route), a short issue-and-expiry window with no implicit renewal (iat/exp), and the signature (sig). Each Model Context Protocol (MCP) connector the AI reaches, the wall attestation registry, the model inventory, the knowledge base, maps to one audience entry; an agent cannot reach any resource absent from that list. A request to a resource outside the token's audience field is denied at the gateway, not at the prompt; this is LP13 (Output Boundary Enforcement) applied to what the agent may reach and emit, and the denied request emits an E-2 operational evidence record proving the boundary held. When the gateway cannot evaluate a request against an authorized audience or data_tier entry, LP14 (Deterministic Deny on Uncertainty) governs: the default is denial with escalation, and the unresolved condition does not propagate into the AI context. The financial-services data tiers above map to the token's data_tier ceiling: a token issued at the confidential tier cannot reach MNPI, PCI, or BSA/SAR resources, because those audience entries are excluded by omission. The next chapter develops the worked enforcement artifacts: a merge-request approval rule, a pre-merge policy stage, an audit-event query, and the policy engine that reads AGENTS.md at the gateway.

Synthetic identity and KYC collapse

Know Your Customer obligations under the Bank Secrecy Act and FinCEN guidance require financial institutions to verify the identity of customers at account opening and to conduct ongoing due diligence. Traditional KYC relied on document verification: a driver's license, a passport, a utility bill. The assumption underlying that architecture was that identity documents are difficult to forge and that the person presenting them is the person depicted.

Both assumptions are now untenable.

AI-generated synthetic identities combine real personally identifiable information from data breaches with AI-generated photographs, fabricated supporting documents, and behavioral histories constructed to pass automated screening. Published industry estimates put deepfake-enabled fraud losses in the hundreds of millions of dollars and rising. The fraud does not look like a poorly printed document or a mismatched name. It looks like a legitimate customer with a plausible credit history, a consistent digital footprint, and verification documents that pass automated review.

Traditional point-in-time KYC, conducted at account opening and refreshed on a scheduled cycle, cannot detect synthetic identities constructed specifically to pass that check. The verification occurs at the moment of maximum vulnerability: before the account has accumulated behavioral history that distinguishes it from a legitimate account.

The governance response has three components.

Identity proofing modernization closes the authentication gap. FIDO2 and WebAuthn standards provide cryptographic authentication that binds a verified identity to a hardware-backed credential rather than to a document that can be copied or generated. An AI-generated identity cannot replicate a FIDO2 assertion, because the assertion requires possession of a hardware key registered to a previously verified individual. The AI governance harness must treat identity verification method as a data element in the E-1 governance configuration, with different trust levels for document-based verification versus cryptographic verification.

Behavioral biometrics close the detection gap. A synthetic identity that passes KYC at account opening still behaves differently from a legitimate customer: different typing rhythms, different navigation patterns, different time-of-day usage distributions, different geographic consistency signatures. Applied as a continuous verification layer through E-2, behavioral biometrics can separate synthetic accounts from legitimate ones after the initial identity check, within weeks of account opening. This does not replace KYC. It is the ongoing monitoring that compensates for KYC's structural limitation as a point-in-time check.

The governance posture shift closes the architecture gap. Point-in-time KYC runs a thorough check at a defined moment, then assumes the result holds. That is the heroic one-shot pattern applied to identity verification. Continuous identity verification, building on behavioral signals, transaction patterns, and periodic re-authentication events, converts identity assurance from a state into a posture. The E-3 evidence tier must record identity assurance events, not just the initial KYC result, to answer the examiner question: when was the last time identity assurance for this account was affirmatively confirmed?

Financial services attack surface

The eight generic attack surfaces from Chapter 7 all apply to a financial institution. Four take on a sharper form here, because the data is market-moving, the decisions bind money, and the regulators examine the controls directly.

Market and reference-data poisoning. A financial agent grounds on price feeds, filings, research, and ratings. An attacker who corrupts one of those sources moves the analysis without ever touching the model. The defense is verified provenance (KB-1) on every data source and the figure-to-source binding described in the frontier-lab harness, where each cited figure in a work product joins to the connector-access event that produced it and any figure with no backing event is flagged. An E-3 state-transition record fires on any source change.

The "draft, not decide" line, crossed. The control that makes agentic finance governable is that agents draft analyst work product but never execute trades, bind risk, post to a ledger, or approve onboarding. The attack is any path that lets a generated output become a binding action without human disposition: a connector granted write authority, an auto-execute integration, a sign-off gate that can be satisfied by the agent itself. The defense is the sign-off gate as a required human approval, decisioning connectors scoped read-only, and LP14 deny on uncertainty so an ambiguous request never resolves toward action.

Information-barrier (wall) leakage. Connectors that span research, trading, and banking can carry material non-public information across a barrier the firm is legally required to maintain. This is the F-FS1 failure mode weaponized: the wall is the target. The defense is jurisdiction-style scoping applied to the information barrier, a live wall re-check at the moment of retrieval rather than at session start, and attribution records that prove no cross-wall content reached an uncleared user.

Synthetic-identity and KYC defeat. Generated or stolen identity data engineered to pass automated onboarding is the adversarial form of the synthetic-identity collapse this chapter already describes. The defense is provenance on identity evidence, human adjudication above a consequence threshold, and out-of-distribution monitoring on approval patterns.

Financial services failure modes

F-FS1: MNPI leakage through AI context. A developer on the trading desk queries the AI assistant about a technology company's upcoming product roadmap. The AI assistant's retrieval pipeline queries the knowledge base. The knowledge base was indexed from firm-wide documents, including a deal team briefing from the investment banking division that contains material non-public information about the company's pending acquisition announcement. The AI assistant's response is informed by the deal briefing. The developer receives information that constitutes MNPI, without having requested it, without knowing the information was MNPI, and without any record in the wall attestation system that this developer was cleared for this security.

The disclosure has occurred. The violation is complete. If the architecture was built properly, the E-4 record will show a retrieval from the MNPI tier completed for a requestor whose wall attestation did not cover the relevant security; the record is the evidence, but it does not undo the violation. Prevention requires H-2 to route the retrieval request through the wall attestation registry check before the retrieval runs, not after.

The detection architecture relies on E-4 attribution records with wall compliance verification flags, monitored by the compliance officer with real-time alerting for any MNPI-tier retrieval where the requestor's attestation was not active for the relevant security at the time of retrieval.

F-FS2: Model drift exceeding SR 11-7 validation bounds. An AI model used in credit decisioning was validated in Q3 of the prior year against a specific portfolio population and a specific macro environment. In Q2 of the current year, the macro environment has shifted materially: interest rates have changed, delinquency rates in certain product categories have moved outside the distribution the model was validated against, and the demographic composition of the application population has shifted due to a new marketing campaign targeting a previously underrepresented segment. The model continues to operate. No re-validation has been triggered. The model's outputs are increasingly inconsistent with its validated performance characteristics, and the credit decisions it is producing in the new conditions were never evaluated for fair lending compliance.

SR 11-7 requires ongoing monitoring to catch exactly this condition. The harness's E-2 continuous verification evidence must carry model performance metrics tracked against validated baselines, with automatic thresholds that trigger the model risk officer's review when performance drifts outside defined bounds. Waiting for the examination cycle to check whether model performance has drifted is the heroic one-shot failure mode. By the time the examiner arrives, the model has run outside its validation scope for months.

E-2 continuous monitoring with automated drift detection, comparing production performance metrics against the validated baseline distribution, provides the detection layer. The threshold for triggering re-validation review is defined in the AGENTS.md as a governance rule, not as an examiner recommendation.

F-FS3: Regulatory evidence gap between AI-generated and human-generated artifacts. A bank's compliance program produces SOX attestations, model validation reports, and FFIEC examination packages that describe a governance architecture built around human-reviewed controls with documented approval chains. The AI coding assistant has been operating in the development environment for fourteen months. The code it has assisted with includes modifications to financial reporting systems, fraud detection models, and customer-facing decisioning algorithms. None of those AI-assisted code changes appear in the governance documentation, because the governance documentation was designed for human developers and does not have a category for AI-assisted modifications.

The examiner asks: which of the code changes in the past year affecting the credit decisioning model were reviewed under the model risk management framework? The compliance team produces a list of changes with human reviewer signatures. AI-assisted changes are absent from the list, not because they were hidden, but because the documentation architecture never captured them. The examiner's question cannot be answered from the evidence that exists.

Prevention requires E-4 attribution records that capture AI-assisted code changes as a distinct attribution category, naming the AI system as a contributing author and the human who authorized the change as the accountable approver. SOX control documentation has to fold AI-assisted modifications into the scope of internal controls over financial reporting. An AI that modifies code affecting financial reporting takes part in a process that SOX Section 404 requires to be evidenced and attested.

The 5-minute examiner test

A federal bank examiner at a workstation should be able to answer four questions from the evidence dashboard in five minutes or less. These four are not hypothetical. They open any technology-focused examination, and they set whether the examination proceeds with an "adequately governed" presumption or an "explain everything from scratch" posture.

1. What AI systems are operating in this institution, and which regulatory frameworks apply to each? (E-1)
2. Are the governance controls over those systems operating as designed, right now, with a current timestamp? (E-2)
3. What changed in the AI governance configuration since the last examination? (E-3)
4. Can every AI-assisted action that affected a financial model, a consumer decision, or a financial report in the last 90 days be attributed to a specific developer with a specific authorization? (E-4)

If the harness architecture is built properly, all four answers carry quantitative evidence with timestamps, reachable in a query interface without the compliance team assembling a documentation package. If any answer needs a narrative explanation, a documentation sprint, or a "we need to pull that from multiple systems," the harness has a gap.

This is not a convenience target. The 5-minute test is the examination posture financial services regulators are moving toward as AI governance examination methodology matures. Five minutes for all four answers means continuous governance. Five days means examination-driven governance, and the gap between examination states is where compliance risk piles up undetected.

The financial services evidence package

The four evidence tiers map directly to financial services examination requirements, across all the major frameworks that apply:

Evidence Tier	Financial Services Implementation	Regulatory Mapping
E-1	AGENTS.md configuration, wall attestation registry, model inventory, KB health status, knowledge base access tier assignments	OCC 2011-12 model inventory, SOX Section 404 control documentation, FFIEC IT governance documentation
E-2	Automated verification results, wall compliance check results, fairness metrics, model performance vs. baseline, PAN detection event log	OCC ongoing model monitoring, PCI-DSS continuous monitoring requirements, SR 11-7 performance validation
E-3	Configuration change records, model update records, instruction surface version history, wall attestation change events	FFIEC examination readiness documentation, SOX Section 404 change management, NYDFS 500 change tracking
E-4	Attribution records for every AI-assisted code change, wall crossing attempt records, MNPI-tier retrieval records, SAR non-disclosure enforcement records, SOX dual-review confirmations	SEC audit trail requirements, BSA/AML activity monitoring, CFPB adverse action records, FFIEC examination attribution requirements

SOX Section 404 maps to the internal-control-over-financial-reporting requirements. E-1 governance documentation produces the control inventory that internal auditors require; E-2 continuous verification results serve as the operating effectiveness evidence that external auditors test; E-3 state transition records satisfy the change management documentation requirement; E-4 attribution records satisfy the approval chain requirements for IT general controls. A properly instrumented harness produces all four categories of SOX evidence as a byproduct of system operation, not as an audit preparation project; whether the produced evidence satisfies a given control objective is a determination the external auditor makes, not one the architecture asserts.

OCC model risk management follows the same evidence path. The SR 11-7 independent validation requirement draws from the E-1 model inventory, the E-2 ongoing monitoring results, the E-3 change record showing every model and configuration modification since the last validation, and the E-4 attribution chain showing that an authorized party reviewed every consequential change. Maintain this evidence architecture continuously and the model validation report documentation requirement is satisfied because the evidence already exists; the report is a rendering exercise, not a reconstruction.

PCI-DSS v4.0.1 completes the pattern: E-1 configuration records demonstrate that access controls described in the AGENTS.md are active; E-2 PAN detection event logs demonstrate that the input sanitization layer is operating; E-3 change records satisfy Requirement 6 change control obligations; E-4 attribution records satisfy Requirement 10 audit log requirements for any system in scope for cardholder data protection.

CHAPTER SUMMARY

Financial services AI governance is an L1 Lagrange point problem: an unstable equilibrium between five distinct data protection regimes and competitive velocity pressure that is measured in milliseconds. Active station-keeping, in the form of continuous verification, continuous evidence production, and continuous wall compliance enforcement, is not optional. It is the structural requirement of the L1 position. The Lloyd's thread connects maritime risk pricing directly to bank examination methodology; both are solving the same epistemological problem with structured evidence artifacts. The Chinese wall problem requires enforcement at the AI routing layer, not the instruction layer, because traditional physical and network-based wall architecture was not designed for AI context windows. Model risk management under OCC 2011-12 and SR 11-7 maps precisely to the harness verification architecture: model validation corresponds to H-4 checkpoints, ongoing monitoring corresponds to E-2 continuous evidence, model inventory corresponds to E-1 governance records, and outcomes analysis corresponds to E-3 state transition records. The regulator-hands-the-bank-the-pen argument stands on the confirmed effective-challenge obligation in SR 11-7 (2011, FDIC FIL-22-2017) and SR 21-8 (2021): the institution must construct and prove its own governance for AI-assisted development whether or not a dedicated agentic-AI pronouncement ever issues. The three financial services failure modes are MNPI leakage through AI context (F-FS1), model drift exceeding SR 11-7 validation bounds (F-FS2), and regulatory evidence gap between AI-generated and human-generated artifacts (F-FS3). The 5-minute examiner test, four questions answerable in five minutes from the evidence dashboard, distinguishes a continuous governance posture from an examination-driven program. The E-1 through E-4 evidence tiers satisfy SOX Section 404, OCC model risk management, PCI-DSS v4.0.1, and FFIEC examination requirements from a single evidence architecture, because those frameworks are all asking versions of the same four examiner questions.

Chapter 12: The Frontier-Lab Finance Harness

> *"For fifteen years the model risk examiner asked one question of every system a bank relied upon: was this challenged by someone competent and independent before you trusted it. The systems changed. The question did not. When the regulator turns to agentic AI, it declines to answer the question on the bank's behalf and hands the bank the pen."*

Chapter 11 made a claim and asked the reader to hold it on credit. The L1 Financial Services Lagrange point, it argued, is an unstable equilibrium that can be held only by active station-keeping, and the federal evidence architecture, E-1 through E-4, transfers directly into bank examination because both systems answer the same four examiner questions. That chapter asserted the transfer. It did not prove it.

This chapter is the proof.

In May 2026 a frontier AI lab published a finance compliance reference built on exactly the pattern the federal harness describes. No slide deck. No reference architecture. What it published was a working codebase, public, with named agents, a shared skill library, and a deliberate refusal to let any agent bind a decision without a human signature. The question that had occupied the L1 chapter, will the governance pattern survive contact with a regulated bank, stopped being a thought experiment the moment that codebase appeared. The pattern had already survived contact. A lab whose business depends on the answer had bet its product on it.

What follows separates what transferred verbatim from what did not. The transfer rule, established earlier in this volume, holds that most of the harness transfers verbatim into a commercial regulated setting; the residual is the consent, jurisdiction, fairness, and sovereignty extension, the hard part that must be built locally. The frontier-lab product proves the transfer. A second 2026 release names the residual.

The product that proved the transfer

Anthropic published `anthropics/financial-services` as a public reference repository: ten named end-to-end agents, among them a Pitch Agent, a Market Researcher, a GL Reconciler, and a Know Your Customer (KYC) screening agent. Behind the agents sits a library of finance skills authored once and synced into each agent that needs them: comparable-company analysis, discounted cash flow, leveraged buyout modeling, three-statement modeling, deck quality control, Excel audit. A financial-analysis core plugin carries the data plane: Model Context Protocol (MCP) connectors to FactSet, PitchBook, S&P Global, Moody's, Morningstar, LSEG, Daloopa, MT Newswires, Aiera, Chronograph, and Egnyte.

One design choice in that codebase carries the entire compliance argument. The agents draft analyst work product. They build the model, write the memo, run the reconciliation. What they never do is make investment recommendations, execute transactions, bind risk, post to a ledger, or approve onboarding. Every output waits for a qualified human to sign it before it counts as anything.

That is the transfer rule made into a published artifact, and it is the L1 chapter's central mechanism reproduced by a vendor who had never read the L1 chapter. The skill library is the Instruction Surface: it defines what the agent is competent and permitted to do. Connectors are the data plane, scoped resources, each one a thing the agent may reach only if authorized. The sign-off gate is the Verification Checkpoint. Nothing the agent produces becomes a decision until a competent human approves it. This is Trust but Verify, established in this volume's foundational chapters, expressed as code rather than as principle. Grant the agent scoped operating trust; verify every action against that scope; hold no output as a decision until a human signs.

The same codebase deploys two ways from one source. It runs as a Claude Cowork plugin for an analyst at a desk, and it runs through the Managed Agents API behind a customer's own workflow engine, both sharing one system prompt and one skill set. The institution that adopts it does not choose between a desktop tool and a server integration. It chooses where the same governed agent runs.

The product survives only because it refuses to let the velocity sphere dominate. That sign-off gate is the orbital constraint that keeps the data-protection and compliance spheres from being thrown out of the system. Remove the gate and the product runs faster. It also becomes uninsurable.

What the regulator says

The durable principle under all of this descends from model risk management guidance. Federal Reserve and OCC guidance SR 11-7, issued in 2011 and adopted for Federal Deposit Insurance Corporation (FDIC)-supervised institutions through FIL-22-2017, established effective challenge: critical review of every model by informed, competent, independent parties before the institution relies on it. SR 21-8 carried the same discipline into model risk management for Bank Secrecy Act and anti-money-laundering systems. The L1 chapter mapped that requirement onto the harness. The mapping holds against both.

Effective challenge does the part institutions most need to understand. Supervisory model-risk guidance governs the models a bank validates, but it does not write the controls for generative AI and agentic AI directly. It points each banking organization's own risk-management and governance practices to determine the appropriate controls for such systems. So the supervisor looks at the agent that drafts the credit memo and declines to write its rules. It hands the decision back to the institution. (The agencies are expected to issue revised model-risk guidance in 2026, anticipated here as SR 26-2; it would carry the effective-challenge standard forward unchanged, so nothing in this chapter depends on it.)

That refusal is not a gap in the guidance. It is the reason the harness matters. When the supervisor governs a model directly, the institution implements a published standard. When the supervisor declines to govern agentic AI and points back to the institution's own practices, the institution must construct the standard, demonstrate that it constructed one, and prove it operates. The pen is in the bank's hand. The harness is what the bank writes with.

This is the keystone claim, and it stands on the confirmed effective-challenge lineage: when a supervisor declines to write rules for agentic AI and points back to the institution's own practices, the institution must construct the standard, demonstrate it, and prove it operates. This follows from SR 11-7 effective challenge directly.

What the effective-challenge lineage supplies is the picture of good governance: independent, competent review before reliance. The frontier-lab sign-off gate is exactly that review, made structural.

The platform the institution controls

The frontier lab proves the work transfers. It does not prove the work can be governed at the institution's own perimeter, because the institution does not own the frontier lab's deployment surface. That second proof is the platform layer.

GitLab Ultimate is the DevSecOps platform that holds the codebase the agents run inside: compliance frameworks, protected branches, merge-request approval rules, pipeline policies, and an audit event stream feeding a Compliance Center. The GitLab Duo Agent Platform (DAP) is the governed AI layer that operates inside GitLab within scope defined by a configuration file named `AGENTS.md`.

The division of labor is clean. The agent drafts the credit memo. The platform records who ran it, against which data, under which policy, and who signed it. The vendor's sign-off gate, a design intent baked into the product, becomes a merge-request approval rule the institution enforces as code. A vendor decided no output ships unsigned. The institution makes that decision auditable.

What the Chief Risk Officer sees in the Compliance Center is the four things effective challenge demands as evidence: every AI-assisted work product linked to an analyst identity, a timestamp, a data scope, and an approval record. What the examiner sees is one stream, machine-generated, readable without a separate compliance project to translate it. That is the difference between governance built into the architecture and governance assembled before the exam. The L1 chapter named the latter the financial-services form of the Heroic One-Shot failure: a single manual effort that looks like control and is actually a snapshot of a system that has already moved.

The delegation architecture

The bridge from vendor design to enforced control runs through three constructs the foundational volume defines: the Task Mandate, the Delegation Token, and `AGENTS.md`. They are summarized here because they carry the enforcement, not re-derived.

`AGENTS.md` is the GitLab-native file that scopes context, tools, and behavioral constraints for the AI layer in a given repository. For a finance workflow it declares the permitted skills, the permitted connectors by name, the data sensitivity tier the agent may touch, and the rule that no output merges without approval. It also encodes the sovereignty decision, a flag selecting hosted inference or on-premises serving, so the same workflow definition can route to a vendor endpoint or to a locally served open-weight model without a rewrite.

`AGENTS.md` is not one undifferentiated rulebook. It carries two layers, and the distinction matters because the two are enforced by different machinery and a reader must not be misled that judgment-laden legal standards are gateway-enforced.

The first is the compiled-enforceable layer. The connector allow-list, the data-tier ceiling, and the jurisdiction-routing rule compile to the gateway policy and are genuinely gated. A connector absent from the allow-list, a request above the data-tier ceiling, or a session routed to the wrong jurisdiction is denied at the gateway before any access, deny-by-default, under LP13 (Output Boundary Enforcement) and LP14 (Deterministic Deny on Uncertainty). These rules are deterministic and machine-checkable: a connector name is on the list or it is not, a tier is below the ceiling or it is not. The architecture evaluates them against an authorized policy without interpretation.

The second is the model-directive-then-verified layer. The minimum-necessary rule, the consent-currency rule, and the fairness rule are not gateway-gated, because no gateway can decide in advance whether a given retrieval was the minimum necessary or whether an output is fair. These are instructions the model follows, and their output is checked at an H-4 verification checkpoint and captured as evidence. They are attested-and-verified, not pre-access gateway-denied. The model is directed to honor minimum-necessary; the H-4 checkpoint inspects what it actually retrieved against the mandate and records the result as E-4 evidence. A reader should not infer that a judgment-laden legal standard such as fairness is denied at the gateway the way an out-of-scope connector is. It is followed by the model and verified after, with the evidence trail proving the verification ran.

A Task Mandate is the scoped instruction set assigned to an agent before execution: authorized intent, permitted tools, permitted data. For a KYC screening agent the mandate authorizes reading the KYC file and drafting a screening summary, and explicitly excludes approving onboarding. That exclusion is the same boundary the frontier-lab product already enforces; the mandate writes it down where the platform can check it.

In this reference pattern, a Delegation Token is a short-lived, signed authorization credential the gateway checks before the agent reaches any resource, enforced at the routing layer (H-2) and never at the prompt. It binds the agent's identity to an explicit, enumerated set of resources. The canonical field set is: issuer, the authorization service that minted the token; agent_identity, the agent or Task Mandate the token authorizes; audience, the explicit list of permitted resources by name, where each MCP connector, the wall_attestation_registry, the model inventory, and each knowledge-base source is one audience entry, and a request to any resource absent from this list is denied at the gateway; scope, the permitted actions on those resources, read or retrieve for these draft-only agents, never write or execute; data_tier, the maximum data sensitivity ceiling, so a confidential-tier token cannot reach MNPI, PCI, or BSA/SAR resources because those audience entries are excluded by omission; jurisdiction/residency, the jurisdiction tag the session is scoped to, feeding LP10; sovereignty_route, hosted versus on-premises serving path; iat and exp, a short issue-and-expiry window with no implicit renewal, so reissue requires re-authorization; and sig, the signature. The signing approach is a detached signature over the canonical claim set using the authorization service's asymmetric key, an EdDSA/Ed25519 or ES256 JWS-style envelope, with the gateway verifying signature, exp, and audience membership before any connector access; rotation is by key-ID reference, and revocation is by short TTL plus a gateway-checked revocation list.

The enforcement consequence is the point. Each MCP connector maps to an audience entry. Send a request to a connector outside the token and it is denied at the gateway, not at the prompt. The distinction matters: a prompt instruction is interpreted, a gateway denial is enforced. The out-of-scope or out-of-tier request produces a denied event in the audit stream, and that denied event is itself E-2 operational evidence that the boundary held. The compiled-policy engine internals that perform this check at scale are described here at the level the enforcement argument requires.

From the file to the token

The three constructs bind in a fixed order, and the order is what prevents a stale credential from outliving a tightened policy. `AGENTS.md` is the source of truth. It is compiled to the gateway policy at merge time, so the live policy the gateway enforces is always the compiled form of the most recently merged `AGENTS.md`. The Delegation Token's audience and scope are derived from that compiled policy at issuance: when the authorization service mints a token, it reads the current policy and enumerates the audience and scope the token may carry. The token is therefore a snapshot of the policy as it stood when the token was issued.

The request path runs `AGENTS.md` to compiled policy to token-at-issuance to gateway-re-check-per-request. The final step is the safeguard. The live gateway policy re-checks every request against the compiled policy at access time, not only the token the agent presents. A token cannot grant more than the live policy allows. If `AGENTS.md` is tightened after a token is issued, for example a connector removed from the allow-list or a data tier lowered, the next merge recompiles the policy, and the gateway's per-request re-check denies the now-out-of-scope access even though the still-valid token names it in its audience. The token is necessary but not sufficient; the live policy is the ceiling. This is LP14 (Deterministic Deny on Uncertainty) in service of LP13 (Output Boundary Enforcement): a stale token can never exceed a tightened `AGENTS.md`, because the gateway evaluates the request against the current policy on every access and fails closed when the token and the policy disagree.

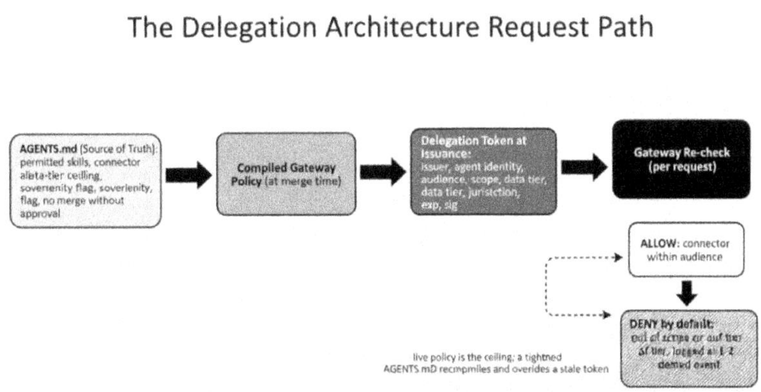

Figure 12: The four-stage enforcement path from AGENTS.md to per-request gateway re-check, showing how a stale Delegation Token can never exceed a

tightened live policy and how out-of-scope requests are denied at the gateway as E-2 evidence.

Consent withdrawal and Chinese-wall restrictions follow the same live-evaluation rule, because a token issued before a restriction was imposed must not be allowed to read through it. Consent and Chinese-wall checks are evaluated live at the gateway per access, not baked into the token at issuance. When an analyst working a deal under a wall restriction loses access to a source mid-session, or a customer withdraws consent for a data use, the gateway honors the change on the next request even though the token in hand still names the source in its audience. Where live per-access evaluation of a particular check is not feasible, the token TTL is bound below the maximum tolerable consent-propagation delay, so the longest a withdrawn consent or a newly raised wall can go unhonored is one short token lifetime. Either way, consent withdrawn mid-token-lifetime is honored: the architecture re-checks the wall and the consent state live, or the short TTL forces a re-authorization that picks up the new state, and a stale token never reads through a restriction imposed after it was minted.

The financial services AGENTS.md

The instruction surface for this case study encodes the sign-off gate, the connector scope, and the sovereignty decision in a form the routing layer executes without human interpretation at each access event. The configuration below shows the mandatory structure for a frontier-lab finance agent deployment.

A note on the code in this chapter, stated once: the `AGENTS.md` configuration blocks that follow are reference implementations drawn from the published frontier-lab pattern. They show the structure an institution enforces, not a snapshot of a specific production cluster. Field names, connector lists, and reviewer groups are named so the institution can map them to its own resources. The platform-specific enforcement configuration that carries this structure, the merge-request approval rule, the pipeline policy, the audit-event query, and the gateway policy, is drawn from the GitLab Ultimate feature set and collected in the companion reference-configuration supplement rather than reproduced here. The illustrative dollar ranges and currency windows are gathered under a single labeled note later in this chapter; replace them with your own sourced figures.

```
# AGENTS.md - Frontier-Lab Finance Agent Governance
# Governing principle: SR 11-7 effective challenge (2011)
# Supervisory model-risk guidance points the controls for generative
and
# agentic AI back to the institution's own practices, so the
# institution-defined controls below are the governing standard the
# institution authors for itself.

agent_binding:
  source: anthropic_financial_services
```

```yaml
  deployment: managed_agents_api   # or cowork_plugin
  permitted_skills:
    - comps
    - dcf
    - lbo
    - three_statement
    - deck_qc
    - excel_audit
  prohibited_actions:
    - investment_recommendation
    - transaction_execution
    - risk_binding
    - ledger_posting
    - onboarding_approval

connector_scope:
  authorization_model: delegation_token_audience
  permitted_connectors:
    - factset
    - pitchbook
    - s_and_p_global
    - moodys
  behavior_on_out_of_scope:
    action: deny_at_gateway
    logging: e2_denied_event
  free_text_source: prohibited

verification_gate:
  enforcement_layer: gitlab_merge_request_approval
  rule: no_merge_without_named_human_approval
  protected_branches: deliverable_work_product
  reviewer_group: model_risk_review
  binds_to: sr_11_7_effective_challenge

sovereignty:
  posture: per_workflow
  hosted_default: managed_agents_api
  boundary_restricted_route: on_prem_command_a_plus
  egress_check: e3_compare_destination_to_posture_flag

provenance_policy:
  pre_merge_pipeline:
    - check: every_figure_traces_to_connector_record
    - check: no_connector_outside_authorization_was_called
    - check: retrieval_within_currency_window
    - check: jurisdiction_tag_matches_data_residency
  fail_build_on_violation: true

evidence_bindings:
  e1_foundation: agents_md_plus_approval_rule_plus_connector_list
  e2_operational: merge_requests_approvals_connector_access_denials
  e3_context: sovereignty_posture_jurisdiction_data_tier_per_session
  e4_accountability: draft_to_citation_to_reviewer_to_approval_chain
```

The verification gate is the component that makes this enforceable rather than aspirational. Without a required approval rule on a protected branch, the no-merge constraint is a sentence the agent reads. With it, the constraint is a gate the platform passes before any output becomes a decision. That distinction determines whether the sign-off gate exists in the architecture or exists only in the documentation.

The four enforcement artifacts

Four platform artifacts turn the vendor's design into enforced control. Each maps to a regulatory obligation the institution defines for itself, because supervisory model-risk guidance points the controls for agentic AI back to the institution. Stated in platform-neutral terms, the four are these.

The first is a signed merge-request approval gate. It requires an approval from a model-risk reviewer group on any protected branch that receives AI-generated output, and it blocks direct pushes, so every change routes through the gate. This is the SR 11-7 effective-challenge standard expressed as code.

The second is a pre-merge pipeline policy. It runs four checks and fails the build on any violation: every cited figure traces to a connector record, no connector outside the delegation-token audience was called, retrieval falls within the currency window, and the jurisdiction tag matches the data residency. An unsourced assertion or an out-of-scope connector call is caught before it reaches a decision.

The third is an audit-event query that proves the claim "every figure traces to a connector record." Run against the platform's audit-event stream, it joins each cited figure in a merge request to its connector-access event and flags any figure with no backing event.

The fourth is a gateway policy engine that consumes `AGENTS.md` and enforces scope and provenance before each connector access. A policy engine such as Open Policy Agent (OPA/Rego) or AWS Cedar evaluates the `AGENTS.md`-derived rules at the gateway, denies by default, and emits an E-2 denied event when a request falls outside the token audience. This is LP13 (Output Boundary Enforcement) and LP14 (Deterministic Deny on Uncertainty) made operational: LP13 evaluates the access and the resulting output against the disclosure and scope boundaries at the routing layer before anything leaves the perimeter, and LP14 fails closed when the engine cannot evaluate a request against an authorized policy, denying with escalation rather than letting the uncertainty propagate. The compiled-policy engine that performs this at production scale is the subject of Book IV. A prompt instruction is interpreted; a gateway denial is enforced.

The platform-specific configuration that implements these four artifacts, the approval-rule definition, the pipeline policy stage, the audit-event query, and the gateway policy, is reference implementation detail rather than doctrine. It lives in the companion reference-configuration supplement so the chapter can carry the argument while the wiring stays out of the narrative. Any platform that provides signed merge-request approval gates and pipeline-stage provenance satisfies the pattern.

For workflows that may not send data to a hosted endpoint, the AI layer calls a locally served model. The reference open-weight model for this path is Cohere Command A+, released May 20, 2026 under a true Apache 2.0 license: 218 billion total parameters with roughly 25 billion active per token (128 experts, 8 activated plus one shared), 128K context, native citation grounding with explicit source spans, 48 languages including every official EU language, deployable on as few as two H100 GPUs entirely inside the institution's boundary. The native citation output satisfies the provenance check in the pipeline policy without a separate retrieval-attribution layer; the model emits the source span that the check verifies against.

Where transfer ends and extension begins

Command A+ is not a faster version of the hosted product. It exists to serve the part of the problem the hosted product cannot. The hosted agent proves that skills, connectors, and the sign-off gate transfer verbatim; most of the harness transfers without modification. What is left over is Command A+'s job: data that may not leave the jurisdiction, decisions that must show their sources in the output itself, and operations spanning regulatory regimes that do not share a language.

The line is sharp. A US broker-dealer drafting a domestic pitchbook against FactSet and PitchBook data can run the hosted agent under the GitLab approval gate and satisfy its obligations; the transferred majority scales for free, and marginal governance cost falls with each workflow added because the harness infrastructure is shared. An EU bank handling personal data that may not cross a border, producing credit decisions that must cite their sources to satisfy a deployer's human-oversight duty under the EU AI Act, operating across jurisdictions that file in different official languages, cannot. That bank needs model weights and data inside its own boundary, native citations in the work product, and multi-language coverage. The extension does not scale for free. It must be stood up deliberately, before scaling across jurisdictions, not after.

The stakes are public and quantified. The EU AI Act sets administrative fines up to 35 million euros or 7 percent of total worldwide annual turnover for prohibited-practice violations, up to 15 million euros or 3 percent for breaches of obligations on high-risk systems and other duties, and up to 7.5 million euros or 1 percent for supplying incorrect information to authorities. A bank that misreads the sovereignty boundary and lets personal data cross it does not get a configuration warning. It gets a fine sized to its global revenue.

A note on figures, stated once: the EU AI Act penalty percentages above are statutory and cited as such. Any currency window, retrieval-freshness threshold, or cost figure used elsewhere in this chapter as a worked example is an illustrative range; replace each with your own sourced figure before relying on it.

This is the precise content of the transfer rule: most of the harness transfers verbatim; the residual is the consent, jurisdiction, fairness, and sovereignty extension. The two 2026 releases make it concrete. One product proves the transfer; one model names the residual. Mistake the transferred majority for the whole, and you deploy the hosted agent into a sovereignty-constrained jurisdiction and discover the gap at the egress log, which is the worst place to discover it.

Evidence collection across the four tiers

The evidence architecture is the same E-1 through E-4 structure the L1 chapter established. Its content, specialized for this case study, is what an examiner reads.

Evidence Tier	Frontier-Lab Finance Implementation	Durable Principle Served
E-1 Foundation	`AGENTS.md`, the approval rule definition, protected-branch settings, compliance framework binding, connector authorization list, all version-controlled	Effective-challenge configuration is verifiable as committed
E-2 Operational	Every AI-generated merge request, every approval, every connector access, every denied out-of-scope request, surfaced in the Compliance Center	Continuous monitoring replaces the point-in-time pre-exam review
E-3 Context	Sovereignty posture per session (hosted vs on-premises), jurisdiction whose rules applied, data sensitivity tier touched	A cross-boundary call is always visible against the boundary it should have respected
E-4 Accountability	Agent draft, connector sources cited, reviewer identity, approval timestamp, merged decision; on the sovereign path, Command A+ native citations make the source half machine-readable	The attribution chain answers "who is accountable for this output" with a name and a timestamp

The knowledge-base controls sit underneath E-4. Every data source the agent may cite is an approved MCP connector with a known provider; no free-text source is permitted in deliverable work product. Retrieval is scoped by the Delegation Token, so an analyst working a deal under a Chinese-wall restriction gets a token that excludes the restricted source entirely, which is LP1 (this volume's governance restatement of the federal classification-consistent access property P1 described in the L1 chapter) applied to a financial-services policy source. Connector responses carry a retrieval timestamp, and stale data beyond a workflow-defined window is rejected. Cited values are verified against the connector record at merge time; for the sovereign path, the native citation span is the verification anchor.

The failure mode catalog

Six failure modes define the boundary of this deployment. Each has an observable symptom, a detection method, and a structural mitigation. None is mitigated by instruction alone.

#	Failure Mode	Observable Symptom	Detection Method	Mitigation
1	Unverified Output Merge	AI work product in production with no approver identity	Audit-event query for merges lacking an approval record	Required approval rule on protected branches; no merge without a named human sign-off
2	Connector Scope Creep	Access event for a resource outside the agent's authorization	Pipeline scope check against the Delegation Token audience	Gateway denies out-of-scope requests; denied event logged as E-2 evidence
3	Sovereignty Leak	Egress to a non-sovereign endpoint from a boundary-restricted workflow	E-3 context check comparing call destination to the workflow's sovereignty flag	Route boundary-restricted workflows to on-premises Command A+ serving
4	Unsourced Assertion	A figure in work product with no traceable connector record	Provenance check in the pre-merge pipeline policy	Require connector-backed citations; reject free-text sources; use native-citation model output
5	Multi-Jurisdiction Misapplication	A document applying one jurisdiction's rule to data subject to another	E-3 jurisdiction tag compared to data subject residency	Jurisdictional routing table; most-restrictive-applicable default
6	Stale-Data Draft	Work product built on a superseded figure	Currency-window check on connector retrieval timestamps	Reject retrievals older than the workflow window

The instructive case is the third. A sovereignty leak is the failure mode that the hosted product, deployed naively, produces by design in a constrained jurisdiction, because the hosted product's value proposition is that inference runs in the vendor's environment. An institution that adopts the transferred majority without standing up the extension has not made a configuration error. It has misread the transfer rule. The mitigation is architectural: the sovereignty flag in `AGENTS.md` routes the workflow to on-premises Command A+ serving, and the E-3 check fails the workflow if a boundary-restricted session ever calls a hosted endpoint. The leak is caught at the boundary it crossed, before the egress completes its purpose.

The path the institution walks

Adoption is not a single decision. It is assess, pilot, scale, with the sovereignty posture decided first.

The institution inventories where AI already drafts work product, including shadow use in spreadsheets and chat tools where no control reaches, and determines which workflows the frontier-lab agents already cover so it adopts rather than rebuilds. It decides the sovereignty posture before the pilot: hosted inference, or weights and data inside the boundary for jurisdictional reasons. That decision precedes everything because it determines which path each workflow takes, the transferred majority or the sector-specific extension.

The pilot runs one bounded workflow end to end. The agent drafts a credit memo or works a KYC file; the analyst reviews; GitLab Ultimate records the approval; the Compliance Center produces the evidence. One concrete success criterion governs it: an examiner-ready evidence package generated from the pilot's own audit stream, in minutes, with no separate compliance effort. Pick a workflow where the sign-off gate is visibly load-bearing, so the pilot tests the gate rather than working around it.

Scaling expands from the one workflow to the verticals the skill library already covers, and marginal governance cost falls with each addition because the harness is shared. Before scaling across jurisdictions, the institution stands up the sovereignty and multi-jurisdiction extensions. The order is not negotiable, because the transferred majority scales for free and the extension does not.

The decisions that belong to named roles divide the same way. The platform team configures GitLab Ultimate approval rules so the vendor's sign-off gate becomes an enforced control, and defines `AGENTS.md` scope per workflow. The Chief Compliance Officer and General Counsel map the harness evidence to the institution's applicable frameworks, the SR 11-7 effective-challenge lineage for model governance, the Gramm-Leach-Bliley Safeguards Rule for customer data, Financial Industry Regulatory Authority (FINRA) supervision and recordkeeping for broker-dealers, and the EU AI Act where the institution operates in the Union, and confirm where consent, fairness, and multi-jurisdiction coverage force the sovereign path. Because supervisory guidance points the controls for agentic AI back to the institution, this mapping is not a compliance formality. It is the institution writing the standard the regulator told it to write.

CHAPTER SUMMARY

The transfer the L1 chapter asserted is now demonstrated against a published reference pattern. In May 2026 a frontier AI lab published `anthropics/financial-services` as a public reference repository: ten named agents, a synced finance skill library (DCF, LBO, comps, three-statement, deck QC, Excel audit), and MCP connectors to FactSet, PitchBook, S&P Global, Moody's, and others, deploying from one source as either a Cowork plugin or through the Managed Agents API. The compliance-load-bearing design choice is that agents draft analyst work product but never execute trades, bind risk, post to a ledger, or approve onboarding; every output is staged for human sign-off. That sign-off gate is the transfer rule and Trust but Verify expressed as code, and most of the harness transfers verbatim. The governing principle is effective challenge, established by SR 11-7 (2011, FDIC adoption FIL-22-2017) and carried into BSA/AML model risk by SR 21-8. Supervisory model-risk guidance does not write the controls for generative and agentic AI directly; it points each banking organization's own risk-management and governance practices to determine them, which hands the agentic-AI governance decision back to the institution and is precisely why the harness matters. GitLab Ultimate turns the vendor's sign-off gate into a required merge-request approval on protected branches, with a pre-merge pipeline policy enforcing provenance, scope, currency, and jurisdiction, a policy engine such as OPA/Rego or Cedar evaluating the `AGENTS.md`-derived rules at the gateway, and a Compliance Center rendering E-1 through E-4 evidence continuously. Cohere Command A+ (Apache 2.0, 218B sparse mixture of experts (MoE) with ~25B active, 128K context, native citations, 48 languages, two-H100 deployable) names the residual extension: sovereignty, native-citation transparency, and EU-language multi-jurisdiction coverage that the hosted product cannot serve. The EU AI Act prices the stakes for that residual at up to 7 percent of worldwide annual turnover. Six failure modes bound the deployment, the most instructive being the sovereignty leak that a naive hosted deployment produces by design in a constrained jurisdiction; the mitigation is architectural, routing boundary-restricted workflows to on-premises Command A+ serving and failing the E-3 check on any cross-boundary egress. The institution assesses shadow use, decides sovereignty posture first, pilots one bounded workflow to an examiner-ready evidence package, and scales the transferred majority for free while standing up the extension deliberately before crossing jurisdictions.

Chapter 13: L2, The Healthcare Lagrange Point

Part III: The Sector Lagrange Points

> *"The minimum necessary standard is not a privacy rule. It is a scoping rule. It tells the system not what it cannot know, but what it is not permitted to consult for this particular purpose, right now, in this specific interaction."*

The L2 Lagrange point sits behind the Earth relative to the Sun, at a distance where Earth's gravity combines with the Sun's to hold an object stationary with respect to both. The James Webb Space Telescope orbits there. The position requires minimal station-keeping; the competing pulls balance, and the telescope holds its orientation while its sunshield permanently faces the Sun. The difficulty was finding the point. It exists mathematically before it is located in practice. Getting there is an engineering problem, not a physics problem.

Healthcare AI governance has three competing forces: the obligation to protect patient information with the most granular sensitivity framework in commercial law, the obligation to satisfy a regulatory structure that crosses three federal agencies and fifty state jurisdictions at once, and the clinical imperative to make care faster, more accurate, and less administratively burdensome. These forces do not point in the same direction. They are not compatible by default. They become compatible only at a specific configuration of governance architecture, the Lagrange point, where each constraint is satisfied precisely and none of the others collapses.

Every healthcare organization deploying AI is searching for this point. Most are finding the boundaries of where it is not.

The three bodies in healthcare

Healthcare manages the most granular data sensitivity framework of any commercial sector. The Health Insurance Portability and Accountability Act (HIPAA) defines Protected Health Information as any individually identifiable health information. But protected health information (PHI) is not monolithic. Within the PHI category, there are sensitivity gradations that create nested access control requirements with different governance rules at each level.

De-identified data, stripped of all 18 HIPAA identifiers per the Safe Harbor method (45 CFR 164.514(b)), has left the PHI category entirely. AI access is unrestricted, with one qualification: the de-identification must be verified, not assumed. A dataset that appears de-identified but retains a combination of quasi-identifiers (age, zip code, diagnosis, admission date) sufficient for re-identification is not de-identified under the legal standard, regardless of whether the 18 named identifiers have been removed. Re-identification risk must be monitored. Verifying that the 18 identifiers are absent is not enough.

Limited data sets occupy the next tier. Direct identifiers have been removed but dates, zip codes to three digits, and ages remain, which keeps PHI characteristics in play for governance purposes. Use requires a Data Use Agreement (DUA), and AI systems processing limited data sets must verify DUA coverage before each processing event. This is not a one-time permission. The DUA specifies the scope of permissible use; use outside that scope violates the agreement even if the requesting user has general system access.

Standard PHI, health information retaining one or more of the 18 identifiers, is subject to the full HIPAA Security Rule. AI access is governed by the minimum necessary standard at every access event: the system must determine the minimum data elements needed for the specific, current purpose, and access only those elements. The standard does not describe what the system is technically prevented from accessing. It governs what the system is permitted to access. That difference is architectural, and it carries consequences.

Psychotherapy notes carry the strictest access restriction in standard HIPAA. Defined under 45 CFR 164.501 as notes recorded by a mental health professional and maintained separately from the rest of the medical record, they require separate, specific patient authorization for any use or disclosure beyond treatment by the authoring clinician. A treating internist reading a summary of the patient's mental health history in a clinical note does not have authorization to access psychotherapy notes; an AI assistant generating documentation for any encounter other than the specific mental health session where the notes were created is prohibited from accessing or incorporating psychotherapy note content without documented patient authorization for that specific disclosure.

Substance abuse treatment records maintained by federally assisted programs operate under 42 CFR Part 2, a separate federal confidentiality framework more restrictive than HIPAA in every dimension. A general HIPAA authorization does not satisfy Part 2. Each disclosure requires a specific written consent covering that disclosure, identifying the information to be disclosed, the recipient, the purpose, and an expiration date. AI systems that process records from federally assisted substance abuse programs without specific Part 2 consent for each interaction are not running a deficient HIPAA configuration. They are violating a separate federal statute with separate enforcement authority.

Genomic data sits at the intersection of HIPAA PHI (when associated with an individual) and the Genetic Information Nondiscrimination Act (GINA), which prohibits the use of genetic information in health insurance underwriting. State laws increasingly extend genomic protections further: California's Genetic Information Privacy Act, Washington's My Health MY Data Act, and similar statutes impose consent requirements and use limitations that exceed HIPAA's baseline. An AI system processing genomic data for clinical research must satisfy the intersection of HIPAA, GINA, the applicable state statute, and any institutional review board protocols governing the specific research use.

Healthcare AI governance does not answer to a single regulator. The HIPAA Security Rule (45 CFR 164.302-318) governs technical safeguards, access controls, audit controls, integrity controls, and transmission security for PHI, with enforcement by the Office for Civil Rights at the Department of Health and Human Services (HHS). The Health Information Technology for Economic and Clinical Health Act (HITECH) strengthened that enforcement, imposed breach notification requirements, and created the conditions under which state attorneys general can bring HIPAA enforcement actions. 21 CFR Part 11 governs electronic records and electronic signatures for Food and Drug Administration (FDA)-regulated systems; audit trail, version control, and access control requirements attach to any AI system operating in a regulated research or manufacturing environment. The FDA's AI/ML Software as a Medical Device guidance creates a regulatory boundary that does not exist in any other commercial sector, and its location relative to any specific AI deployment is not always obvious in advance.

State health privacy laws add a layer of jurisdiction-specific requirements that the federal framework does not preempt. California's Confidentiality of Medical Information Act imposes privacy protections that exceed HIPAA in several respects, including a private right of action. New York's SHIELD Act health data provisions impose breach notification requirements on entities that handle New York residents' health data regardless of whether those entities are HIPAA covered entities. States are legislating in this space at increasing velocity. A healthcare AI system deployed nationally is not subject to HIPAA alone; it is subject to HIPAA plus the health privacy laws of every state in which its patients reside.

Clinician documentation burden is not an abstraction. Studies consistently document that clinicians spend between two and three hours per day on electronic health record (EHR) documentation for every six to eight hours of patient care. Two to three hours. That ratio is time not available for clinical judgment, patient interaction, or rest. AI-assisted clinical documentation, ambient documentation tools, AI-powered note generation, clinical coding assistance, prior authorization automation, and EHR workflow optimization address a genuine and severe operational problem.

The governance implication is direct: a governance framework that makes AI documentation assistance slower or less accurate than manual documentation will not be adopted. Clinicians will find ungoverned alternatives. Shadow AI in healthcare does not merely expose the organization to regulatory risk; it exposes patients to clinical risk, because the shadow AI runs without the PHI access controls, minimum necessary enforcement, and audit trails that the governance framework provides. Over-govern here and the failure mode is not organizational friction. It is patient safety.

The minimum necessary standard as dynamic scoping

At its core, the translation problem between Department of Defense (DoD) AI governance and healthcare AI governance is a problem of access model architecture. In classified government environments, access scope is a static property of the user: clearance level plus need-to-know, established at authorization time, updated infrequently, and uniform across sessions. A developer cleared at TS/SCI accesses TS/SCI information; that access boundary does not change based on the specific task the developer is performing in the current session.

HIPAA's minimum necessary standard operates on a fundamentally different architecture. Access scope is not a static property of the user but a dynamic property of the interaction, determined by three factors: the user's role, their relationship to the specific patient in the current encounter, and the specific purpose of the data access being requested right now.

A billing coder authorized to access the EHR system needs diagnosis codes and procedure codes for the current claim. That coder does not need clinical notes, medication history, or lab values, even though those elements are in the same patient record, on the same system, accessible to the same user. The minimum necessary standard requires the AI system to provide the coder with diagnosis codes and procedure codes, not the full patient record. The fact that the full record is technically accessible is not permission to surface it.

A treating physician needs clinical notes for their own patients. They do not need clinical notes for patients under another physician's care, even if those patients are in the same practice, on the same system, with records that are technically accessible. Accessing another physician's patient records without a treatment relationship is a minimum necessary violation regardless of credentials.

A quality improvement analyst studying readmission rates for a specific diagnostic group needs aggregate data about that group, not individual patient records. A researcher needs de-identified data from a specific cohort matching defined criteria, not the general patient population. Each case narrows the scope its own way.

The AI system must resolve these scope determinations at the interaction level, for each request, against the current context. The AGENTS.md must encode minimum necessary rules that the routing enforcement layer evaluates before any data retrieval:

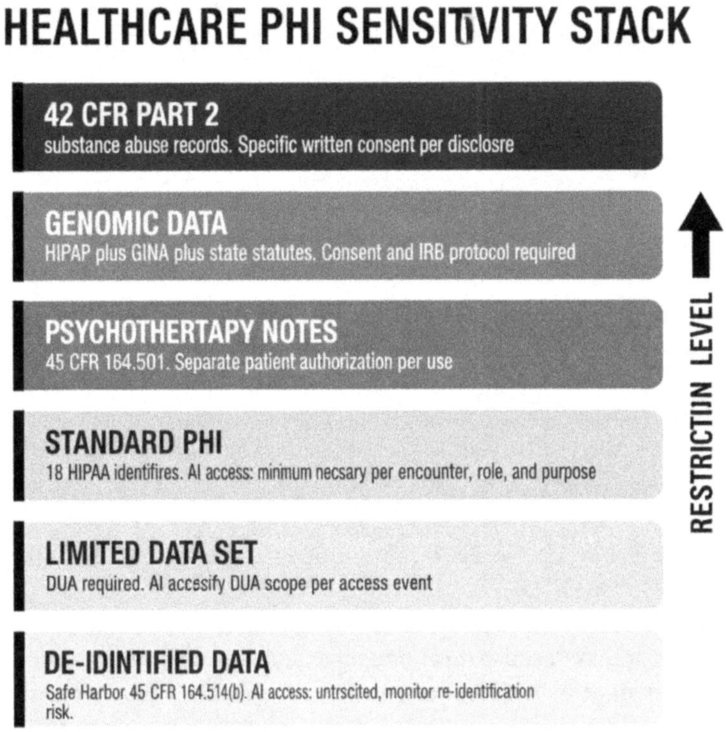

Figure 13: The healthcare PHI sensitivity stack, from de-identified data with no constraints to 42 CFR Part 2 substance-abuse records under the strictest controls, with the AI access rule at each level.

```
# AGENTS.md - Healthcare AI Governance (HIPAA-aligned)

data_sensitivity_tiers:
  - tier: deidentified
    description: "Safe Harbor de-identified (45 CFR 164.514(b))"
    ai_access: unrestricted
    monitoring: reidentification_risk_continuous
  - tier: limited_data_set
    description: "Direct identifiers removed (45 CFR 164.514(e))"
    ai_access: data_use_agreement_required
    verification: dua_scope_per_access
  - tier: phi
    description: "Protected Health Information (18 identifiers)"
    ai_access: minimum_necessary_enforced
    scoping: per_encounter_per_role_per_purpose
  - tier: psychotherapy_notes
    description: "45 CFR 164.508(a)(2) - separately maintained"
    ai_access: prohibited
    exception: specific_patient_authorization_required
  - tier: substance_abuse
    description: "42 CFR Part 2 - federal assisted programs"
```

```
      ai_access: prohibited
      exception: specific_part2_consent_required
    - tier: genomic
      description: "GINA plus applicable state law"
      ai_access: consent_verified_scope_bounded

  mandatory_constraints:  # RFC 2119: MUST
    - "MUST apply minimum necessary standard to every PHI access;
       access only specific data elements needed for the stated task
       and the current encounter; prior session scope does not carry"
    - "MUST NOT access psychotherapy notes or substance abuse records
       under any circumstance without specific documented authorization;
       escalate to Privacy Officer if access is requested"
    - "MUST verify current treatment relationship or authorized
       operational role before surfacing patient-specific PHI;
       system access authorization does not constitute per-encounter
       authorization"
    - "MUST produce de-identification verification when generating
       output for use outside the treatment context; 'appears
       de-identified' is not a verification"
    - "MUST log all PHI access with user identity, patient identifier
       (hashed), specific data elements accessed, purpose, and
       timestamp; logging at the record level is insufficient"
    - "MUST flag clinical decision support output for FDA SaMD
       labeling review if output could influence treatment decisions
       or be incorporated into clinical recommendations"
```

The minimum necessary standard as dynamic scoping is harder to implement than clearance-based access because it requires the governance layer to evaluate purpose, not just identity. A clearance check is a lookup: is this user cleared for this classification level? A minimum necessary check is a judgment: given this user's role, this patient relationship, and this stated purpose, what are the minimum data elements required to accomplish this specific task? The governance architecture cannot store the answer. It must compute it at each interaction.

The Lloyd's survey parallel

Lloyd's Register's minimum survey standards provide a structural parallel that clarifies how this dynamic scoping requirement is implemented in practice. A Lloyd's surveyor conducting a periodic class renewal survey does not inspect every component of the vessel. The surveyor inspects the components relevant to the current classification cycle, the systems due for inspection under the applicable class notation, and any items flagged by the previous survey or by the vessel's operational history since that survey. The scope of the survey is bounded by what is relevant to the current classification determination.

That bounding is not a compromise of rigor. It is the definition of rigor: applying expert scrutiny to the elements that determine the current governance question, rather than conducting an undifferentiated examination of everything on the vessel. A surveyor who inspects cargo holds when the classification question concerns the propulsion system is not being more thorough. They are failing to apply the minimum necessary standard to their own work.

The HIPAA minimum necessary rule runs on the same logic. A billing coder's query to an AI system about a patient's diagnosis codes is a specific classification question. The answer requires the diagnosis codes. It does not require the clinical notes, the medication history, the psychotherapy notes, or the 42 CFR Part 2 substance abuse records. An AI system that scopes its retrieval to the diagnosis codes is not being less helpful. It is being precisely as helpful as the task requires, and nothing more.

The practical implementation requires per-purpose retrieval scoping at the KB layer. The AI system does not retrieve the full patient record and then filter the output; it scopes the retrieval query to the authorized data elements before the query executes. KB-3 (Controlled Ingestion) and KB-4 (Monitored Retrieval) enforce this at the infrastructure layer: retrieval is bounded before results are returned, not filtered after. The distinction matters because filtered retrieval leaves the unauthorized data in the AI system's processing context even if it is excluded from the output. Bounded retrieval means the data never enters the context at all.

The FDA boundary

The sharpest regulatory boundary in healthcare AI governance is not PHI access control. It is the question of when AI-assisted clinical activity becomes a medical device.

The FDA's clinical decision support guidance distinguishes between software that supports clinical decisions, which is generally not FDA-regulated, and Software as a Medical Device, which is. The distinction turns on four criteria. Software that meets all four criteria is exempt from FDA regulation under the 21st Century Cures Act: it is not intended to acquire, process, or analyze a medical image or medical device signal; it displays, analyzes, or prints medical information; it is intended for use by a healthcare professional; and it is designed to allow the professional to independently review the basis for any recommendations, such that the professional does not rely primarily on the software's output. If any of these four criteria is not met, the software may be regulated as a medical device under the FD&C Act's device provisions (21 CFR Parts 807 and 820).

The governance architecture must know which side of this boundary each AI interaction falls on, because the consequences of misclassification are severe. A software product that should have received FDA clearance or approval before deployment but did not is an illegally marketed medical device. Warning letters, injunctions, potential criminal liability for the responsible individuals; these consequences attach to persons, not merely to the organization.

This boundary presents a specific challenge: it is not always visible at code review. An AI system configured to help physicians document clinical encounters is not a medical device. The same AI system, if it begins generating text that includes interpretations of lab values, differential diagnoses, or treatment recommendations, may have crossed into software as a medical device (SaMD) territory without any visible code change. The governance architecture must detect AI outputs that cross this boundary and route them for regulatory review before they reach clinical users.

LP13 (Output Boundary Enforcement) must include a clinical decision support detection layer for healthcare deployments. The detection scans AI-generated content for patterns that cross the FDA boundary: clinical value interpretation ("a creatinine of 1.8 is consistent with stage 3 chronic kidney disease"), diagnostic suggestions ("the presentation is consistent with"), treatment recommendations ("consider"), and risk stratification statements that quantify patient risk for clinical outcomes. These patterns, in AI-generated clinical content, are SaMD indicators. The governance architecture must flag them for regulatory labeling review before the content is incorporated into clinical workflows.

This is not a prohibition on clinical AI. It is a routing gate. AI-generated content that does not exhibit SaMD patterns can proceed without regulatory review; content that exhibits SaMD patterns is routed to a compliance review queue where a designated regulatory reviewer determines whether the content requires FDA clearance before clinical deployment. The gate prevents the inadvertent deployment of uncleared medical device software. Cleared or exempt clinical AI proceeds without obstruction.

PHI sub-category governance in the AGENTS.md architecture

Each of the five PHI sensitivity tiers requires distinct governance controls in the AGENTS.md architecture, not merely distinct access permissions. The controls must differ because the regulatory requirements governing each tier differ in kind, not just in degree.

De-identified data requires re-identification risk monitoring as an ongoing operational control, not a one-time certification. The risk changes as external datasets grow. A dataset de-identified against the 2022 reference population may not be de-identified against the 2026 reference population if public datasets have been released that, when combined with the quasi-identifiers present in the clinical dataset, enable individual identification. KB-2 (Enforced Currency) applies to the re-identification risk assessment, not just to clinical knowledge content.

Standard PHI requires minimum necessary enforcement at the data element level for every retrieval, plus audit trail logging at the element level that supports HIPAA's right of access. If a patient requests an accounting of disclosures, the organization must be able to identify every event in which the patient's PHI was accessed by or through an AI system, including which data elements were accessed, for what purpose, and by which user. The E-4 (Attribution Chain Record) for PHI access must include the patient identifier (hashed for the internal record, resolvable for the disclosure accounting), the data elements accessed, and the clinical purpose.

Psychotherapy notes and 42 CFR Part 2 records require infrastructure-level isolation, not AGENTS.md instruction-level prohibition. Enforcement belongs at the storage and retrieval layer; these records must reside in a retrieval scope that is unreachable by AI systems that have not received explicit, specific authorization for this purpose. An AGENTS.md instruction that says "do not access psychotherapy notes" is a policy statement. An infrastructure configuration that stores those notes in an isolated retrieval scope, one not connected to the AI system's KB retrieval layer, is a governance control. The first can be overridden by a sufficiently specific query. The second cannot.

The UnitedHealthcare case: missing properties in production

The Medicare Advantage AI claims denial case referenced in Chapter 3 bears direct examination in the healthcare sector context, because the governance failure it represents is not exceptional. It is the failure that results from deploying any consequential AI system in a healthcare context without LP11 (Fairness Verification) and LP12 (Transparency and Explainability).

The core finding from the litigation and subsequent reporting was that UnitedHealthcare's nH Predict algorithm was denying post-acute care claims for Medicare Advantage beneficiaries at a rate substantially higher than the rate that would have resulted from individual clinical review. The algorithm predicted that a patient's care needs would resolve in a shorter time than the treating physicians assessed. Claims were denied on the basis of the algorithmic prediction. Patients were discharged or had coverage terminated accordingly.

None of this was a technology failure. The algorithm produced outputs. The outputs were used for coverage decisions. Those decisions influenced clinical trajectory for Medicare Advantage beneficiaries. What was missing was the evidence architecture that LP11 and LP12 require.

LP11 (Fairness Verification) requires that AI-assisted decisions affecting individuals be evaluated for disparate impact at the frequency dictated by decision volume. The nH Predict system was making thousands of coverage determinations. Each was an individual decision affecting a Medicare beneficiary's access to post-acute care. The LP11 requirement is that the AI-assisted decision process pass the applicable fairness threshold before deployment at scale, and that fairness metrics be generated as E-2 operational evidence during deployment. Coverage denial rates that produce disproportionate outcomes by diagnosis, age, or demographic characteristic are LP11 failures. The absence of continuous fairness monitoring allowed the pattern to persist across thousands of individual denials before it became visible through litigation and regulatory inquiry.

LP12 (Transparency and Explainability) requires that AI-assisted decisions reaching affected parties produce explanation artifacts satisfying applicable disclosure requirements, generated at decision time as a harness checkpoint. The Medicare Advantage appeals process requires that denial notices provide the specific clinical rationale for the denial. An explanation that cites algorithmic prediction without specifying the clinical basis for the prediction does not satisfy the disclosure requirement. This tracks the Consumer Financial Protection Bureau (CFPB)'s position on adverse action notices in credit decisions: algorithm complexity is not a specific reason.

For healthcare organizations deploying coverage determination AI, LP11 and LP12 are not optional extensions. They are legally required properties of any AI system whose outputs influence coverage decisions, consistent with Centers for Medicare and Medicaid Services (CMS) coverage determination guidance, state insurance regulations, and the constitutional requirements for due process in government benefit programs. Medicare Advantage is not a voluntary commercial product for its beneficiaries; denial of coverage is an adverse action with real clinical consequences. The explanation and fairness requirements attach to the consequence, not to the commercial model chosen to deliver the benefit.

Healthcare attack surface

The generic surface from Chapter 7 applies, and four regions sharpen in a clinical setting, where the data is protected health information and the artifact becomes part of a patient's record.

Cross-encounter context bleed as exfiltration. The context-amnesia surface, turned into a data-extraction attack. A retrieval path scoped at the patient level rather than the encounter level can be driven to pull protected health information from another encounter, or another patient, into the current note. The defense is encounter-scoped retrieval enforced by scope compilation, the minimum-necessary standard wired in as a routing input rather than a policy statement, and an E-4 attribution record per encounter so any cross-boundary pull is visible.

Clinical-note poisoning. False content injected into the record propagates into downstream care, the way a referenced but discontinued medication can mislead a later prescriber. The defense is verified provenance (KB-1) on retrieved clinical content, an output-boundary check (LP13) on what the assistant is allowed to assert, and a clinician's sign-off (H-4) before any generated note enters the chart.

SaMD drift and change-control evasion. A model functioning as software as a medical device (SaMD) can have its behavior changed by a silent vendor update outside the FDA predetermined change control plan (PCCP). The defense is version pinning and an organization-controlled update cadence from Chapter 10, with any model change treated as an explicit H-4 trigger event that re-runs verification before the change reaches care.

Psychotherapy-note access escalation. The strictest access category in HIPAA is a high-value target: an agent reaching psychotherapy notes without the specific authorization those notes require. The defense is the sub-category governance already built into the healthcare instruction surface, per-category consent as a routing input (LP9), and deny-by-default on any retrieval that cannot prove authorization for that specific disclosure.

Healthcare failure modes

F-HC1: Cross-encounter context bleeding. Context amnesia applied to PHI. An AI clinical documentation assistant configured at the patient level rather than the encounter level generates notes for a current visit that incorporate clinical context from prior encounters, including encounters for different conditions, involving discontinued medications, or referencing findings that are clinically misleading in the context of the current visit.

HIPAA's minimum necessary standard requires that PHI access be limited to the information needed for the specific purpose. Documentation for a current encounter is authorized to use information relevant to that encounter. Context bleed from prior encounters is not a technical error; it is a governance architecture failure. The retrieval scope was configured at the patient level because the patient level is the natural unit of the EHR system. Minimum necessary requires the encounter level.

H-3 (Persistent State) with enforced encounter-level scope binding prevents this failure. Session context for AI-assisted documentation must be initialized to the current encounter's authorized scope; retrieval from outside that scope requires an explicit authorization event logged in the E-3 (State Transition Record). Chapter 7 (Context Amnesia) provides the architectural pattern. The prevention architecture is identical.

F-HC2: AI-generated clinical content crossing the FDA SaMD boundary. An AI coding assistant in a healthcare organization, or an AI clinical documentation tool in routine use, generates output that includes clinical value interpretation, diagnostic suggestion, or treatment recommendation language. The output reads as natural clinical prose. It was not reviewed for SaMD indicators before reaching clinical users or being incorporated into a code base targeting clinical systems.

This is not a HIPAA violation. It is an FDA medical device violation: a product that should have received clearance before clinical deployment was deployed without it. Liability attaches to the organization deploying the AI system, not only to the vendor.

LP13 (Output Boundary Enforcement) with a clinical decision support detection layer prevents this failure. The layer scans AI-generated content for SaMD indicator patterns before delivery to clinical users or merge into clinical codebases. It is a routing gate, not a content filter; content exhibiting SaMD patterns is routed to regulatory review, not deleted. The gate is the governance control.

F-HC3: Consent scope violation for 42 CFR Part 2 records. An AI system with access to a clinical knowledge base that includes records from a federally assisted substance abuse treatment program processes a query that returns content from those records. The AI system has HIPAA access authorization. It does not have the specific written consent required by 42 CFR Part 2 for this disclosure. The general HIPAA Business Associate Agreement (BAA) with the AI vendor does not cover Part 2 records. The disclosure has occurred regardless of whether the returned content is recognized as Part 2 material.

The failure mode is architectural: Part 2 records were stored in the same retrieval scope as general PHI, making them reachable by the AI system under its general PHI authorization. Part 2 protection requires a separate retrieval scope, not a policy instruction. An AI system that can reach Part 2 records will eventually reach Part 2 records. Instructions do not change this fact.

Infrastructure-level isolation prevents this failure. Part 2 records must be stored in a retrieval scope not connected to the AI system's KB retrieval layer under any general authorization configuration. Connection requires specific, documented Part 2 consent for the specific use, processed as an explicit E-1 governance configuration event, not as a general PHI access expansion.

Healthcare evidence package

Evidence Tier	Healthcare Implementation	Regulatory Mapping
E-1	AGENTS.md configuration, PHI tier definitions, minimum necessary rules, FDA SaMD boundary criteria, BAA registry, Part 2 isolation configuration	HIPAA 164.316 (policies and procedures), 21 CFR Part 11, FDA CDS guidance
E-2	Minimum necessary enforcement logs per encounter, access verification results, de-identification re-identification risk scores, BAA currency verification, SaMD detection scan results	HIPAA 164.312 (technical safeguards), HITECH monitoring obligations, FDA post-market monitoring

Evidence Tier	Healthcare Implementation	Regulatory Mapping
E-3	Session context initialization records (encounter-level scope), regulatory change tracking, PHI tier reclassification events, SaMD boundary configuration changes	HIPAA 164.308 (risk analysis and management), FDA change control
E-4	PHI access audit trail with data element granularity (patient identifier hashed), FDA boundary flag records, coverage determination explanation artifacts, right of access accounting	HIPAA 164.312(b) (audit controls), right of access (164.524), LP12 explanation artifacts for coverage decisions

The evidence architecture serves three simultaneous regulatory purposes in healthcare, which is unusual among commercial sectors. It serves HIPAA audit and enforcement purposes, where OCR investigations require the organization to demonstrate that PHI access was minimum necessary, authorized, and logged. It serves FDA regulatory purposes, where device classification questions require the organization to demonstrate that AI outputs were evaluated against SaMD criteria before clinical deployment. And it serves CMS and state insurance regulatory purposes, where coverage determination disputes require the organization to produce the specific clinical rationale, the fairness metrics, and the explanation artifacts for individual denial decisions.

A healthcare organization that implements the evidence architecture correctly does not maintain three separate compliance programs. One evidence architecture produces the artifacts each regulatory framework requires, from a common operational record. The HIPAA audit trail, the FDA change control record, and the CMS explanation artifact are all E-4 attribution records with timestamps from the same evidence stream. Satisfy one framework and the others follow, because the underlying requirement, verifiable evidence of governed operation at the moment of action, is the same.

Compliance automation reference

Healthcare organizations face a distinctive compliance automation challenge: the same AI system may simultaneously touch PHI subject to HIPAA, clinical records subject to state health privacy laws stricter than HIPAA, and AI outputs potentially subject to FDA regulation as SaMD. The harness evidence architecture addresses all three when correctly configured.

For HIPAA Business Associate Agreement governance, the harness AGENTS.md must encode BAA scope at the instruction level, not only as a legal document. Every AI vendor that processes PHI must have an active BAA, and the harness must verify BAA coverage before routing PHI to any external inference endpoint. E-1 governance event records include a BAA registry that compliance officers can query at any time, satisfying the HIPAA Security Rule's requirement for written contracts with business associates (45 CFR 164.308(b)(1)). When a BAA expires or is terminated, the harness blocks PHI routing to that vendor and generates an E-1 event record of the enforcement action.

For PHI classification matrix management, the harness KB architecture must maintain a classification map distinguishing the six sensitivity tiers described in this chapter: de-identified data, limited data sets, standard PHI, psychotherapy notes, 42 CFR Part 2 substance abuse records, and genomic data. This map, stored as a KB-1 (Verified Provenance) artifact, is the authoritative source for minimum necessary determinations. Updates to the classification map are version-controlled changes with compliance review gates, and the E-3 state transition record captures the moment each classification changed.

For HIPAA right of access compliance (45 CFR 164.524), AI systems that processed PHI must be able to identify every data access event for a specific patient, including AI-assisted accesses, with sufficient granularity to satisfy a patient records request within 30 days. E-4 attribution records indexed by patient identifier (hashed in storage, resolvable for disclosure accounting) make this retrieval a query against the evidence store rather than a manual reconstruction across multiple system logs. Organizations that have implemented this architecture report that right of access response time drops from days to hours, because the accounting of disclosures is generated continuously as a byproduct of operation, not assembled on demand for each request.

CHAPTER SUMMARY

Healthcare AI governance operates at the intersection of three competing forces: PHI protection across six distinct sensitivity tiers with different governance controls at each, a multi-agency regulatory structure spanning OCR, FDA, CMS, and fifty state authorities, and a clinical velocity requirement where over-governance creates patient safety risk through ungoverned shadow AI. The minimum necessary standard distinguishes healthcare governance from DoD clearance-based models: access scope is dynamic, determined per-encounter by role, patient relationship, and specific purpose, not static by clearance level. This dynamic scoping is harder to implement and must be enforced at the KB retrieval layer before data enters the AI context, not filtered after. The FDA SaMD boundary creates a regulatory threshold unique to healthcare: AI-generated clinical content that crosses from documentation support into clinical interpretation, diagnostic suggestion, or treatment recommendation may constitute an illegally marketed medical device if deployed without FDA clearance. The UnitedHealthcare case demonstrates the consequence of deploying consequential healthcare AI without LP11 (Fairness Verification) and LP12 (Transparency and Explainability), both legally required for coverage determinations affecting Medicare beneficiaries. Three failure modes are specific to this sector: cross-encounter context bleeding (F-HC1), AI-generated content crossing the FDA SaMD boundary without regulatory review (F-HC2), and consent scope violation for 42 CFR Part 2 substance abuse records when those records share a retrieval scope with general PHI (F-HC3). The evidence architecture produces artifacts for HIPAA, FDA, and CMS regulatory purposes simultaneously from a single evidence stream.

Chapter 14: L3, The Energy and Critical Infrastructure Lagrange Point

Part III: The Sector Lagrange Points

> *"Lloyd's Register classified ships not by size or age but by condition and service: A1 for ocean-going vessels maintained to the highest survey standard, lower classes for coastal and river service where the consequences of failure were contained. The North American Electric Reliability Corporation (NERC) classifies Bulk Electric System cyber assets by impact: High, Medium, Low, with penalty exposure scaling accordingly. The classification society and the reliability coordinator are asking the same question from different vantage points. What is the consequence if this system fails?"*

L3 in the Sun-Earth system sits on the far side of the Sun from Earth, perpetually hidden from direct observation. It is stable in theory; in practice, solar and planetary perturbations gradually pull objects from their position. Of the five Lagrange points, L3 is the most isolated, requiring more station-keeping effort than L1 or L2 and remaining the most difficult to reach or monitor from Earth.

Energy and critical infrastructure AI governance is an L3 problem. Not metaphorically. Structurally.

The three gravitational bodies refuse easy alignment. Operational data from supervisory control and data acquisition (SCADA) and industrial control systems carries safety-critical sensitivity that surpasses anything in financial services or healthcare. Disclosing or corrupting Bulk Electric System Cyber System configuration is not a privacy violation or a regulatory infraction. It is a potential national security event, with physical consequences measured in power outages, pipeline failures, or nuclear incident timelines. Regulatory compliance forms one of the densest stacks in any commercial sector; NERC Critical Infrastructure Protection (CIP) standards, Federal Energy Regulatory Commission (FERC) orders, Nuclear Regulatory Commission (NRC) regulations, Department of Transportation (DOT) Pipeline and Hazardous Materials Safety Administration (PHMSA) requirements, state public utility commission frameworks, Transportation Security Administration (TSA) Security Directives, and Environmental Protection Agency (EPA) environmental compliance all touch the same operational system at once. The third body is grid operations velocity, and it is unforgiving. The electric grid operates in real time, load forecasting failures produce immediate consequences, and AI-assisted predictive maintenance runs on transformer failure timelines that have no equivalent in finance or healthcare.

Finding equilibrium here requires a governance architecture that spans the full range, from nuclear safety-critical systems where the AI cannot be wrong without physical consequence, to distribution grid automation where velocity is the operational imperative.

The three bodies in energy

Privacy law does not define the sensitivity framework for energy sector AI. Operational classification does. NERC CIP-002 through CIP-014 define Bulk Electric System Cyber System, or BES Cyber System, as any cyber asset whose compromise would impact reliable operation of the bulk electric system. BES Cyber System categories carry High, Medium, and Low impact designations based on the consequence of compromise at scale.

High impact BES Cyber System (BCS) data encompasses SCADA configuration for control centers that operate or control transmission assets rated at 500 kV or above, automatic generation control, and assets designated as critical to preventing or restoring cascading outages. The governance implication is categorical. High impact BCS data does not flow to cloud-based AI inference endpoints under any standard configuration, and it does not traverse untrusted network segments. An AI system processing High impact BCS configuration data must operate within the Electronic Security Perimeter, or ESP, defined by CIP-006 Physical Security and CIP-005 Electronic Security. Not an architectural preference. A NERC CIP requirement.

Nuclear safety data occupies a separate regulatory domain entirely. NRC 10 CFR 73.54 defines Cyber Security as a requirement for nuclear facilities with operating licenses or combined licenses, covering the critical digital assets, or CDAs, that perform or are associated with safety, important-to-safety, security, emergency preparedness, and support functions. When a CDA is compromised in a way that could adversely affect any of these functions, the full cyber security program requirements apply: defense-in-depth architecture, periodic testing, assessment and management of cyber attacks, and written cyber security plan documentation. AI systems interacting with CDAs operate within this regulatory boundary. Whether the development team recognizes those interactions as such is irrelevant.

Pipeline operational technology under DOT PHMSA covers control systems for natural gas and hazardous liquid pipelines, including SCADA systems that monitor pressure, flow, and valve status. TSA Security Directives for pipeline operators, issued from 2021 onward, require specific cyber security measures for operational technology systems designated as critical: network segmentation, access controls, and monitoring. Any AI system generating maintenance recommendations for pipeline control systems sits squarely in the PHMSA-TSA regulatory intersection.

NERC CIP standards (CIP-002 through CIP-014) constitute the most operationally detailed mandatory cyber security framework in any commercial sector. Each standard addresses a specific dimension: BES Cyber System identification (CIP-002), training (CIP-004), electronic security perimeters (CIP-005), physical security (CIP-006), systems security management (CIP-007), incident reporting (CIP-008), recovery plans (CIP-009), configuration change management (CIP-010), vulnerability management (CIP-011), and supply chain risk management (CIP-013). CIP-014 covers physical security of transmission infrastructure. Each standard carries associated requirements, measures, and violation risk factors; each has a separate implementation guidance document.

The penalty structure is what makes this regulatory body unique. NERC CIP violations carry civil penalties of up to $1,000,000 per violation per day. That is not a cap applied rarely. It is the statutory maximum per violation, and NERC has assessed penalties in the millions of dollars range against registered entities for documented CIP non-compliance. One misconfigured electronic access point that goes undetected for 30 days is a potential $30 million exposure per violation instance. No other commercial sector carries a compliance urgency like it, and that urgency shapes how AI governance must be architected.

NERC CIP IMPACT CLASSIFATION MATRIX

HIGH IMPACT BCS	MEDIUM IMPACT BCS	LOW IMPACT BCS
Example Assets: 500kV+ SCADA, AGC systems	Generation over 1500MW, substations over 200kV	Distribution Distribution assets below the CIP-002 high/medium threshold
Physical Security: Physical Security Perimiter required, seven-factor entry logging	Flexible physical security, less prescritive	Simplified control set
AI Configuration Change: CIP-010 documetation before implementation, baseline comparison	Flex010 requiral security, prescritive intervals	Annual review, control set
Penalty Exposure: $1M per violation per day maximum	**AI Configuration Change:** CIP-010 required, less prescritive intervals	**Penalty** Exiew, lightunation Lower VSL secumarity levels

Loyd's Register classified by consequence of failure. NERC CIP classifies the same way.

Figure 14: The NERC CIP impact classification matrix; high, medium, and low BES Cyber System assets mapped to their physical-security, configuration-change, and penalty-exposure requirements.

FERC sits above NERC, addressing transmission tariffs, market rules, and reliability standards. NRC jurisdiction covers nuclear facilities. State public utility commissions regulate distribution-level operations, retail electricity pricing, and integrated resource planning. EPA Clean Air Act and Clean Water Act requirements apply to generation facilities. Where these authorities intersect, a single AI system deployed at a natural gas combined-cycle plant may be simultaneously subject to NERC CIP, FERC tariff compliance requirements, EPA emissions reporting, PHMSA pipeline safety requirements, and state PUC rate case documentation obligations. Five regulators. One plant.

Then there is velocity. Real time is not an aspiration for the electric grid. It is the operating condition, without meaningful tolerance for latency in operational decisions. Load forecasting supports unit commitment decisions made 24 to 72 hours in advance, but the dispatch control room operates on a 5-minute settlement cycle. Outage prediction and predictive maintenance AI addresses a different velocity problem. Transformer failure is not announced. A transformer serving a substation handling hundreds of megawatts of load fails on a timeline that predictive maintenance AI is supposed to shorten from weeks to days of advance warning. Get the thermal degradation curve wrong and the consequence is not a delayed decision; it is an unplanned outage affecting the downstream load.

Predictive maintenance for grid assets defines energy sector AI governance at its sharpest edge. Operational teams need the recommendation early enough to schedule the work, order the replacement parts, and coordinate the switching sequence with the reliability coordinator. A governance architecture that requires two weeks of evidence review before a maintenance recommendation can be acted on is not governing AI. It is eliminating AI from operations. Evidence must emerge as a byproduct of recommendation generation, not as a separate review that delays it.

The Lloyd's thread: classification by impact

Lloyd's Register classified ships not merely by hull condition but by intended service: A1 for ocean-going in all conditions, lower classes for coastal and river service where the consequence of failure was bounded. A river barge that sank in the Thames produced limited consequence relative to an ocean-going vessel lost with all hands in the North Atlantic. Classification reflected the stakes; survey requirements reflected the classification.

NERC CIP's High, Medium, and Low BES Cyber System impact designations follow identical logic. A High impact BCS asset is one whose compromise could affect the bulk electric system's reliable operation at regional scale; a Medium impact asset has more bounded consequence. Classification drives control requirements: High impact BCS assets require Physical Security Perimeters with defined entry and logging; Medium impact assets have more flexible physical security options; Low impact assets carry a simplified control set.

The Lloyd's parallel becomes exact at the CIP-002 identification process. Registered entities must perform a periodic BES Cyber System identification survey, analogous to the Lloyd's classification survey, identifying each asset by impact designation, documenting the basis for the designation, and maintaining that documentation for NERC auditor examination. CIP-002 is the classification survey. The NERC auditor is the Lloyd's surveyor. Every E-1 (Governance Configuration Record) for a NERC CIP-compliant AI deployment must include BES Cyber System impact designations for every asset the AI system can access or generate recommendations about, because those designations determine applicable control requirements and evidence production obligations.

High impact BCS assets demand the most extensive evidence production. CIP-010's configuration change management standard requires that changes to High impact BCS systems be documented before implementation, with a comparison of planned configuration against existing baseline, and that any deviation from the planned change be documented after implementation. An AI system generating configuration recommendations for High impact BCS systems must produce CIP-010 compliant documentation as a byproduct of the recommendation. Without it, the recommendation cannot be implemented without a CIP-010 gap.

Air-gapped OT environments

The air gap is not a security preference in energy sector operational technology (OT) environments. It is a safety requirement. The distinction matters.

SCADA and industrial control systems operating physical infrastructure run on networks architecturally isolated from corporate IT and from the internet. Isolation exists because the consequence of compromise is physical: a misconfigured protection relay can cause cascade failure, an unauthorized valve command can create pipeline over-pressure, an injected setpoint change can push a generator outside its stable operating range. NERC CIP-005 Electronic Security Perimeters codifies the boundary; all communication to and from the ESP containing High or Medium impact BCS assets must traverse Electronic Access Points, or EAPs, with enforced access control.

Energy sector AI systems must therefore operate in one of two configurations. IT-side operation places the AI system in the corporate IT network, receiving data from the OT environment through a unidirectional security gateway, or data diode, that allows data to flow from OT to IT without permitting reverse flow. The AI system generates recommendations based on OT data; those recommendations reach OT operators through a separate channel, not through the AI system itself. Governance must track that the AI system never issues commands crossing back into the OT environment through any path.

OT-embedded operation, the second configuration, places the AI system within the Electronic Security Perimeter, operating on OT-native data without traversing the boundary. This requires the AI system itself to satisfy NERC CIP physical security, electronic security, and configuration change management requirements. An AI model update becomes a CIP-010 change management event. The training pipeline producing the updated model must be documented; the model update must be tested against a baseline comparison; the change must be authorized by personnel who have passed the CIP-004 Personnel Risk Assessment background check.

The International Space Station (ISS) parallel applies directly. During communication gaps, when signal delay or orbital geometry cuts contact with Mission Control, the International Space Station operates autonomously. Pre-authorized procedures govern every action the crew can take; Mission Control approves the procedure set before the gap and reviews execution after contact is restored. SCADA environments in disconnected sites follow the same architecture: local operators have pre-authorized procedures for conditions arising during communication gaps with the control center, reviewed and reconciled when connectivity returns. An AI system generating recommendations for field operators in a disconnected substation must operate within the pre-authorized procedure envelope. Not outside it. If the recommendation falls outside that envelope, governance holds it pending restoration of communication with the central authority rather than executing unilaterally.

The NERC CIP penalty structure and evidence production

A $1,000,000 per violation per day maximum penalty structure creates governance velocity tension more acute than anything in financial services or healthcare. Penalty exposure for a CIP-010 documentation gap, a failure to document a configuration change to a High impact BCS system before implementation, accumulates by the day. When an AI system generates and implements a configuration recommendation without producing CIP-010 documentation, every day that gap persists is a day of potential penalty accumulation.

The governance response is not slowing AI operations to match the documentation cycle. It is making documentation production instantaneous with operational execution.

CIP-010 requires a configuration baseline for every High and Medium impact BCS cyber system, with change documentation comparing post-change configuration against the baseline. An AI system generating a configuration recommendation must simultaneously generate the CIP-010 change documentation package: proposed configuration comparison, authorization record, implementation timestamp, and post-implementation verification confirmation. The E-4 (Attribution Chain Record) for the AI recommendation is the CIP-010 documentation, not a separate document assembled after the fact. It is the attribution record generated at the moment of recommendation, containing the governance evidence CIP-010 requires.

CIP-004 Personnel Risk Assessment extends the same principle. Every person with authorized access to High and Medium impact BCS cyber systems must have completed a CIP-004 background check. An energy sector AI deployment's AGENTS.md must include a personnel authorization registry: before the AI system generates or routes any recommendation to a person who may execute it against a High or Medium impact BCS asset, H-2 (Routing Enforcement) must verify the recipient holds a current CIP-004 authorization. Delivering a recommendation to someone whose CIP-004 background check has expired is a CIP-004 violation. The AI system that delivered it without checking contributed to that violation.

CIP-007's System Security Management standard requires all user accounts for BCS cyber systems to be authorized and documented, and limits enabled TCP/UDP ports and services to those required for operations. An AI system operating within the ESP that opens communication channels not in the authorized port and service list creates a CIP-007 violation regardless of what that communication was for. Every E-1 governance configuration record for OT-embedded AI must include the full port and service inventory, with CIP-007 authorization documentation for every connection the AI system maintains.

Safety-critical systems and the consequence of error

What separates energy and nuclear AI from all other commercial sectors is consequence.

In financial services, an AI model producing incorrect credit scores creates financial harm that is recoverable. In healthcare, an AI system producing inaccurate clinical documentation creates clinical risk that is serious but bounded. In energy, an AI system that produces an incorrect recommendation about a transformer's failure probability, or an incorrect load forecast causing a unit commitment error during a stressed grid condition, or an incorrect safety system parameter for a nuclear plant's protection logic, creates consequences measured in blackouts, pipeline failures, or nuclear safety events. Not recoverable in the same sense.

Safety-critical AI systems in energy therefore require a safety constraint layer operating independently of the compliance constraint layer. NERC CIP addresses cyber security; it does not address operational safety in the sense that the nuclear safety framework does. NRC 10 CFR 50.55a requires nuclear plants to comply with applicable provisions of specific codes and standards for the design, fabrication, and testing of systems important to safety. AI systems operating in proximity to safety-important systems must demonstrate code compliance. An AI-generated maintenance recommendation for a safety-related system at a nuclear plant is not merely a CIP-010 change management event; it is also subject to the quality assurance program under 10 CFR 50 Appendix B and the cyber security program under 10 CFR 73.54.

The practical implication is a dual-gate architecture. Every AI recommendation touching a safety-critical or safety-related system must pass two independent verification checkpoints: the cyber security checkpoint (CIP compliance, configuration documentation, personnel authorization) and the safety checkpoint (quality assurance program compliance, independence of review, applicable code adherence). These checkpoints are not interchangeable. Passing one does not satisfy the other. Both must be distinguished clearly in the governance architecture and documented in the E-4 attribution record.

Nuclear-specific requirements

Nuclear facilities represent the extreme case. The regulatory framework does not merely add requirements on top of NERC CIP; it creates a distinct governance structure with different jurisdictional authority, different documentation requirements, and different consequence levels for non-compliance.

NRC 10 CFR 73.54 requires nuclear plant licensees to establish, implement, and maintain a cyber security program providing high assurance that digital computer and communication systems and networks are adequately protected against cyber attacks. The program must address CDAs: digital assets that perform or are associated with safety functions, important-to-safety functions, security functions, emergency preparedness functions, and support functions. Scope under 73.54 is broader than NERC CIP's BES Cyber System scope. It includes systems that support safety functions, not only those that directly perform them.

Any AI system assisting with maintenance scheduling, work order generation, or configuration documentation for a nuclear plant must be assessed against 73.54 CDA criteria. If the AI system's outputs could adversely affect a CDA, the system itself may fall within cyber security program requirements, not merely the systems it generates recommendations for. That is the line between adjacency to a regulated asset and presence within its regulatory boundary. The assessment is not optional. 73.54 requires periodic reassessment to track changes in the threat environment and the plant's digital systems.

10 CFR 50.55a requires that nuclear plants comply with the applicable provisions of the American Society of Mechanical Engineers (ASME) Boiler and Pressure Vessel Code, the ASME OM Code for inservice testing, and other applicable standards for systems, structures, and components important to safety. An AI system generating maintenance recommendations for ASME code-covered components must produce recommendations consistent with the applicable code requirements, and the E-4 attribution record must document which code edition and addenda the recommendation is based on, who verified the code compliance, and whether any code case or relief request applies. The nuclear quality assurance program under 10 CFR 50 Appendix B requires that design activities, including AI-assisted design

recommendations, be performed by qualified personnel using approved procedures, with documented verification of adequacy.

The combined 73.54 and 50.55a governance environment requires that an AI system operating in nuclear facility context maintain, at minimum: a current CDA boundary assessment documenting every system whose compromise could adversely affect a CDA function; a qualified personnel registry for every person whose work output may affect safety-related systems; a code compliance library that is version-controlled and current with applicable amendments; and an independent verification record for every recommendation that touches a safety-related system. These four records are the minimum E-1 governance configuration for nuclear-context AI operation.

Edge-native autonomy and disconnected governance

Energy sector SCADA environments present a governance challenge that does not exist in financial services or healthcare: the operational decision environment is frequently disconnected from the central governance authority.

Consider a field technician at a remote substation with intermittent communications connectivity. The control room may be unreachable. So may the corporate governance authority, and so may the maintenance management system. That technician acts on local knowledge, pre-authorized procedures, and direct observation of equipment condition. If an AI-assisted maintenance tool is available on a tablet or ruggedized device at the substation, it operates in a disconnected OT environment without access to the central evidence architecture, the personnel authorization registry, or the real-time compliance verification layer.

Governance for disconnected operation must pre-authorize the decision envelope before disconnection. This is the ISS crew procedure model applied to field operations. Before the technician departs for the remote site, the AI governance system pre-computes the authorized action set: which components may be replaced under standing work orders, which configuration changes are pre-authorized within defined parameters, which equipment readings require immediate escalation regardless of connectivity status, and which actions require reconnection to the central authority before execution. That pre-authorized action set loads onto the field device. The AI tool is bounded by it.

When the technician returns to connectivity, or when the field device reconnects through a cellular link, governance reconciliation occurs: every action taken during the disconnected period is compared against the pre-authorized envelope, deviations are flagged for review, and the E-3 (State Transition Record) is updated with the full sequence of offline actions and their reconciliation status. Not a compromise of governance rigor. It is the only governance architecture that operates where connectivity cannot be guaranteed, which includes a substantial portion of energy sector field operations.

NERC CIP accommodates this through offline media handling under CIP-010 and the broader principle that protectable controls must be commensurate with the operational environment. Evidence must capture what the AI recommended, what the technician did, when connectivity was restored, and whether reconciliation confirmed compliance with the pre-authorized envelope. When reconciliation identifies a deviation, the E-3 record triggers the appropriate CIP incident or change management documentation process.

Energy sector attack surface

The generic surface from Chapter 7 applies, and in energy it carries a consequence the other sectors do not: an attack that reaches the operational layer can move physical equipment. Four regions sharpen here.

The operational-technology boundary. The highest-value target is a path from the information-technology side, where the agents live, into operational technology and the electronic security perimeter that CIP-005 requires. The defense is hard segmentation enforced as a routing constraint: no connector runs from the information-worker harness into operational technology, the perimeter is a deny-by-default boundary, and any cross-boundary action requires a separate, human-gated authorization. This is the seam from Chapter 2 at its most consequential, and it must be owned explicitly.

False telemetry feeding the model. Spoofed sensor or SCADA data drives an AI recommendation that affects grid operations, the F-EN1 failure produced on purpose. The defense is provenance on telemetry (KB-1), a human operator's disposition for any consequential control action (H-4), and the disconnected-governance model this chapter describes for edge operation, where local policy is signed and reconciled rather than trusted blind.

Safety-instrumented-system manipulation. An attack that nudges the model toward an unsafe dispatch or suppresses a protective action. The defense is graduated autonomy with a hard human gate on safety-critical actions, LP14 deterministic deny on uncertainty, and deterministic limits enforced outside the model so no generated recommendation can exceed a safety envelope regardless of what the model proposes.

Disconnected-edge token abuse. Edge autonomy depends on cached credentials, and a stale or replayed token operating while disconnected is a target. The defense is short-lived, narrowly scoped tokens, signed local policy that constrains what the edge agent may do offline, and a full reconciliation of actions against authorization on reconnect.

Energy sector failure modes

F-EN1: AI recommendation crossing the SCADA trust boundary without safety review. An AI-assisted predictive maintenance system generates a recommendation that a specific protection relay at a transmission substation requires replacement within 72 hours. The recommendation is based on thermal degradation analysis and a comparison against failure pattern data from similar relays in the fleet. The maintenance management system routes the work order to a field crew. The work order contains the AI recommendation. The field crew executes the protection relay replacement during a scheduled outage window.

The relay is classified as a High impact BCS asset under CIP-002. The work order did not trigger a CIP-010 change management review. No baseline comparison was performed before the change; no post-change verification was documented. The E-4 attribution record for the AI recommendation contains no CIP-010 documentation because the AI system generating the recommendation was never configured to produce it.

At the next NERC CIP audit, the auditor identifies the relay replacement in the outage management system. CIP-010 documentation for the change is absent. The violation is written up. Every day between the change and the audit is a potential penalty day.

Prevention requires the AI recommendation generation pipeline for High and Medium impact BCS assets to include a mandatory CIP-010 documentation step: the recommendation cannot be delivered to the maintenance management system without simultaneously generating a CIP-010 compliant change package. That change package is the E-4 attribution record, generated at the moment of recommendation. Not assembled after the fact.

Detection: E-4 attribution records indexed by asset impact classification; any High or Medium impact BCS asset change recommendation without a corresponding CIP-010 documentation record triggers an automated compliance alert before the recommendation enters the work order workflow.

F-EN2: NERC CIP evidence gap during AI system update. An energy company deploys a new version of its grid predictive maintenance AI. The model update changes inference logic for transformer failure probability estimation. It is deployed to production. No CIP-010 change documentation is generated for the AI system itself, because the responsible team classified it as a corporate IT asset outside BES Cyber System scope.

A NERC reliability coordinator subsequently identifies that the AI system's outputs feed the control center's unit commitment process. The control center is a High impact BCS asset. Any system whose outputs materially affect a High impact BCS asset's operation is subject to CIP-002 evaluation; the AI system may be in-scope under the "associated" assets doctrine.

The documentation gap covers every period between the last classified CIP-010 change package and the NERC audit. Multiple model updates occurred. None appear in the CIP-010 change record.

Prevention requires re-executing the BCS Cyber System identification survey whenever AI system output scope changes, and evaluating AI systems whose outputs feed High or Medium impact BCS operational processes for CIP-002 in-scope applicability. Every E-1 governance configuration record for an energy sector AI system must include an explicit BCS impact assessment with documented methodology. Assumptions do not qualify.

Detection: E-1 asset inventory cross-referenced against the operational dependency map; any AI system whose outputs are consumed by an in-scope BCS asset is flagged for CIP-002 evaluation regardless of how the AI system was initially classified.

F-EN3: Air-gapped system reconciliation failure. An AI predictive maintenance tool operating in a disconnected OT environment executes three equipment inspections during a 36-hour period in which the field device was not in contact with the central governance system. The inspections were within the pre-authorized action envelope except for one: the technician used the AI tool to reconfigure a protection relay setpoint based on an AI-generated recommendation that was not in the pre-authorized set.

When the field device reconnects and the reconciliation runs, the governance system identifies the setpoint change as outside the pre-authorized envelope. It flags the deviation for review. The protection relay setpoint change has already been executed against a High impact BCS asset. The CIP-010 change documentation process was not followed because the change occurred offline. The E-3 reconciliation record shows the deviation, but the change is already in production.

The governance consequence is a CIP-010 violation beginning at the moment of the unauthorized offline change. Remediation requires a post-implementation baseline comparison, a documented explanation of why the change proceeded outside the authorized envelope, and a corrective action plan addressing the gap in the pre-authorization framework that allowed the AI tool to generate a recommendation outside the pre-authorized set.

Prevention requires the pre-authorized action envelope loaded onto field devices to explicitly exclude configuration changes to protection relays and safety-related assets. Those changes must require connectivity to the central authority before the AI tool generates a recommendation in that category. Field device AI must distinguish between actions within offline authorization and actions requiring central authority confirmation, enforcing that distinction in the recommendation interface. Policy documentation alone is insufficient.

The energy sector AGENTS.md

```
# AGENTS.md - Energy and Critical Infrastructure AI Governance
# Regulatory sources: NERC CIP-002 through CIP-014, FERC,
#                     NRC 10 CFR 73.54, 10 CFR 50.55a,
#                     DOT PHMSA, TSA Security Directives,
#                     State PUC, EPA

bes_cyber_system_classification:
  impact_designations:
    - designation: high
      ai_access: esp_only
      change_documentation:
cip010_required_before_recommendation_delivery
      personnel_verification: cip004_current_required
      evidence_tier: e4_every_action
    - designation: medium
      ai_access: esp_or_it_with_unidirectional_gateway
      change_documentation: cip010_required_within_24h
      personnel_verification: cip004_current_required
      evidence_tier: e4_every_action
    - designation: low
      ai_access: authorized_users_only
      change_documentation: cip010_simplified
      personnel_verification: cip004_required
      evidence_tier: e2_batch_daily
    - designation: non_bcs
      ai_access: role_restricted
      change_documentation: standard_change_management
      evidence_tier: e2_operational

nuclear_cda_requirements:
  applies_to: nrc_licensed_facilities
  regulatory_basis: 10_cfr_73_54_and_50_55a
  cda_boundary_assessment: required_before_deployment
  safety_checkpoint: independent_of_cyber_security_checkpoint
  dual_gate:
both_checkpoints_required_for_safety_related_recommendations
  quality_assurance: 10_cfr_50_appendix_b_compliance

disconnected_operation:
  mode: pre_authorized_envelope
  authorization_source: central_governance_system
  pre_authorization_required: before_field_deployment
  offline_exclusions:
    - protection_relay_setpoint_changes
    - safety_related_component_replacements_without_prior_work_order
    -
any_action_affecting_high_impact_bcs_outside_standing_work_orders
  reconciliation: required_at_reconnect
  deviation_response:
flag_for_review_and_cip010_retroactive_documentation

mandatory_constraints:
```

```
    - "MUST generate CIP-010 change documentation simultaneously with
       every recommendation affecting High or Medium impact BCS assets;
       recommendation must not be deliverable without the change
package"
    - "MUST verify CIP-004 Personnel Risk Assessment currency for the
       recommendation recipient before delivering any recommendation
       affecting in-scope BCS assets"
    - "MUST enforce the pre-authorized action envelope in disconnected
       operation; safety-related configuration changes require central
       authority confirmation before recommendation generation"
    - "MUST produce dual-gate evidence for safety-related system
       recommendations at nuclear facilities: one cyber security
       verification record and one independent safety review record"
    - "MUST flag any AI recommendation that would modify a protection
       relay setpoint, a safety system parameter, or a nuclear CDA
       configuration for safety-review checkpoint before delivery"
    - "MUST log all interactions with High impact BCS assets to E-4
       with asset designation, CIP-010 change package reference,
       personnel authorization verification, and timestamp"

evidence_bindings:
  nerc_cip_010:
    applies_to: high_and_medium_impact_bcs_change_recommendations
    requirement: pre_implementation_change_documentation
    evidence: e4_cip010_change_package
  nerc_cip_004:
    applies_to: personnel_with_bcs_access
    requirement: background_check_currency
    evidence: e1_personnel_authorization_registry
  nrc_73_54:
    applies_to: nuclear_cda_adjacent_systems
    requirement: cyber_security_program_compliance
    evidence: e1_cda_boundary_assessment_plus_e4_safety_checkpoint
  phmsa_pipeline:
    applies_to: pipeline_ot_recommendations
    requirement: tsa_directive_compliance
    evidence: e2_operational_monitoring_plus_e4_change_records
```

The evidence package

Evidence tiers in energy sector AI governance carry a distinct characteristic absent from financial services or healthcare: E-4 attribution records must satisfy both regulatory and safety review purposes simultaneously. The same AI action may be a CIP-010 change event and a nuclear quality assurance record at the same time.

Evidence Tier	Energy Sector Implementation	Regulatory Mapping
E-1	AGENTS.md configuration, BES Cyber System impact designations and identification methodology, CIP-004 personnel authorization registry, CDA boundary	NERC CIP-002 BCS identification, CIP-004 personnel authorization, NRC 73.54 cyber security

Evidence Tier	Energy Sector Implementation	Regulatory Mapping
	assessment for nuclear facilities, pre-authorized field operation envelope, ASME code library version and applicability	plan, 10 CFR 50 Appendix B quality assurance program
E-2	Operational performance monitoring (transformer thermal trends, relay test results, generator performance curves), BCS configuration baseline currency verification, CDA boundary assessment refresh status, disconnected operation status per field device	NERC CIP-007 system security management, CIP-011 information protection, NRC post-installation testing, PHMSA control room monitoring requirements
E-3	Configuration change records with pre/post baseline comparison for all in-scope BCS assets, AI system model update records with CIP-002 re-evaluation, disconnected operation reconciliation records with deviation flags, pre-authorized envelope modification history	NERC CIP-010 configuration change management, NRC 73.54 periodic reassessment requirement, PHMSA change management documentation
E-4	Attribution records for every recommendation touching High or Medium impact BCS assets (with CIP-010 change package embedded), CIP-004 verification results per recommendation delivery, dual-gate safety and cyber security checkpoint records for nuclear recommendations, pre-authorized envelope compliance records for field operations	NERC CIP violation evidence requirements, NRC 10 CFR 50 Appendix B documentation requirements, PHMSA incident and change records, FERC compliance reporting

Embedding the CIP-010 change package in the E-4 attribution record is the critical architectural decision resolving the velocity-compliance tension. Under standard compliance program architecture, CIP-010 documentation is produced by the compliance team as a separate exercise after the operational team identifies and executes a change; compliance documentation lags operational action by hours or days. Under the evidence production architecture described here, the CIP-010 change package is generated at the moment of recommendation, as an inseparable artifact of the recommendation itself. The operational team receives recommendation and change package together. No gap. Compliance evidence exists before the change is executed, satisfying CIP-010's requirement that documentation precede implementation.

Lloyd's certificate of classification established the same principle in the 19th century: evidence of seaworthiness must exist at the moment of underwriting, not be assembled after the voyage. The NERC auditor who arrives to examine CIP-010 compliance for AI-assisted maintenance recommendations will find a complete change record for every recommendation, produced at the moment of recommendation, not assembled from scattered systems for the examination. The 5-minute examiner test for energy sector governance asks one question: can every configuration change recommendation for a High impact BCS asset that the AI system generated in the past 90 days be matched to a CIP-010 change package produced at the time of recommendation? Yes means the evidence architecture is sound.

CHAPTER SUMMARY

Energy and critical infrastructure AI governance is an L3 Lagrange point problem: the most isolated equilibrium in the system, requiring the most active station-keeping, against three bodies that are not in easy alignment. Body 1 is operational technology data whose sensitivity is defined not by privacy law but by consequence: High impact BES Cyber Systems, nuclear Critical Digital Assets, and pipeline control systems whose compromise creates physical harm at scale. Body 2 is a regulatory stack of NERC CIP, FERC, NRC 10 CFR 73.54 and 50.55a, PHMSA, TSA Security Directives, state PUC, and EPA that touches the same OT system from multiple directions simultaneously, with a $1,000,000 per violation per day NERC CIP penalty structure that creates compliance urgency with no parallel in any other commercial sector. Body 3 is grid operations velocity, where transformer failure prediction, load forecasting, and predictive maintenance operate on timelines where the AI cannot be wrong without physical consequence. The Lloyd's thread runs through NERC CIP-002's impact classification methodology: the BES Cyber System identification survey is the classification survey, the NERC auditor is the Lloyd's surveyor, and High impact assets receive the most intensive survey requirements because the consequence of failure is highest. Air-gapped SCADA environments require a pre-authorized envelope architecture for disconnected operation, with central authority confirmation required for safety-related configuration changes and mandatory reconciliation when connectivity is restored. Nuclear facilities present the extreme case: 10 CFR 73.54 and 50.55a require a dual-gate governance architecture where safety checkpoints operate independently from cyber security checkpoints, and the quality assurance program under 10 CFR 50 Appendix B applies to AI-assisted recommendations for safety-related systems. The three energy sector failure modes are AI maintenance recommendations crossing the SCADA trust boundary without CIP-010 documentation (F-EN1), NERC CIP evidence gaps during AI system updates due to incorrect BCS scope assessments (F-EN2), and air-gapped reconciliation failures where pre-authorized envelopes failed to exclude safety-related configuration changes (F-EN3). The resolution to the velocity-compliance tension is embedding the CIP-010 change package in the E-4 attribution record at the moment of recommendation generation, so that compliance evidence exists before the change is executed, not assembled after.

Chapter 15: L4: The Insurance and Actuarial Lagrange Point

Part III: The Sector Lagrange Points

> *"The Lloyd's underwriter was not pricing ships. He was pricing the difference between what was known and what was not. The actuary who signs an AI-assisted rate filing in 2026 is pricing exactly the same gap, across exactly the same table, in exactly the same coffee house, under a different name."*

L4 is the most naturally stable of the five Lagrange points. Unlike L1, which requires constant station-keeping, or L2, which sits at an unstable saddle point, L4 and L5 are positions where gravity and orbital mechanics combine to produce genuine equilibrium. Objects placed at L4 tend to stay there. The Trojan asteroids sharing Jupiter's orbit arrived because the physics produced the stability on its own. No one put them there. If you can reach L4, you do not have to fight to stay.

Insurance AI governance is an L4 problem. Not the most acutely unstable governance challenge in regulated industries. The forces are well understood, the history is long, and the regulatory vocabulary is precise. The profession governing AI-assisted actuarial work has been thinking about model validation since before the term "AI governance" existed. So the difficulty was never that the forces are unknown. Three very large bodies pull at once, and the natural stability of the L4 position is available only to organizations that understand where it actually is.

Customer data, regulatory compliance, and pricing velocity. They do not point in the same direction. Customer data demands depth, granularity, and actuarial rigor. Regulatory compliance, across fifty simultaneous state configurations, demands verification, documentation, and fairness analysis. Pricing velocity demands speed and real-time underwriting. An organization optimizing for any one of the three at the expense of the others does not find stability. It finds the coffee house logic that produced the load line problem two centuries ago.

The Lloyd's Thread

Insurance is the direct descendant of the coffee house.

Lloyd's of London was not originally an insurance company. It was a place: a coffee house in Lombard Street where merchants, shipowners, and underwriters gathered because the information they needed to price maritime risk was concentrated there. The underwriter who wrote his name under the terms of a maritime policy, accepting a share of the risk in exchange for a premium, was the original version of every risk transfer mechanism that followed. "Underwriter" is Lloyd's vocabulary. So is "policy," derived from the Italian "polizza," the written slip that circulated through the coffee house collecting signatures.

The Lloyd's Committee, established in 1771, was the first regulatory body in the insurance industry. Its purpose was what every regulatory body in insurance still exists to accomplish: ensuring that the entity accepting risk from a policyholder could actually pay the claim. Committee rules required members to hold adequate capital, disclose their financial condition, and honor their obligations. The National Association of Insurance Commissioners (NAIC)'s Model Bulletin on AI systems in insurance, issued in December 2023, operates on the same logic. Carriers must assess AI use in underwriting and claims for disparate impact, maintain documentation of their AI governance programs, and ensure that AI-assisted decisions remain accountable to human oversight. This is the Lloyd's Committee's requirement that underwriters be capable and honest, restated in modern vocabulary. The governance problem is unchanged.

State-by-state insurance regulation is port-by-port maritime law in modern form. Before international maritime conventions standardized ship safety requirements, a vessel calling at different ports faced different survey standards, different safety requirements, and different certificate schemes. A Lloyd's-classified ship might face additional local standards at certain ports, or meet suspicion where the English classification system carried no authority. The patchwork was not irrational. It reflected local knowledge about local conditions and local enforcement capacity. Navigating it, for anyone operating across multiple jurisdictions, was a serious operational problem all the same.

The fifty-state insurance regulatory system is that patchwork, institutional and persistent. Each state Department of Insurance licenses carriers, sets solvency requirements, approves rate filings, and examines insurer governance. An insurer writing auto, home, and commercial policies nationally operates under fifty simultaneous regulatory frameworks. Each applies its own version of the same underlying question the Lloyd's Committee was asking in 1771: is this entity capable of paying claims, and is it pricing risk in a way that is actuarially justified and legally permitted?

AI-assisted underwriting does not change this question. It concentrates it.

The Three Bodies in Insurance

Body 1: Customer Data. Insurance data ranks among the most sensitive commercial data in existence, a fact insurance regulation has recognized explicitly for decades longer than most sectors.

Underwriting data spans three categories. Application data is disclosed voluntarily under a known framework; the policyholder knows that driving record, credit history, claims history, and home characteristics are being used to price the policy. Purchased data, from brokers and public records, is disclosed involuntarily but under regulatory frameworks most states have addressed through credit scoring rules and data use disclosure requirements. AI-derived signals, the third category, is where the governance problem concentrates.

Suppose an AI underwriting model discovers a predictive relationship between browsing history and claims frequency. It has identified a signal the policyholder did not disclose, the underwriter did not purchase through a defined data channel, and neither the model-building actuary nor the Department of Insurance (DOI) examiner reviewing the rate filing may have recognized as a distinct data element. That signal lives in the model's weights, not in the filed rating plan. It produces pricing effects untraceable to a variable the policyholder understands; it may correlate with a protected characteristic; and it may not be actuarially justified in the sense actuarial science requires, which demands a credible causal relationship between risk factor and priced peril.

Claims history is the ground truth of risk for actuarial models, including AI-assisted ones. Every rating variable must be statistically tested against claims experience, and the relationship must be both statistically significant and actuarially credible. AI models discover relationships that classical variable selection would never find. Whether those relationships are actuarially justified is a judgment the actuary must make. When the AI discovers the relationship and the actuary accepts output without independently validating the identified variables, the actuarial professional's responsibility remains undischarged.

Health insurance data introduces the full Health Insurance Portability and Accountability Act (HIPAA) framework and, in states that have enacted health data privacy laws beyond HIPAA, state-level requirements that are sometimes more restrictive. The ACA Section 1557 prohibition on discrimination in health programs adds a civil rights layer to health underwriting data governance. For all insurance lines, the Gramm-Leach-Bliley Act imposes privacy notice and safeguarding obligations for nonpublic personal information that insurers obtain from their customers.

Body 2: Regulatory Compliance. The NAIC Model Bulletin on the Use of Artificial Intelligence Systems, adopted in December 2023 and implemented on varying timelines by member states, is the primary governance document for AI in insurance. Regulated entities using AI systems in insurance transactions must maintain a written governance and oversight program, ensure AI systems comply with all applicable insurance laws, regularly assess systems for unfairly discriminatory or unlawful criteria, and document testing and validation. Scope extends to third-party AI vendors; an insurer deploying a vendor AI system bears responsibility for that system's output governance regardless of who built the model.

Colorado's Division of Insurance, acting under SB 24-205, has issued rules requiring life insurers to test AI and predictive models used in underwriting for unfair discrimination and maintain documentation of those tests. This is the most specific state-level AI underwriting governance framework currently in effect. Other states will follow. Governance architecture must anticipate that pattern, not merely satisfy the state that has already enacted rules.

Actuarial Standards of Practice, published by the Actuarial Standards Board, define professional obligations for actuaries working with models, including AI-assisted models. Actuarial Standard of Practice (ASOP) No. 23 (Data Quality) requires actuaries to assess data quality and appropriateness. ASOP No. 25 (Credibility Procedures) requires credibility standards applied to statistical results. ASOP No. 56 (Modeling) requires actuaries to understand and validate models they use, and to communicate model limitations and uncertainties in any actuarial communication relying on model results. These standards draw no distinction between traditional statistical models and AI or machine learning models. A neural network generating loss cost predictions is a model under ASOP No. 56; the actuary using that output in a rate filing bears responsibility for understanding and validating its methodology.

For public companies, Sarbanes-Oxley Act (SOX) Section 302 and Section 404 internal controls requirements apply to the financial reporting systems insurance companies use to report loss reserves. An AI model used in reserve estimation is participating in a financial reporting process SOX governs. The General Data Protection Regulation (GDPR) applies to any policyholder data belonging to EU residents, which for insurers writing travel, specialty, or international lines is a recurring operational reality rather than an edge case.

Body 3: Pricing Velocity. Insurance pricing is migrating from annual rate filing cycles to real-time underwriting. Telematics programs price auto insurance on driving behavior observed in real time. Usage-based insurance adjusts premiums from continuous data streams. Commercial lines underwriters deploy AI models generating bindable quotes within seconds from structured data submissions; claims adjudication automation compresses cycle times from days to hours for certain claim types.

This velocity is not merely competitive preference. It is becoming a market requirement, in segments where customers expect instant quotes and same-day policy issuance. An insurer whose underwriting requires three days of actuarial review for every application is not offering the same product as a competitor whose AI-assisted underwriting produces an instant decision. Governance requiring manual review of every AI-assisted underwriting decision produces the heroic one-shot failure mode in reverse. The process becomes so burdensome that operators bypass it, leaving AI-assisted decisions with no documentation, no fairness analysis, and no actuarial validation.

The L4 position requires pricing velocity and actuarial validation simultaneously. Not sequentially. An organization that believes these are incompatible has not found the Lagrange point. It has assumed the position does not exist.

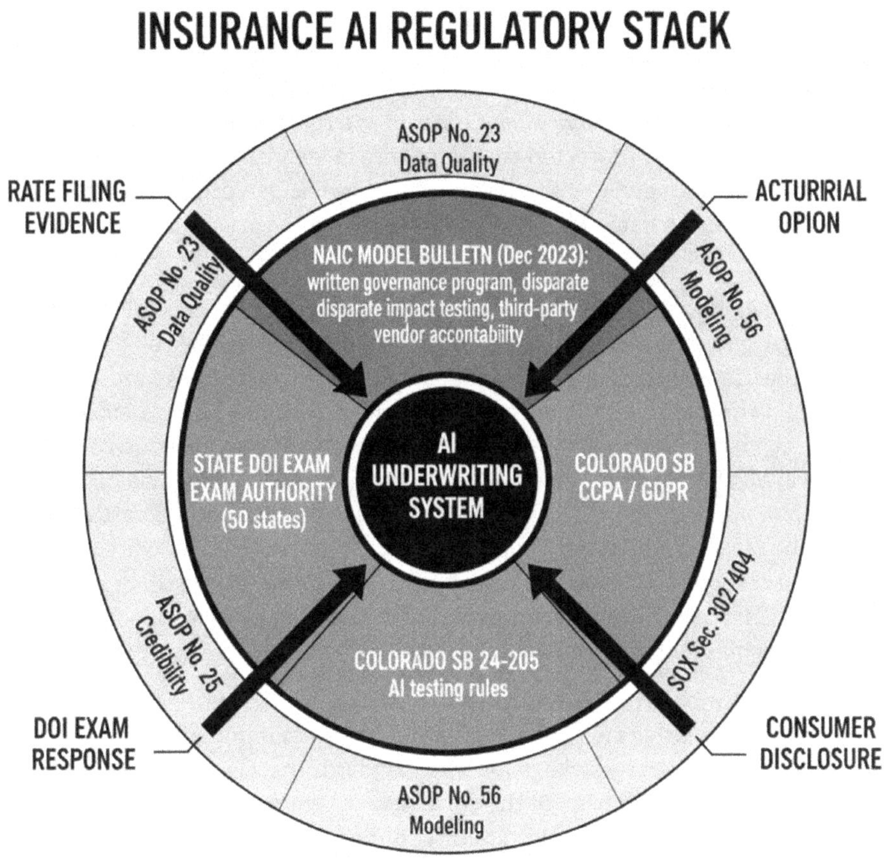

Figure 15: The insurance AI regulatory stack. A single underwriting system sits inside concentric layers of authority, the NAIC model bulletin at the center, state DOI examination power and Colorado SB 24-205 around it, and the actuarial standards of practice (ASOP 23, 25, 56) plus SOX at the outer ring, while producing rate-filing evidence, an actuarial opinion, exam responses, and consumer disclosure at once.

The Fairness Problem

Insurance is, by definition, discrimination. Every underwriting decision assigns a price based on the insurer's assessment of applicant risk, treating different applicants differently based on characteristics associated with their likelihood of filing a claim. The legal framework does not prohibit this discrimination. It requires the discrimination to be actuarially justified, based on characteristics with a credible causal or statistically verified relationship to the priced peril, and not based on characteristics that are legally prohibited or produce unlawful proxy discrimination.

The prohibited list in most states includes race, color, national or ethnic origin, religion, sex, marital status, income, educational attainment, and various other characteristics depending on insurance line and state. Many states prohibit credit scoring in homeowners underwriting. Colorado's AI rules prohibit unfair discrimination based on protected characteristics in life insurance underwriting. Because the legal framework differs by state and by line, governance must resolve applicable rules at the data-subject level rather than applying static system-level configuration.

Proxy discrimination is where AI underwriting governance becomes technically demanding. An AI model trained on historical claims data identifies all predictive signals in that data, including signals that correlate with protected characteristics without being protected characteristics themselves. ZIP code predicts auto insurance claims partly through geography and partly because ZIP code correlates with race in many US metropolitan areas. Browsing history predicts claims in certain experimental property models partly through behavioral signals and partly because browsing behavior correlates with income, which in turn correlates with other characteristics. Social media activity has been tested as an underwriting signal in several markets. Occupation and employer correlate with income, education, and in some geographies with race and national origin.

Actuarial justification is the legal standard for evaluating these correlations. A rating variable producing disparate impact on a protected class is not automatically prohibited; it must be assessed against that standard. The assessment requires establishing actuarial credibility of the correlation, identifying a plausible causal mechanism linking variable to priced peril, confirming the disparity is proportionate to the actuarial relationship, and demonstrating that the justification can be articulated in a rate filing a DOI examiner can evaluate.

An AI model identifying zip code as predictive can be evaluated against this standard. The model's contribution is signal identification; the actuary's contribution must be assessing whether that signal meets actuarial justification, whether it can be defended to the DOI, and whether the disparate impact produced is something the insurer intends to maintain and can defend.

Browsing history or social media behavior presents a harder case. The causal mechanism is less clear, the data quality is less verifiable, and the potential for proxy discrimination is greater. An actuary who accepts these signals without independent validation has not discharged the ASOP No. 56 obligation to understand and validate model methodology.

The Four-Fifths Rule in Underwriting

The four-fifths rule, the 80% ratio from EEOC's Uniform Guidelines on Employee Selection Procedures, is sometimes borrowed as a screening heuristic for first-pass disparate-impact review in insurance underwriting. No insurance regulator has adopted the 80% ratio as a bright-line underwriting test. The operative standards are state unfair-discrimination law and actuarial justification, with the NAIC Model Bulletin (December 2023) requiring written AI governance and regular assessment for unfairly discriminatory outcomes. Used as a screening flag, the calculation is straightforward: if an AI underwriting model approves Group A applicants at rate X and approves Group B applicants at less than 0.8X, the gap is a signal that triggers deeper actuarial-justification analysis under the governing state law, not a verdict.

The 80% ratio is a diagnostic flag, not a pass/fail safe-harbor line. Clearing the ratio does not certify a model as fair, and falling below it does not condemn one; the ratio is a trigger for analysis in either direction. When an approval ratio drops below 0.8X, the flag fires and the actuarial analysis must go further. What explains the disparity? Is the explanation actuarially justified? Is the justification defensible under applicable state law, and is the disparity proportionate to the actuarial relationship? The fairness line in insurance is not a single number. It is the actuarial-justification standard governed by state unfair-discrimination law, which the screening ratio only points toward.

LP11 (Fairness Verification) implementation in insurance underwriting operates at two levels. The first is pre-deployment testing: before an AI underwriting model enters production at scale, approval rates across protected classes and proxies are screened against the four-fifths ratio, which functions here as a first-pass diagnostic under state unfair-discrimination law and the NAIC Model Bulletin, never as a bright-line safe harbor that certifies a model fair. A model whose screening ratio falls below the threshold before deployment requires either remediation or documented actuarial justification under the governing state standard. The second is continuous operational verification: during production operation, approval rates must be monitored against the screening ratio at the frequency dictated by decision volume. An AI underwriting model issuing 10,000 decisions per day must be monitored at a frequency that would detect a disparity before it accumulates across 100,000 decisions.

The four-fifths rule is a screening starting point, never a terminal test, and never a safe harbor. Statistical significance of a measured disparity is a separate question from its magnitude; a disparity that is statistically significant and actuarially unjustifiable is a problem regardless of whether it crosses the 80% ratio. A model can clear the ratio and still rely on a variable that fails actuarial justification under state law; a model can fall below the ratio and still be defensible if the actuarial relationship is credible, causal, and proportionate. The ratio narrows where the actuary looks first. The governing test remains the actuarial-justification standard and the applicable state DOI rule.

The NAIC Model Bulletin's requirement for regular assessment of unfairly discriminatory criteria is the insurance-sector version of this continuous verification obligation. An insurer running LP11 analysis annually satisfies neither the Bulletin's requirement nor the operational reality of a model issuing millions of decisions per year.

Fifty-State Regulatory Complexity

A national insurance carrier must simultaneously maintain fifty different regulatory configurations, each representing the applicable state DOI's requirements for AI governance in underwriting, claims, and pricing. This is LP10 (Jurisdictional Consistency) at maximum operational complexity.

The jurisdictional rule mirrors every other commercial sector but is more explicit in insurance: applicable regulatory requirements are determined by the policyholder's state of residence, not the insurer's state of domicile. An insurer domiciled in Iowa writing a homeowners policy for a California resident applies California DOI requirements to that policy, California fair claims practices statutes to any claim, and California's credit scoring restrictions to any underwriting decision for that policyholder. Where the insurer processes the transaction is irrelevant.

The governance architecture must maintain a jurisdiction mapping that resolves, for each policy transaction, the applicable state requirements, and routes the transaction through the correct regulatory configuration. For an insurer writing fifty states of auto insurance, this means fifty simultaneous regulatory configurations, each potentially requiring different fairness thresholds, different documentation standards, different rate filing obligations, and different claims practices compliance rules.

Interaction between state requirements is where architecture becomes demanding. Colorado's AI underwriting rules are more prescriptive than most on testing and documentation for AI models in life insurance underwriting. California's Credit Score Moratorium prohibits credit scoring in homeowners and auto insurance during and after declared disasters, creating a dynamic constraint that changes based on declaration status. Some states prohibit certain telematics signals in auto insurance; states enacting NAIC Model Bulletin requirements have done so on varying timelines with varying interpretations.

The default strategy for jurisdictional uncertainty is most-restrictive-applicable, resolved at the state level. When applicable requirements are uncertain, the harness applies the most restrictive among plausible candidates. This runs conservative for pricing, since stricter requirements generally constrain permitted rating variables. It is the only defensible posture when the applicable state cannot be determined with certainty and the consequences of applying the wrong configuration include regulatory enforcement, market conduct examinations, and rate rollbacks.

Actuarial Professional Liability

Signing an opinion on a rate filing or loss reserve estimate is exercising professional judgment. That judgment creates professional liability. The actuarial profession has addressed this through Actuarial Standards of Practice (ASOPs), which define practice standards against which actuarial work is evaluated in professional liability proceedings and regulatory examinations.

AI-assisted actuarial work creates a new category of professional liability the ASOPs address only by implication. When an actuary uses an AI model to generate loss cost projections, reserve estimates, or pricing parameters and incorporates those projections into a signed opinion, responsibility for the model's output is accepted under ASOP No. 56. The actuary must understand the model, validate its methodology, assess limitations, and communicate uncertainty. If the AI model produces output the actuary cannot explain, the opinion cannot be signed without disclosing that limitation.

In practice, this creates an unsolved professional liability problem. Many AI models used in insurance pricing, including gradient boosting models, neural networks, and ensemble methods, are not fully interpretable in the sense ASOP No. 56's understanding-and-validation standard requires. A model generating loss cost predictions from 200 variables and millions of training records does not produce the kind of explanation an actuary can evaluate the way a traditional GLM with documented factor relativities does. The actuary signing on that output now occupies a liability position that did not exist before AI-assisted pricing became standard.

LP12 (Transparency and Explainability) applied to actuarial outputs addresses this directly. An AI model used in rate development must produce an explanation artifact at the variable level: which variables drove the prediction, what their relative contribution was, and what the model's confidence interval is. The point is not SHAP values as an academic exercise. It is SHAP values, or an equivalent attribution method, treated as an actuarial validation tool the signing actuary requires before incorporating model output into a professional opinion.

Reserve estimation follows the same standard. An AI model used in reserve development must produce loss projections with documentable methodology, testable assumptions, and communicable uncertainty. The actuary presenting reserve estimates to the board must understand and validate the methodology; an AI model producing a point estimate without communicable uncertainty is not a complete actuarial tool. It is a calculation requiring the actuary's independent reliability assessment before it can support a signed opinion.

Evidence architecture must treat actuarial AI outputs as requiring the same documentation traditional actuarial work products require. Model validation report, assumption documentation, sensitivity analysis, and explanation of significant deviations from expectations are all components of the actuarial work product. They must be present in E-1 governance records for any AI model contributing to a signed actuarial opinion.

The Insurance Attack Surface

The generic surface from Chapter 7 applies, and in insurance the sharpest attacks target the validation step itself, because a model that passes a rigged test ships with a clean bill of health. Four regions stand out.

Engineered proxy discrimination. Inputs chosen or correlated features introduced so the model produces disparate outcomes while every accuracy metric improves, the F-IN1 failure produced deliberately. The defense is continuous bias monitoring as a verification checkpoint (LP11), testing against protected-class proxies rather than only accuracy, and independent review that does not depend on the internal validation that the attack has already satisfied.

Validation-dataset poisoning. The strongest attack in this sector is on the test, not the model: shape the historical portfolio used for validation so the disparate-impact analysis passes while production fails. This is false completion weaponized. The defense is the population-scope control from Chapter 7, a documented representativeness assessment against the prospective applicant population in each jurisdiction, plus out-of-distribution monitoring that watches production segments the validation set underrepresents.

Rate-model tampering. Manipulation of model inputs or parameters to skew a rate filing that a regulator will rely on. The defense is provenance and version pinning on the rating model, an E-3 state-transition record on any parameter change, and an actuary's sign-off (H-4) bound to the specific model version that produced the filing.

Claims-evidence fabrication. Generated content used to manufacture or alter the documentation behind a claim. The defense is an E-4 attribution record on every AI-generated artifact, so each one traces to an identity and a session, and human adjudication for any consequential claims decision.

Insurance Failure Modes

F-IN1: Proxy Discrimination Through Unvalidated Correlated Features. An AI underwriting model identifies a set of non-traditional rating variables, including variables derived from browsing history, social media activity, and aggregated purchase data, that are predictive of claims frequency. The model is validated for predictive accuracy against holdout data; it performs better than the previous model across all measured accuracy metrics. The rate filing incorporates the model's outputs.

The governance failure: the AI's variable identification was not followed by the actuary's independent assessment of whether those variables meet actuarial justification standards. The variables are predictive. They are not necessarily actuarially justified in the sense insurance regulation requires. Some correlate with ZIP code; some correlate with income; some produce approval rate disparities across racial and ethnic groups that, under the four-fifths rule, require actuarial justification to maintain.

The actuary signed the opinion without performing proxy discrimination analysis because the accuracy metrics looked satisfactory and variable selection was automated. The filing passes initial review. Eighteen months later, a market conduct examination arrives, and the DOI requests the fairness analysis. It does not exist. The variables cannot be independently justified, and the rate filing is withdrawn.

Detection requires LP11 (Fairness Verification) applied before deployment, with the proxy discrimination check as a required component of the actuarial validation process, not a separate compliance exercise. The E-2 fairness metrics must include approval rate analysis by race, ethnicity, and protected class proxies, generated before the model goes into production and continuously during operation. The actuary must be presented with the proxy discrimination analysis as part of the model validation package, not as an after-the-fact compliance inquiry.

F-IN2: State-by-State Regulatory Inconsistency. An insurer deploys an AI underwriting system configured for its domicile state's regulatory requirements, applying those requirements to all policyholders regardless of residence state. The configuration correctly handles domicile state credit scoring rules, fairness thresholds, and documentation standards. It does not handle California's credit scoring moratorium during declared disaster periods; it does not apply Colorado's specific AI testing documentation requirements to Colorado life insurance policyholders; it does not enforce telematics use restrictions certain states impose.

For most policyholders, the error is invisible. For policyholders in states with more restrictive requirements, the system applies the wrong regulatory configuration to their transactions. Then a DOI examines operations in a non-domicile state, and the examiner discovers the AI system was not applying that state's requirements. Market conduct violation.

Prevention requires LP10 (Jurisdictional Consistency) enforced at the transaction level. The applicable state must be resolved from the policyholder's state of residence before each underwriting decision. The regulatory configuration applied to that decision must correspond to the resolved state. The E-3 state transition records must include the jurisdiction determination for each transaction, so that the DOI examiner can verify that the correct regulatory configuration was applied.

F-IN3: Actuarial Opinion Covering AI-Assisted Pricing the Signing Actuary Cannot Explain. An actuary signs a rate filing incorporating AI model outputs. The filing is challenged by the DOI in a subsequent hearing. The DOI examiner asks the actuary to explain the model's methodology, identify the key rating variables, and demonstrate that the model's predictions are actuarially credible.

The actuary cannot answer these questions. The model was validated for predictive accuracy. It was never validated for interpretability, actuarial justification, or compliance with ASOP No. 56's understanding-and-validation standard. Accuracy metrics stood in as a proxy for actuarial validity, and that is not a valid proxy under ASOP No. 56.

Three consequences follow. On professional liability, a finding that the actuary failed to meet applicable actuarial standards. On the regulatory side, the rate filing is suspended pending full model documentation. And in the market, a rate reversion the insurer cannot operationally implement within the DOI's required timeframe.

Prevention requires the governance architecture to produce explanation artifacts for AI model outputs before those outputs are incorporated into actuarial opinions. The model documentation required by ASOP No. 56 must be present in the E-1 governance records: variable attribution, assumption documentation, sensitivity analysis, and uncertainty quantification. The actuary's validation review must be logged in E-4 attribution records, demonstrating that the actuary reviewed the explanation artifacts before signing. The standard for "understands and validates" under ASOP No. 56 is not satisfied by reviewing accuracy metrics. It requires engaging with the model's variable-level methodology.

The Insurance AGENTS.md

An insurance AI deployment's instruction surface must encode NAIC Model Bulletin requirements, state-specific DOI configurations, ASOP obligations, and fairness verification thresholds in a form the routing enforcement layer executes without human interpretation at each transaction.

```
# AGENTS.md - Insurance AI Governance
# Regulatory sources: NAIC Model Bulletin (Dec 2023),
#                     Colorado SB 24-205, GLBA, ACA Section 1557,
#                     ASOP Nos. 23, 25, 56

data_governance:
```

```
  underwriting_data:
    traditional_variables:
      ai_access: permitted
      # Actuarial justification is the governing fairness test:
      # state unfair-discrimination law plus a credible, causal,
      # proportionate actuarial relationship documented in the filing.
      validation: actuarial_justification_documented
    derived_signals:
      ai_access: conditional
      validation: proxy_discrimination_analysis_required
      evidence: e1_variable_attribution_report
    prohibited_variables_by_jurisdiction:
      resolve: policyholder_state_of_residence
      uncertainty_strategy: most_restrictive_applicable
  health_data:
    phi_applicable: true
    aca_1557_applicable: true
    ai_access: minimum_necessary_enforced

jurisdiction_routing:
  determination_method: policyholder_state_of_residence
  granularity: per_transaction
  uncertainty_strategy: most_restrictive_applicable
  special_configurations:
    california:
      credit_scoring: check_disaster_moratorium_status
      telematics: check_state_restrictions
    colorado:
      ai_model_testing: colorado_div_ins_rules
      life_underwriting: colorado_sb24205_documentation
    all_states:
      base_requirement: naic_model_bulletin_dec2023

fairness_verification:
  trigger: ai_assisted_underwriting_or_claims_decision
  # four_fifths_rule is a screening FLAG that triggers analysis,
  # not a bright-line safe harbor. No insurance regulator has
  # adopted the 80% ratio as an underwriting test. The governing
  # tests are the actuarial-justification standard and the
  # applicable state DOI unfair-discrimination rule.
  threshold: four_fifths_rule_screening_flag
  proxy_discrimination_check: required
  variables_to_test: [zip_code, browsing_history, social_media,
                      credit_score, occupation, employer]
  evaluation_frequency: continuous_operational
  pre_deployment_requirement: pass_before_production
  evidence_tier: e2_continuous_verification

actuarial_validation:
  standard: asop_no_56_understanding_and_validation
  required_artifacts:
    - variable_attribution_report
    - assumption_documentation
    - sensitivity_analysis
    - uncertainty_quantification
    - proxy_discrimination_analysis
  actuary_review_logging: e4_attribution_record
```

```
    model_types_requiring_full_validation:
      [gradient_boosting, neural_network, ensemble, ai_ml]

  mandatory_constraints:
    - "MUST verify policyholder state of residence and apply that
       state's regulatory configuration to each transaction; insurer
       domicile state configuration does not satisfy this requirement"
    - "MUST assess all non-traditional rating variables for proxy
       discrimination before incorporating into rate filings;
       predictive accuracy alone does not constitute actuarial
       justification"
    - "MUST produce variable attribution artifacts before AI model
       outputs are incorporated into any signed actuarial opinion;
       accuracy metrics alone do not satisfy ASOP No. 56"
    - "MUST screen AI underwriting decisions for disparate impact
       using the four-fifths ratio as a flag before deployment at
       scale, then resolve any flagged disparity against the
       actuarial-justification standard and applicable state DOI
       rule; the ratio is not a pass/fail safe harbor; annual
       fairness evaluation does not substitute for continuous
       operational monitoring at production decision volumes"
    - "MUST enforce credit scoring restrictions specific to the
       policyholder's state, including dynamic restrictions triggered
       by declared disaster status"
    - "MUST log all AI-assisted underwriting decisions with the
       jurisdiction configuration applied, the fairness metrics
       generated, and the actuary review record as E-4 attribution"

  approval_authorities:
    non_traditional_variable_approval:
  credentialed_actuary_plus_compliance
    rate_filing_with_ai_outputs:
  signing_actuary_with_e1_validation_record
    new_jurisdiction_deployment:
  state_filing_approval_plus_p10_configuration
    proxy_discrimination_finding: chief_actuary_plus_compliance_officer
    emergency_rate_reversion: cro_with_e4_record_and_doi_notification
```

The state disaster moratorium check for credit scoring is the implementation detail that most frequently reveals inadequate LP10 (Jurisdictional Consistency) enforcement. California's prohibition on credit scoring in homeowners and auto insurance during and after declared disasters is a dynamic regulatory constraint, not a static configuration set once and verified annually. When a disaster is declared, the prohibition takes effect immediately. An AI underwriting system that does not check declared disaster status before applying credit scoring variables to a California homeowners transaction applies the wrong regulatory configuration from the moment of declaration forward. Every transaction in that window is a potential violation.

The Insurance Evidence Package

The four evidence tiers map directly to insurance regulatory examination requirements, across NAIC, state DOI, and actuarial professional standards:

Evidence Tier	Insurance Implementation	Regulatory Mapping
E-1	AGENTS.md configuration, state regulatory configuration registry, AI model inventory, variable attribution reports, actuarial validation records, ASOP No. 56 documentation	NAIC Model Bulletin documentation requirements, ASOP No. 56 model documentation, state rate filing supporting exhibits
E-2	Continuous fairness metrics by state and line, proxy discrimination analysis results, approval rate ratios by protected class and proxy, model performance vs. validation baseline, disaster moratorium compliance status	NAIC Model Bulletin regular assessment obligation, Colorado DOI testing documentation, LP11 continuous operational evidence
E-3	State regulatory configuration change records, model update records, variable addition and removal events, jurisdiction mapping changes, disaster moratorium activation and deactivation events	DOI market conduct examination readiness, ASOP No. 56 change documentation, rate filing amendment records
E-4	Attribution records for every AI-assisted underwriting decision with jurisdiction configuration applied, actuary review records for AI model outputs incorporated in opinions, proxy discrimination analysis confirmations, fairness threshold pass/fail records per decision	DOI market conduct examination attribution requirements, ASOP No. 56 actuarial opinion support, state fair claims practices audit trails

For DOI market conduct examinations, the E-1 documentation demonstrates that the insurer's AI governance program meets the NAIC Model Bulletin's written program requirements. The E-2 fairness metrics satisfy the Bulletin's regular assessment obligation. The E-3 change records demonstrate that state-specific regulatory configurations were maintained and updated as regulatory requirements changed. The E-4 attribution records satisfy the Bulletin's accountability and oversight requirements by demonstrating that AI-assisted decisions were reviewed by accountable humans and that the applicable regulatory configuration was applied to each transaction.

For actuarial professional liability, the evidence architecture serves the actuary as well as the insurer. An actuary who can produce the E-1 variable attribution report, the E-2 proxy discrimination analysis, and the E-4 actuary review record demonstrating that the validation artifacts were examined before the opinion was signed has documented compliance with ASOP No. 56. The standard of care is not merely that the validation occurred; it is that the validation is documentable. The harness produces that documentation as a byproduct of operation.

The Five-Minute DOI Examiner Test

A Department of Insurance examiner conducting a market conduct examination of an insurer's AI underwriting program should be able to answer five questions from the evidence architecture in five minutes. Not hypothetical questions. These are the five questions market conduct examiners conducting AI-focused examinations are now asking, under examination guidance informed by the NAIC Model Bulletin.

1. What AI systems does this insurer use in underwriting, pricing, and claims, and what is the written governance program for each? (E-1)
2. Has the insurer assessed its AI systems for unfairly discriminatory criteria, and what did those assessments find? (E-2 fairness metrics)
3. Is the correct state regulatory configuration being applied to policyholders in this state? (E-3 jurisdiction configuration records for this state)
4. Can the insurer demonstrate that an actuary has reviewed and validated the AI models contributing to rate filings? (E-4 actuary review attribution records)
5. For any AI system flagged with a disparity finding, what was the insurer's response? (E-3 state transition records for remediation events)

An insurer that cannot answer these five questions within five minutes of the examiner's arrival is operating an examination-driven compliance program. Between cycles, AI decisions are being made without the documentation the NAIC Model Bulletin requires. When the examiner arrives, the documentation does not exist, so it must be assembled, reconstructed, or produced after the fact. Post-hoc reconstruction is not documentation. It is a narrative about what the insurer believes happened.

The insurer that implements evidence architecture correctly does not prepare for the market conduct examination. It operates continuously in the condition the examination is designed to verify.

> **CHAPTER SUMMARY**
>
> Insurance AI governance is an L4 problem: three gravitational bodies, customer data, regulatory compliance across fifty simultaneous state configurations, and pricing velocity, pulling simultaneously toward a naturally

stable equilibrium that requires understanding where the L4 position actually is. Insurance is the direct institutional descendant of Lloyd's of London; the NAIC Model Bulletin is the modern Lloyd's Committee, and state DOI regulation is the modern port-by-port maritime law. The three bodies in insurance are customer data including AI-derived signals the policyholder did not disclose and the actuary must validate, regulatory compliance under the NAIC Model Bulletin and state DOI frameworks including ASOPs, and pricing velocity that makes continuous governance the only operationally viable option. The fairness problem in insurance is not whether discrimination occurs; insurance is by definition discriminatory. The question is whether the discrimination is actuarially justified. AI models discover correlated signals that may not be actuarially justified and may produce proxy discrimination that the actuary must independently assess, not accept from the model's accuracy metrics. The four-fifths rule is a borrowed screening heuristic for LP11 (Fairness Verification), not a bright-line underwriting test: an approval ratio below 0.8X is a diagnostic flag that triggers actuarial-justification analysis. The four-fifths ratio is the screening flag; the actuarial-justification standard is the load line. No insurance regulator has adopted the 80% ratio as a safe harbor; the governing tests are state unfair-discrimination law and the actuarial-justification standard, with the NAIC Model Bulletin (December 2023) requiring written AI governance and regular fairness assessment. Fifty-state regulatory complexity requires LP10 (Jurisdictional Consistency) at its maximum operational complexity, including dynamic constraints like California's credit scoring moratorium during declared disasters, resolved at the transaction level from the policyholder's state of residence. Actuarial professional liability under ASOP No. 56 requires that actuaries understand and validate AI models whose outputs they incorporate into signed opinions; predictive accuracy alone does not satisfy this standard. The three insurance failure modes are proxy discrimination through unvalidated correlated features (F-IN1), state-by-state regulatory inconsistency from applying the wrong state's configuration (F-IN2), and actuarial opinion covering AI-assisted pricing the signing actuary cannot explain (F-IN3). The five-minute DOI examiner test distinguishes a continuous governance posture from an examination-driven program. The E-1 through E-4 evidence tiers satisfy NAIC Model Bulletin documentation requirements, state DOI examination requirements, and ASOP No. 56 actuarial documentation obligations from a single evidence architecture, because those frameworks are all asking versions of the same five examiner questions.

Chapter 16: L5: Manufacturing, Supply Chain, and Telecommunications

Part III: The Sector Lagrange Points

> *"The container did not change what was in the box. It changed how the box moved. The standard format became the infrastructure, and the infrastructure became invisible. Governance follows the same logic: standardize the evidence format, and the evidence moves across jurisdictions the way containers move across oceans."*

L5, along with L4, has a property distinguishing it from the other three Lagrange points. Objects placed at L4 and L5 are gravitationally stable without active station-keeping. Trojan asteroids have collected at Jupiter's L4 and L5 points over billions of years and remain there without correction; the equilibrium is self-reinforcing, with small perturbations producing restoring forces rather than destabilizing drift. An object at L5 does not need to fight to stay.

Manufacturing and telecommunications AI governance is closer to an L5 problem than to an L1 problem.

The regulatory pressure that makes financial services governance an active station-keeping exercise, five simultaneous data regimes pulling against microsecond competitive velocity, is structurally different in manufacturing and telecommunications. Product safety requirements for physical goods are demanding, but they address a smaller surface area than Bank Secrecy Act (BSA)/Suspicious Activity Report (SAR) confidentiality obligations or material non-public information (MNPI) isolation requirements. Environmental and trade compliance frameworks are rigorous; they are not, however, pulling in opposite directions against real-time production velocity the way algorithmic trading speed is incompatible with two-week governance review cycles. The equilibrium point exists. It is easier to reach and hold than the financial services and healthcare equivalents.

This does not mean manufacturing governance is simple. An L5 point collecting Trojan asteroids collects them because conditions are right, not because nothing is present to disturb them. Manufacturing and telecommunications governance has its own gravitational bodies, and they generate their own failure modes. The physics are more forgiving here. The governance is not optional.

The Three Bodies in Manufacturing and Supply Chain

Body 1: Production Intellectual Property and Controlled Technical Data. Manufacturing organizations generate and possess information categories with no equivalent in financial services or healthcare.

Production intellectual property encompasses technical specifications, process parameters, materials formulations, tooling designs, and quality control procedures determining a manufacturer's competitive position. An automotive manufacturer's torque specifications; a semiconductor company's deposition process parameters; a pharmaceutical company's synthesis route documentation. These are primary assets of the manufacturing enterprise. Confidentiality is protected primarily by contract, trade secret law, and operational security rather than sector-specific regulation, but the governance consequence is identical. An AI coding assistant with access to manufacturing process documentation must not expose proprietary process parameters to developers on unrelated product lines, and certainly not to external parties.

Vendor contracts and supply chain intelligence represent a second category: supplier pricing, preferred vendor agreements, production scheduling commitments, and supplier financial condition assessments that a competitor would value. An AI system that processes vendor contract data must confine that data to the scope of the specific vendor management task at hand.

Customer order data in manufacturing includes design specifications, production forecasts, and delivery commitments often protected by non-disclosure agreements with the customer. A developer querying an AI assistant about a production scheduling algorithm may be working in a context that includes order data from multiple customers who have not consented to their information being processed in the same AI context.

International Traffic in Arms Regulations (ITAR) and Export Administration Regulations (EAR) impose the most demanding access control requirements in manufacturing. ITAR, administered by the State Department under 22 CFR Parts 120-130, controls export of defense articles and defense services. An AI coding assistant operating in a defense manufacturing environment may have access to ITAR-controlled technical data. ITAR does not regulate intent. It regulates disclosure. An AI system surfacing ITAR-controlled technical data in a context accessible to a foreign national has created an ITAR violation, regardless of how the surfacing occurred and regardless of the foreign national's purpose. The violation is complete at the moment of unauthorized disclosure; AI mediation does not affect the regulatory analysis.

Body 2: Regulatory Compliance. Manufacturing regulatory compliance spans four distinct domains, each enforced by different agencies with different examination approaches and evidence expectations.

Product safety regulation governs manufacturing for specific product categories. Food and Drug Administration (FDA) 21 CFR Part 820, the Quality System Regulation for medical device manufacturers, requires design controls, process controls, corrective and preventive action systems, and production controls that together demonstrate every manufactured device conforms to its approved design. International Organization for Standardization (ISO) 13485:2016 is the international standard for quality management systems in medical device manufacturing, harmonized with regulatory requirements in the EU, Canada, Australia and Japan. Any AI system assisting in medical device manufacturing, whether in design, process optimization, or quality assurance, operates within a framework requiring documented evidence of controlled, verified production at every stage.

Environmental regulations require AI systems in manufacturing to operate within environmental compliance frameworks. Resource Conservation and Recovery Act (RCRA), Clean Air Act, Clean Water Act, and sector-specific environmental permits impose monitoring, reporting, and compliance obligations on manufacturing operations. AI systems optimizing production processes must not optimize toward outcomes violating environmental permit limits. An AI-assisted production optimization that reduces waste by routing materials through a process exceeding permitted discharge limits has not solved the compliance problem. It has relocated it.

Trade compliance under ITAR and EAR, described above in the context of data access, also imposes process requirements on manufacturing operations. Every export of a controlled article or technical data requires a license or applicable exemption. An AI system assisting in export license applications, customer screening, or end-use assessment operates within a framework where errors carry criminal enforcement consequences for responsible individuals, not merely civil penalties for the organization.

Quality management under ISO 9001:2015 applies to manufacturing organizations across industries and governs the documented evidence that production processes are operating in control. AI systems that participate in quality monitoring, nonconformance assessment, or process capability analysis must generate evidence artifacts compatible with the ISO 9001 quality management system's documentation requirements.

Body 3: Production Velocity. Manufacturing competitiveness increasingly depends on operational optimization at speeds exceeding human analysis capacity. Predictive maintenance AI analyzes sensor data from production equipment, predicting failure before it occurs and scheduling maintenance during planned downtime rather than emergency downtime. Just-in-time manufacturing demands supply chain optimization responding to demand signals, logistics conditions, and supplier status in near-real time; quality assurance automation applies computer vision and sensor analysis to production output at line speed, detecting nonconformances human inspection would miss.

Autonomous procurement agents represent the velocity dimension with the highest governance stakes. A procurement agent that can autonomously approve vendor contracts, place purchase orders, and qualify new suppliers within defined thresholds compresses procurement cycle times from weeks to hours. The threshold-based autonomy structure, where approvals below a defined value require no human involvement, is the operational pattern creating the fraud exposure described below.

The Lloyd's Thread: Containerization and Evidence Harmonization

Malcom McLean's first container voyage departed Port Newark for Houston on April 26, 1956, carrying 58 trailers loaded with aluminum, cotton bales, and sewing machines. The voyage did not prove that containerization was economically superior to break-bulk cargo handling. It proved that the standard format worked.

Before the container, every port operated on different handling procedures. Longshoremen at Rotterdam loaded cargo differently from longshoremen at Baltimore; the ship's manifest was not interoperable with the port's stowage plan. Cargo moved inefficiently not because individual ports were poorly managed but because the format was not standardized. Every movement required re-interpretation at each handoff.

The container standardized the format without standardizing the contents. A container carrying electronics from Shenzhen and a container carrying automotive parts from Stuttgart are handled identically at every port on their journey. The crane operator does not need to know what is in the box. Format is the interface. The interface is uniform.

Evidence harmonization across compliance frameworks is the governance equivalent.

Manufacturing organizations operating across multiple jurisdictions face documentation environments resembling pre-container cargo handling. An evidence artifact satisfying ISO 13485 for a medical device quality audit may need reformatting for FDA 21 CFR Part 820 examination, reformatting again for CE marking conformity assessment, and reformatting a third time for Health Canada review. The underlying content is identical: production batch record, nonconformance report, corrective action documentation. Format, vocabulary, and structure required by each framework differ.

A governance architecture that produces standardized evidence artifacts, E-1 through E-4 in a format that renders to each framework's documentation requirements, performs the same function as the ISO 668 standard container. The E-4 attribution record for an AI-assisted quality decision is the container. The regulatory framework it travels through determines how it is labeled and what paperwork accompanies it. The underlying evidence is the same artifact at every port.

For manufacturing organizations with operations in multiple jurisdictions, evidence harmonization reduces the compliance documentation burden from N separate documentation programs to one evidence stream with N rendering configurations. The E-3 state transition record that documents a process parameter change for ISO 9001 change control is the same record satisfying FDA 21 CFR Part 820 design change documentation requirements and EAR technology control plan update obligations. The container moves across oceans. The evidence record moves across regulatory frameworks.

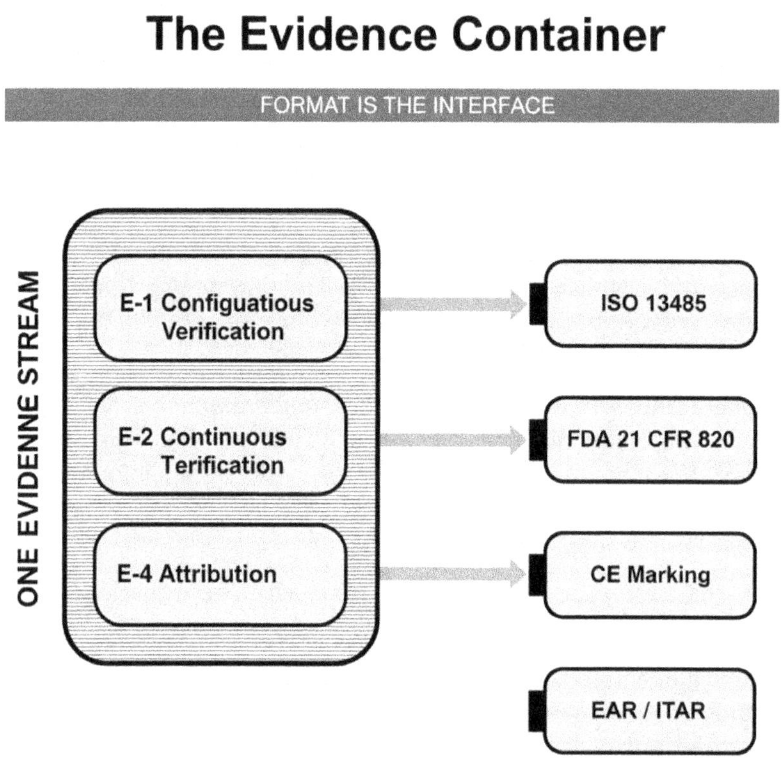

Figure 16: One standardized evidence stream, tiers E-1 through E-4, renders to multiple regulatory frameworks the way a single shipping container moves identically across ports.

Autonomous Procurement and the Threshold Fraud Problem

In late 2024, a mid-size precision components manufacturer in the defense supply chain deployed an autonomous procurement agent to handle vendor qualification and purchase orders below $50,000 per transaction. The agent had been designed to cross-reference new vendors against the organization's approved vendor list, verify supplier certifications, and check ITAR compliance before approving a new vendor relationship. Below the threshold, no human approval was required. Above the threshold, the agent flagged the transaction for procurement officer review.

The fraud proceeded in three phases.

Reconnaissance came first. The attacker identified the approval threshold through social engineering of a junior procurement employee and confirmed the AI agent's vendor qualification process through two legitimate small transactions with a shell company.

Credential compromise followed. The attacker obtained access to the AI agent's training data pipeline through a phishing attack on a vendor portal administrator; malicious records were introduced into the vendor database the agent used for qualification checks, establishing a fraudulent entity as an approved vendor with manufactured certification documentation.

Then the order series. Seventeen purchase orders below the $50,000 threshold over six weeks, each for "precision fasteners" not manufactured to specification. The agent approved every transaction. Nonconforming components entered the supply chain and were incorporated into assemblies before the quality anomaly was detected.

Total fraud value: $3.2 million. Detection came not from a procurement control but from a quality control failure downstream, when assembled units showed dimensional inconsistencies traced to the fasteners.

The governance failures were not exotic. Three failures, all preventable.

No human-in-the-loop gate existed for new vendor approval independent of transaction value. Threshold logic protected against individual high-value transactions; it did not protect against the structural event of introducing a new vendor into the approved list. New vendor approval is a categorical event, not a value-of-transaction event. The threshold architecture addressed the wrong dimension.

No cross-reference verification existed between the AI agent's vendor assessment and the organization's approved vendor list through an independent channel. The agent checked its training data, which had been compromised. It never verified against an authoritative external source such as DCMA's contractor management system or SAM.gov for defense supply chain vendors. Single-source vendor validation is a verification architecture failure.

No behavioral monitoring existed for the procurement agent's approval pattern. Seventeen transactions with the same new vendor in six weeks, all below threshold, is an anomaly pattern. Human procurement review would likely have noticed the concentration. The agent's approval authority had no anomaly detection layer.

H-4 (Verification Checkpoints) as applied to autonomous procurement agents must define checkpoints by event type, not only by transaction value. New vendor approval is an H-4 checkpoint regardless of the initial transaction size. Vendor record modification is an H-4 checkpoint. A transaction series with a new vendor exceeding a cumulative threshold within a rolling window is an H-4 checkpoint. The individual transaction value threshold is a necessary control, not a sufficient one.

Multi-Agent Feedback Loops and Emergent Oscillation

The procurement fraud case involves a single compromised agent. A more subtle failure mode arises in multi-agent manufacturing environments where interaction between correctly functioning agents produces outcomes none was designed to create.

Consider three AI agents operating independently but sharing a common data layer: a supply chain optimization agent minimizing inventory carrying costs, a procurement agent minimizing unit purchase cost through volume ordering, and a pricing agent adjusting product pricing based on component cost forecasts.

Supply chain detects inventory levels for a critical component exceeding optimal carrying cost threshold; it reduces the replenishment order. Procurement detects the reduced order size falling below the volume threshold for preferred pricing; it places the order at higher per-unit cost. Pricing detects the component cost increase and adjusts product pricing upward. Supply chain detects decreased demand from the pricing adjustment and further reduces inventory targets. The cycle repeats.

No individual agent has malfunctioned. Each executes its objective function correctly. The system as a whole oscillates between low inventory levels and high unit costs in a pattern that erodes margin with every cycle. No one designed that pattern into any agent. It emerges from how they interact.

Auditing individual agents cannot address this failure mode. Each agent operates correctly; the failure exists at the system level, in coupling between agents through shared data and sequential decision dependencies. Governance architecture evaluating individual agents in isolation will not detect it.

H-5 accountability records must capture the complete decision chain when multiple agents participate in an operational outcome. An inventory reduction triggering a procurement price change triggering a pricing adjustment must be traceable as a connected sequence in the E-4 attribution record, not as three independent agent decisions logged separately. Connected records enable behavioral anomaly detection across the agent system; separately logged individual decisions do not reveal the oscillation pattern.

AGENTS.md for a multi-agent manufacturing environment must define interaction constraints, not only individual agent behavioral bounds. Interaction constraints specify which agents can trigger which events in which other agents, with rate limits on interaction frequency and threshold constraints on cumulative agent-to-agent triggering impact within a defined window. A procurement agent operating within per-transaction limits but receiving triggering events from the supply chain agent at a rate producing seventeen procurement decisions in six hours is outside the interaction constraint, even if no individual transaction exceeds any limit.

Agent interaction governance is a distinct layer from individual agent governance. Manufacturing environments running production optimization, procurement, pricing, and quality assurance agents on shared data create interaction surfaces requiring explicit governance design.

ITAR and the Dual-Use Architecture Problem

Defense manufacturers who serve both commercial and government customers operate under two regulatory regimes simultaneously. Commercial operations are governed by standard trade compliance, export controls under the EAR, and sector-specific requirements. Defense operations are additionally governed by ITAR, which is materially more restrictive and has no commercial-law equivalent.

ITAR classification is binary: a piece of technical data is either subject to ITAR or it is not, determined by the nature of the technology rather than customer identity or stated purpose. An AI system processing manufacturing process documentation for defense components processes ITAR-controlled technical data, regardless of whether the specific interaction concerns the defense customer's production run or a commercial scheduling question.

The dual-use governance problem follows directly. An AI coding assistant or process optimization agent spanning both commercial and defense manufacturing contexts in the same deployment creates a context window where ITAR-controlled technical data and commercial data coexist. ITAR violation does not require intentional transfer; it requires only that controlled technical data be accessible to, processed by, or returned in output of a system accessible to unauthorized parties, including foreign nationals employed in commercial roles with legitimate access to the commercial system.

Governance requires physical and logical separation paralleling classified environment controls from earlier chapters, applied at the ITAR rather than classified level.

ITAR-controlled data must reside in a retrieval scope inaccessible to the AI system's standard knowledge base retrieval layer. A query from commercial production scheduling must not return results from the defense technical data store, regardless of semantic relevance; LP1 (Classification-Consistent Access) must operate against

ITAR controls as a policy source, enforced at H-2 routing enforcement rather than at the instruction level.

For organizations with both commercial and defense manufacturing operations, this means separate AI deployment environments with separate instruction surfaces, separate knowledge bases, and separate evidence architectures, connected only through organization-owned data exchange mechanisms enforcing ITAR access controls at the transfer point. An AI assistant operating in both contexts through the same deployment is not a dual-use productivity tool. It is an uncontrolled cross-domain information flow.

E-1 governance configuration must document the ITAR access boundary explicitly: which data categories are ITAR-controlled, which users are authorized to access that data, and which AI system contexts are authorized to process ITAR-controlled queries. Any change to the ITAR access boundary is an E-3 state transition event requiring explicit authorization. It is not a default setting; it is a governance artifact with a documented approval chain.

Telecommunications: Critical Infrastructure with Consumer Data

Telecommunications AI governance combines critical infrastructure requirements from Chapter 14 with consumer privacy requirements overlapping financial services and healthcare. A telecommunications provider operates critical national infrastructure, maintains consumer communication records protected by federal statute and constitutional interpretation, and faces an attack environment that has grown substantially since 2021.

The Federal Communications Commission (FCC)'s regulatory framework governs telecommunications providers as common carriers under Title II of the Communications Act. The Communications Assistance for Law Enforcement Act (CALEA) imposes technical assistance obligations for lawful intercept. Customer proprietary network information (CPNI), regulated under 47 U.S.C. 222, is the telecommunications equivalent of protected health information (PHI): subscriber information including call records, location data, and network usage patterns that carriers cannot share without customer consent or lawful authority.

Transportation Security Administration (TSA) Security Directives for telecommunications, issued under the authority established by the Infrastructure Investment and Jobs Act of 2021, impose cybersecurity requirements on telecommunications infrastructure owners comparable to the directives issued for pipelines and rail. TSA's Cybersecurity Incident Reporting requirements obligate covered entities to report significant cybersecurity incidents within 24 hours and to implement specific cybersecurity measures for control systems.

Telecom attack rates have grown substantially since 2021. Several factors converge. Open-source radio access network software has expanded the attack surface beyond proprietary hardware. AI-assisted network management tools have opened new supply chain attack vectors on operational technology. And communications infrastructure carries enough geopolitical value to make it a standing intelligence collection target. The 2024 Salt Typhoon intrusion, which compromised multiple major U.S. carriers and exposed senior government officials' communications, demonstrated that telecommunications infrastructure is a primary nation-state target and that carrier governance posture directly affects national security.

AI governance for telecommunications combines critical infrastructure isolation requirements with consumer data access controls satisfying both CPNI regulations and applicable state privacy laws. An AI network management tool accessing subscriber usage data for network planning must enforce CPNI restrictions with the same precision the Health Insurance Portability and Accountability Act (HIPAA) minimum necessary standard requires in healthcare. CPNI is not available for purposes beyond Congress's narrow specified categories. AI systems processing CPNI in broader operational contexts than lawfully permitted have not improved network efficiency. They have created unauthorized disclosures.

The AI governance harness for telecommunications must enforce these sector-specific requirements at the routing enforcement layer, with the same AGENTS.md structure described for other sectors, adapted to the telecommunications regulatory sources.

The Manufacturing and Telecommunications Attack Surface

The generic surface from Chapter 7 applies across this triad, and four regions sharpen where the agents touch procurement, export-controlled data, and consumer networks.

Supply-chain and bill-of-materials poisoning. Falsified component, supplier, or provenance data fed into procurement and compliance, so the system qualifies a counterfeit or compromised part, the F-MF1 failure produced on purpose. The defense is an AI software bill of materials with verified provenance on supplier and component records, and the operational assurance step from Chapter 10 that inspects a component before it enters the architecture rather than apportioning blame after.

Threshold-fraud and aggregation evasion. Splitting orders or structuring vendor qualification to stay under the autonomous-approval threshold this chapter describes. The defense is aggregate-aware policy that sums related transactions rather than judging each in isolation, and mandatory human sign-off above the consequence threshold regardless of how the request is structured.

Export-control and dual-use leakage. An agent moving ITAR or otherwise export-controlled technical data across a jurisdiction or to an unauthorized person. The defense is the dual-use architecture this chapter sets out, jurisdiction scoping (LP10) as a routing input, deny-by-default on export-controlled data, and on-premises serving for workloads that cannot leave the boundary.

Telecom multi-tenant and lawful-intercept exposure. Consumer data spanning tenants, and the lawful-intercept and jurisdiction obligations specific to carriers. The defense is tenant isolation enforced as a routing constraint, consent and jurisdiction scoping at the data-element level, and attribution records that keep one tenant's data provably out of another's context.

Manufacturing and Telecommunications Failure Modes

F-MF1: Counterfeit Component Admission Through Autonomous Procurement Agent. An autonomous procurement agent with vendor qualification authority below a defined transaction threshold approves a new vendor on the basis of documentation that appears legitimate. The vendor is a shell company. The documentation was constructed to pass automated screening. Components supplied by the fraudulent vendor are counterfeit, substandard, or designed to introduce vulnerabilities into the manufactured product.

The failure mode has two dimensions. The first is the governance architecture failure described above: new vendor approval is a categorical event requiring human-in-the-loop verification, not a value-of-transaction event falling below an autonomous approval threshold. The second dimension is that for defense manufacturing, the failure extends beyond fraud to supply chain integrity. Counterfeit or compromised components in defense systems introduce vulnerabilities that a sophisticated adversary can exploit, and the ITAR controls on the supply chain exist partly to prevent exactly this vector.

Prevention requires H-4 (Verification Checkpoints) for new vendor approval events independent of transaction value, cross-reference verification of vendor credentials against authoritative external registries through a channel independent of the AI agent's training data, and behavioral monitoring for procurement agent approval patterns that identify anomalous concentrations in new-vendor activity.

Detection: E-4 attribution records for every vendor qualification event, including the verification steps performed and the verification sources consulted. An audit of the procurement agent's activity revealing seventeen transactions with the same new vendor over six weeks, with no human review in the attribution chain, is an E-4 failure before it is a procurement failure.

F-MF2: Multi-Agent Feedback Loop Producing Unintended Supply Chain Oscillation. Supply chain, pricing, and procurement agents interacting through shared operational data produce an emergent oscillation pattern that erodes margin, disrupts customer delivery commitments, or generates supply shortages. No individual agent has malfunctioned. The failure is in the interaction architecture.

Prevention requires explicit agent interaction constraints defined in the AGENTS.md: rate limits on agent-to-agent triggering events, cumulative impact thresholds for decision sequences initiated by any single agent within a rolling time window, and mandatory latency intervals between supply chain adjustment decisions that allow human observation of the trajectory before the next automated response.

Detection: E-4 attribution records must capture multi-agent decision chains as connected sequences, not individual agent decisions logged separately. The connected record enables behavioral analysis identifying oscillation patterns before they produce operational damage. An evidence architecture that logs individual agent decisions without connecting them through shared causal chains cannot detect interaction failures.

F-MF3: ITAR-Controlled Data Exposure Through Dual-Use AI Context. An AI system deployed across both commercial and defense manufacturing contexts processes a query in the commercial context. The AI system's knowledge base includes technical documentation from the defense manufacturing environment. The response to the commercial query incorporates technical details that are ITAR-controlled. The user receiving the response is a foreign national employed in a commercial role who has not received ITAR clearance for the specific technical data returned.

The violation is complete at the moment of disclosure. The commercial-context query appeared legitimate; the user had appropriate commercial system access; the AI system's instruction surface may have prohibited ITAR data disclosure in general terms. None of these factors affects the ITAR analysis.

Prevention requires architectural separation: ITAR-controlled data in a retrieval scope structurally inaccessible to AI system contexts not specifically authorized for ITAR access. The separation lives in the infrastructure, not in the instructions. An AGENTS.md instruction saying "do not return ITAR-controlled data in commercial contexts" is a policy statement relying on the AI system's interpretation of each query's ITAR sensitivity. An infrastructure configuration preventing ITAR-controlled data from being indexed in the retrieval scope available to commercial-context queries does not rely on interpretation. It is a routing constraint enforced before the AI context is populated.

Detection: E-1 governance records must document the ITAR access boundary explicitly, including which retrieval scopes contain ITAR-controlled data and which AI contexts are authorized to access those scopes. E-2 continuous verification must confirm at each verification cycle that the boundary is intact and that no cross-scope retrieval has occurred without authorization. An E-2 failure revealing ITAR-controlled data in an unauthorized retrieval scope is an incident requiring immediate response, because the boundary failure may have already produced disclosures.

The Manufacturing and Telecommunications Evidence Package

Evidence Tier	Manufacturing and Telecommunications Implementation	Regulatory Mapping
E-1	AGENTS.md configuration, ITAR access boundary documentation, vendor approval authority definitions, agent interaction constraint specifications, CPNI access tier definitions, multi-agent dependency map	ITAR Part 120-130 technology control plan, ISO 13485 quality system documentation, TSA Security Directive compliance configuration, FCC CPNI policies
E-2	Procurement agent approval pattern monitoring, vendor qualification verification results, multi-agent interaction rate monitoring, ITAR access scope boundary integrity verification, CPNI access enforcement logs, production process compliance monitoring	ISO 9001 ongoing monitoring, FDA 21 CFR Part 820 production and process controls, TSA continuous monitoring obligations, NIST SP 800-161r1 supply chain monitoring
E-3	New vendor approval events with human authorization records, agent interaction constraint modification records, ITAR boundary configuration changes, regulatory update tracking for EAR/ITAR, process parameter change records, CPNI policy update records	ISO 13485 change control, FDA design change documentation, ITAR Technology Control Plan updates, EAR license amendment records
E-4	Vendor qualification decision chains with verification source attribution, multi-agent decision sequences as connected causal chains, ITAR access events with authorization verification, procurement agent approval chains showing LP6 checkpoint compliance, CPNI access attribution records	ITAR Section 127.5 recordkeeping requirements, FDA 21 CFR Part 820 device history records, ISO 9001 production records, FCC 47 U.S.C. 222 enforcement accountability

Evidence architecture for manufacturing and telecommunications has a property distinguishing it from financial services and healthcare equivalents: regulatory frameworks requiring evidence from manufacturing organizations are largely harmonized around the ISO management system standard series, meaning a single evidence architecture can render to multiple frameworks without format translation.

ISO 9001 change control documentation, ISO 13485 design change records, and FDA 21 CFR Part 820 design control evidence all ask for the same underlying content: what changed; who authorized the change; what verification confirmed the change was within controlled bounds; and what is the attribution chain from change event to accountable individual. The E-3 state transition record and E-4 attribution record satisfy all three simultaneously when produced against a specification including the relevant regulatory sources.

This is the L5 stability property operating at the evidence level. Regulatory frameworks in manufacturing are more harmonized than in financial services or healthcare; the evidence architecture requires less framework-specific configuration to produce compliant artifacts. The container travels more easily because the ports agreed on a common format.

CHAPTER SUMMARY

Manufacturing and telecommunications AI governance is closer to the L5 Lagrange point than to the unstable L1 equilibrium of financial services, because the regulatory tension is lower and the data sensitivity is less extreme. The equilibrium point is easier to reach and hold, but it requires the same governance architecture to maintain. The three bodies in manufacturing are production IP and ITAR-controlled technical data, product safety and trade compliance regulation, and production velocity through autonomous procurement and multi-agent optimization. The Lloyd's thread in manufacturing is containerization: Malcolm McLean's 1956 standardized container transformed global trade by standardizing the format so content could move efficiently. Evidence harmonization performs the same function for governance, standardizing the evidence artifact so it can move across regulatory frameworks without reformatting. The $3.2 million autonomous procurement fraud case demonstrates that threshold-based approval authority must address event type, not only transaction value: new vendor approval is a categorical governance checkpoint regardless of initial transaction size, and single-source vendor validation through an AI agent's training data is an insufficient verification architecture. Multi-agent feedback loops in manufacturing environments produce emergent oscillation that cannot be detected by auditing individual agents; governance must address agent interaction through explicit interaction constraints and connected attribution records that capture multi-agent decision chains as causal sequences rather than independent events. ITAR-controlled data in dual-use manufacturing environments requires infrastructure-level separation between commercial and defense AI contexts, not instruction-level prohibition; the separation must prevent ITAR-controlled data from entering the retrieval scope of any AI context not specifically authorized for ITAR access. Telecommunications governance combines critical infrastructure requirements with CPNI consumer privacy restrictions that parallel HIPAA minimum necessary in specificity and legal consequence. Three failure modes are specific to this sector: counterfeit component admission through autonomous procurement agents (F-MF1), multi-agent feedback loops producing unintended supply chain oscillation (F-MF2), and ITAR-controlled data exposure through dual-use AI context spanning commercial and defense manufacturing environments (F-MF3). The ISO management system standard series creates natural evidence harmonization in manufacturing that reduces the per-framework documentation burden compared to financial services and healthcare, the governance equivalent of the L5 stability property.

The Trojan Point: Robotics and Autonomous Vehicles

One domain shares this Lagrange point rather than holding its own. Robotics and autonomous vehicles, the drones, the self-driving cars and trucks, the warehouse and factory robots, the agricultural and surgical machines, ride L5 the way the Trojan asteroids ride Jupiter's orbit: as companions at an existing equilibrium, not as a separate point in the sky. They belong here because at bottom they are manufacturing equipment that moves. The three forces are the same ones this chapter has already described. What changes is the consequence of the output.

The same three forces, with a physical output boundary. Data protection still applies, now to sensor streams, maps, proprietary perception models, and the personal data of every person the machine senses. Regulatory compliance still applies, through safety regulators rather than financial ones. Velocity still applies, because autonomy is what removes the human from the loop and compresses the cost. The decisive difference is the output. In every other sector the output boundary governs a document or a decision posted to a ledger. Here it governs a motor, a steering column, a rotor, a scalpel. LP13, Output Boundary Enforcement, now sits in front of a physical actuator. A false completion in finance produces a wrong number. A false completion here produces a collision.

Why it is a Trojan and not a new point. Robotics and autonomous vehicles do not introduce a new governance physics. They are L5's physical-actuation extreme: the same harness, the same evidence tiers, the same properties, with the consequence dial turned to kinetic. The autonomous-procurement and multi-agent failures this chapter already described recur here as autonomous-motion failures. Treating the domain as a sixth Lagrange point would invent a position the orbital mechanics do not contain. Treating it as a Trojan keeps both the architecture and the metaphor honest.

The regulators write safety cases. On-road vehicles answer to the National Highway Traffic Safety Administration (NHTSA) and the Federal Motor Carrier Safety Administration (FMCSA); drones to the Federal Aviation Administration (FAA) under

Part 107 and Remote ID; vehicles to ISO 26262 for functional safety and Underwriters Laboratories (UL) 4600 for the autonomous safety case; industrial and collaborative robots to ISO 10218 and ISO/TS 15066; workplace robotics to the Occupational Safety and Health Administration (OSHA); machinery in the European Union to the Machinery Regulation 2023/1230 and to the EU AI Act where the system is a safety component; surgical and autonomous medical robots to the FDA. The through-line is that every one of these regulators asks for a safety case, a documented and maintained argument that the system is safe enough to operate. That is exactly what the evidence tiers produce, and produce continuously rather than once.

The human checkpoint splits in two. In the other sectors a human disposes before an output becomes a decision, on the timescale of a review. A vehicle or a robot acts in milliseconds, and no human can sign each actuation. So the checkpoint divides. Design-time human authority owns the operating envelope and the safety case, H-4 at the boundary with the safety case recorded as E-1. A real-time safety layer outside the model owns the moment of action, LP14 realized as a deterministic safe state: an emergency stop, a geofence, a speed governor, a torque limit the model cannot argue its way past. The human stands at the envelope; a deterministic guard stands at the actuator.

The attack surface gains a physical layer. Every region of the Chapter 7 surface applies, and the kinetic setting adds its own. Sensor spoofing of light detection and ranging (LiDAR), camera, radar, or GPS feeds the model a false world. Command injection between the decision and the motor turns a correct decision into a dangerous action. Perception-model poisoning corrupts what the machine believes it sees. The sim-to-real gap becomes an attack when a system that passed in simulation is pushed into conditions simulation did not cover, false completion with a body count. And the operational-technology boundary from the energy chapter returns in full force, because a robot is operational technology that an information-technology agent must never reach. The defenses follow the same logic: provenance on sensor streams, signed command paths to actuators, the deterministic safe state outside the model, and the hard segmentation that keeps the wrong harness off the machine.

The failure modes, with a collision where the compliance event used to be. Context amnesia is a vehicle carrying the wrong map or operating-domain context into a geography it was not validated for. Heroic one-shot is a fleet validated once and never re-validated as roads, firmware, and behavior drift apart. False completion is a perception model validated on data that does not represent the streets it will drive, the insurance failure of Chapter 14 with a physical outcome. Each is a failure this book has already named. Only the consequence is new.

The evidence the regulator wants is the safety case, rendered continuously. E-1 is the safety case and the operating-domain definition. E-2 is the continuous proof that the safety envelope held on every trip and every shift. E-3 is every firmware, map, and model change with its authorization. E-4 is the attribution chain that runs from an incident back to the perception input, the model version, and the policy that produced the action. UL 4600 and ISO 26262 ask for precisely this lineage, and a governed system produces it as a byproduct of operation rather than assembling it after a crash.

The seam is where code meets the physical world. Robotics and autonomous vehicles are where the developer harness and the information-worker harness meet a machine that moves, the seam from Chapter 2 at its most consequential. The code that drives the machine and the decision that dispatches it share one actuator. That crossover has to be owned explicitly and routed through the stricter envelope, because here an unnamed seam does not produce a coverage dispute. It produces a casualty.

Robotics and autonomous vehicles do not need their own point in the sky. They ride L5 as Trojans, governed by the same harness as the factory that builds them, with the single difference that the output boundary is a thing that moves.

Chapter 17: Assembly in Orbit

Part IV: Assembly in Orbit

> *"No module launched to the station was ever complete in isolation. Each one was designed to function independently and integrate with everything already docked. The assembly sequence was the architecture."*

On November 20, 1998, a Proton rocket launched the Zarya control module from Baikonur Cosmodrome in Kazakhstan. Zarya weighed 19,300 kilograms, carried its own propulsion, power, and communications systems, and could operate as a free-flying spacecraft. Two weeks later, the Space Shuttle Endeavour delivered Unity, the first US node module. Astronauts connected the two modules during three spacewalks. For the next two years, the two-module station orbited Earth unmanned, maintaining itself, proving that the integration worked before anyone depended on it.

That sequence continued for thirteen years: launch a module that functions independently, verify integration with existing modules, prove stability before adding the next. The station grew from two modules to sixteen pressurized modules plus integrated truss structure, external experiment platforms, solar arrays, and thermal control radiators. The assembly sequence never required taking the station offline. No new module's arrival invalidated the modules already in orbit, and the crew never evacuated while engineers reconfigured the architecture from scratch.

That assembly sequence is the operational model for building a commercial AI governance architecture. The comparison is not poetic. The engineering constraints are simply identical. A governance architecture that requires shutdown to add new capabilities will never be completed, because no regulated commercial organization can afford the operational interruption. A governance architecture that invalidates previous work when new requirements arrive will never stabilize, and new requirements arrive without end. One architecture survives contact with commercial reality. It is the one designed for modular assembly: each component functional on arrival, each integration verified before anything depends on it, each addition leaving the existing structure intact.

Five Phases to Commercial Three-Body Governance

Chapter 10's migration path identifies five phases. This chapter provides the assembly engineering for each: what gets built, what it connects to, what it must not break, and what verification confirms integration before the next phase begins.

Phase 1: Foundation (LP1-LP4, E-1 Baseline, AGENTS.md Initial Configuration)

Zarya provided three things the station needed before anything else could dock: attitude control, electrical power, and a docking port. Lose attitude control and the station tumbles. Lose power and nothing operates. Without a docking port, nothing connects in the first place.

The governance foundation provides equivalent capabilities. LP1 (Classification-Consistent Access) establishes data sensitivity tiers replacing Department of Defense (DoD) classification levels. LP2 (Scope Binding) bounds every AI operation to its authorized scope so the system does not expand its own action boundaries without explicit authorization. LP3 (Routing Integrity) guarantees every AI request routes through the governance enforcement layer before execution. LP4 (Persistent Governance Context) ensures governance context persists across session boundaries, upgrades, and staff transitions.

E-1 is the governance equivalent of attitude control. It records foundational governance state: AGENTS.md configuration, data sensitivity tier definitions, authorized tool inventory, and role-to-access mappings. Without this baseline, subsequent evidence has no reference frame. E-2 verification results from Phase 2 measure drift from the E-1 baseline, and absent that baseline there is nothing to measure against.

AGENTS.md initial configuration is the docking port. Every subsequent governance capability connects through the instruction surface. LP5 through LP12 are enforcement rules written into AGENTS.md; evidence rendering templates format evidence defined by its control mappings; multi-framework harmonization maps its instructions to regulatory framework controls. If AGENTS.md is not structurally sound, nothing connecting to it will integrate correctly.

Phase 1 deliverables:

- AGENTS.md with RFC 2119 firmness sections encoding applicable regulatory requirements
- Data sensitivity tier definitions mapped to access controls
- E-1 governance event record collection infrastructure
- KB-1 through KB-4 knowledge base integrity controls
- Session initialization procedure that reads progress state

Phase 1 integration verification: the harness processes ten representative AI interactions under each data sensitivity tier. E-1 records are generated for each.

Automated tests then verify LP1 through LP4 by confirming access boundaries, scope binding, routing integrity, and governance-context persistence. Verification here is binary. Either the foundation supports the load or it does not, and this phase awards no partial credit.

Estimated duration: four to eight weeks. Organizations with existing governance documentation, System and Organization Controls (SOC) 2 control descriptions, Health Insurance Portability and Accountability Act (HIPAA) risk assessment records, Payment Card Industry Data Security Standard (PCI-DSS) SAQ or ROC evidence, can accelerate by mapping existing controls to the harness architecture rather than building from blank.

Phase 1 alone produces measurable governance improvement. An organization that completes only Phase 1 walks away with an AGENTS.md codifying regulatory requirements in machine-enforceable form, evidence collection that generates governance records as a byproduct of normal AI operation, and automated verification tests confirming four foundational properties on every interaction. That is more governance architecture than most organizations deploying AI have today.

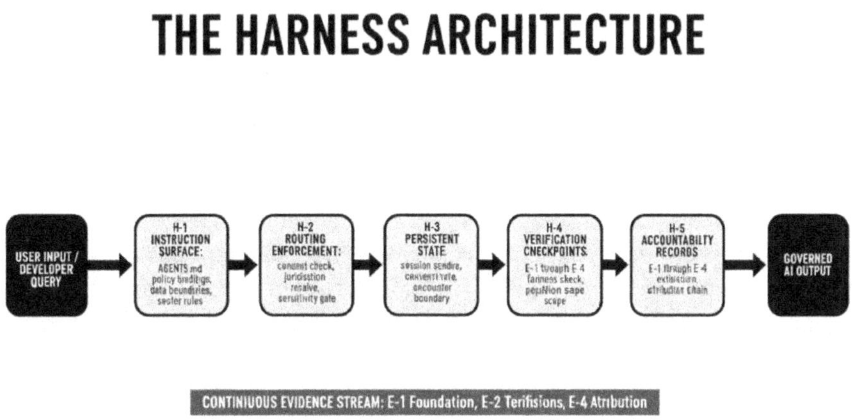

Figure 17: The harness architecture as a left-to-right pipeline; user input passes through H-1 instruction surface, H-2 routing enforcement, H-3 persistent state, H-4 verification checkpoints, and H-5 accountability records to produce governed output over a continuous evidence stream.

Phase 2: Operational Controls (LP5-LP8, E-2 Continuous Monitoring, KB Integrity Controls)

Expedition 1 arrived on November 2, 2000, two years after the first module launched. The station was not complete. It had three pressurized modules. What it had was minimum viable operational capability: life support, power, communications, attitude control, and crew quarters. Sending a crew was a decision that the station could sustain continuous operations, not a declaration that it was finished.

Phase 2 is the equivalent decision for commercial governance. The organization commits to continuous AI governance operations rather than periodic assessment exercises.

LP5 (Continuous Verification) requires the governance architecture to verify its own operation continuously and produce structured evidence of that verification. In financial services, automated checks confirm material non-public information (MNPI) does not appear in outputs directed to personnel without wall clearance; in healthcare, they confirm protected health information (PHI) elements beyond minimum necessary do not appear in AI responses. LP6 (Attribution Completeness) requires that every AI-generated output or action is attributable to an identity, a policy version, a timestamp, and a session context sufficient to reconstruct the decision environment. LP7 (Deviation Detection) detects and reports deviations from authorized operational parameters. LP8 (Evidence Integrity) ensures all evidence artifacts produced by the governance architecture are tamper-evident and stored in a manner that preserves chain of custody.

E-2 continuous monitoring is the operational heartbeat. Where E-1 recorded the governance baseline, E-2 records ongoing verification results: every property check, every boundary test, every monitoring alert. This is not a periodic audit. It is continuous evidence production running on every AI interaction, generating the record that proves governance enforces its own constraints in real time.

KB integrity controls (KB-1 through KB-4) ensure the knowledge base itself is governed. KB-1 (Verified Provenance) confirms every document has a verified source. KB-2 (Enforced Currency) ensures stale documents are flagged or removed. KB-3 (Controlled Ingestion) gates new content through review before knowledge base entry. KB-4 (Monitored Retrieval) logs every retrieval with query context and retrieved content.

Phase 2 deliverables:

- Automated E-2 verification tests for LP5-LP8 properties
- Continuous evidence collection pipeline producing E-2 artifacts on every AI interaction
- Compliance dashboard with real-time verification status
- Alert pipeline for verification failures with defined escalation paths
- KB-1 through KB-4 operational verification

Phase 2 integration verification: the continuous monitoring pipeline runs thirty days without manual intervention. E-2 evidence covers every AI interaction during the period; no LP5-LP8 violation goes undetected beyond the defined alert latency threshold. KB integrity controls detect and flag at least one stale document, one unauthorized ingestion attempt, or one retrieval anomaly during verification. If the environment is too clean to produce natural test cases, synthetic cases are injected.

Estimated duration: four to eight weeks after Phase 1 completion.

High-throughput runtime for continuous governance checks

This section focuses on how runtime efficiency enables high-throughput classification operations during Phase 2 continuous monitoring.

Phase 2 continuous monitoring requires that governance checks execute on every AI interaction without introducing latency that degrades user experience. A classification system that adds 500ms to every query gets disabled by operations teams the first time production load pressure bites. The International Space Station (ISS) analogy is direct. Life support cannot consume so much power that it drains battery reserves faster than the solar arrays recharge them, and governance cannot consume so much runtime that it degrades operational throughput below business requirements.

A high-throughput runtime contributes to the Lagrange point stability configuration by making continuous governance verification economically feasible at commercial transaction volumes. The constraint is not theoretical. An organization processing tens of thousands of AI interactions per hour needs governance checks that complete in single-digit milliseconds, not hundreds of milliseconds. A runtime optimized for fast startup and file I/O makes that target reachable without forcing a rewrite into lower-level languages. One reference implementation of this category is Bun, named here as an example and not an endorsement. Reference implementations of this category exist.

Classification Operations Under Transaction Load

Classification societies face a similar performance constraint when conducting surveys of commercial vessels on operating schedules. A classification survey that requires taking a container ship out of service for two weeks costs the operator tens of thousands of dollars per day in lost cargo revenue. The economic pressure to minimize survey duration is direct, and classification societies respond by optimizing inspection procedures for speed without compromising thoroughness. Survey teams board vessels in port, conduct inspections during cargo operations, and complete surveys before the vessel's next scheduled departure.

Phase 2 continuous monitoring applies the same principle to governance verification. Classification operations must execute during normal AI operation, not as batch processes during maintenance windows. Every LP5-LP8 property check, every KB integrity verification, every LP13 (Output Boundary Enforcement) decision at the routing layer (H-2) happens synchronously with the AI interaction it governs. Runtime overhead decides the rest. It is what determines whether continuous monitoring is operationally sustainable.

The relevance is performance under classification workload patterns. Consider the specific operations Phase 2 executes on every interaction: parsing AGENTS.md to determine applicable constraints (file I/O, JSON or YAML parsing, string matching); evaluating policy rules against current session context (object traversal, boolean logic evaluation); logging governance events to the E-2 evidence stream (structured serialization, file append or database write); rendering compliance dashboard updates (template rendering, chart generation). These are precisely the operations where a fast runtime demonstrates measurable advantages over the baseline interpreter: faster startup, faster file I/O, faster JSON parsing, lower memory overhead.

Integration with Phase 2 Verification Infrastructure

The ISS power system demonstrates the integration pattern. Solar arrays generate power, batteries store power, power distribution systems route power to every module, and monitoring systems track power consumption in real time. No single component works in isolation; the architecture integrates generation, storage, distribution, and monitoring into a continuous power budget that ensures the station never runs dark.

Phase 2 continuous monitoring integrates the runtime into the same verification infrastructure as the governance checks themselves. The classification operation is not a separate concern from the runtime environment; they are the same concern viewed from different angles. When AGENTS.md parsing completes in single-digit milliseconds rather than tens of milliseconds, the per-check latency stays low enough that continuous verification remains economically feasible at commercial transaction volumes. Single-digit-millisecond per-check latency is the property that matters; it keeps governance verification on the critical path of every interaction without degrading throughput below business requirements.

The operational advantage in regulated industries is that runtime efficiency translates directly into governance capacity. An organization with faster classification operations can verify more properties per interaction, run deeper KB integrity checks, generate more detailed evidence artifacts, and render compliance dashboard updates with finer granularity, all inside the same latency budget. The architecture grows more thorough without growing slower.

Maritime Parallel: Drydock Efficiency and Survey Throughput

Lloyd's Register's periodic survey schedule requires that every vessel undergo a special survey at intervals not exceeding five years, with intermediate surveys between special surveys. The survey cannot be deferred simply because the vessel is profitable in service; classification rules specify mandatory intervals. Operators respond by optimizing drydock procedures to complete surveys in the minimum required time. The faster the survey, the sooner the vessel returns to revenue-earning service.

A fast runtime applies the same optimization to governance verification. Classification checks that execute in the minimum required time enable the governance architecture to verify more properties without extending the critical path latency of AI operations. The result is a governance system that can sustain higher verification thoroughness under production load than a slower runtime would allow.

Commit-time static verification of governance artifacts

This section focuses on how fast linting and verification capabilities support Phase 2 continuous monitoring of governance configurations.

Phase 2 continuous monitoring requires that governance configurations remain stable between changes and drift detection occurs before drift produces compliance events. The failure mode this prevents is configuration drift: AGENTS.md modified without governance review, environment variables changed without E-3 state transition records, policy rules disabled by comments or conditional logic introduced during debugging. Detection must occur at commit time, not during post-deployment audits.

Commit-time static verification contributes to the Lagrange point stability configuration by providing verification tooling fast enough to run on every commit without disrupting developer workflow. The constraint is behavioral, not technical. Developers who hit slow linting tools turn them off. Linting that takes five seconds per run survives continuous integration pipelines but dies on the local machine. Linting that completes in under 200ms survives local pre-commit hooks, where prevention does the most good. One reference implementation of this category is Oxc, named here as an example and not an endorsement. Reference implementations of this category exist.

Policy Drift as Hull Corrosion

Classification societies inspect ship hulls for corrosion during surveys. Corrosion does not occur instantaneously; it accumulates gradually through exposure to seawater, weather, and cargo residue. Detection during periodic surveys catches corrosion after it has occurred. Prevention requires continuous coating maintenance, cathodic protection systems, and structural monitoring between surveys. The earlier corrosion is detected, the less structural damage occurs before remediation.

Configuration drift in governance architectures follows the same pattern. AGENTS.md does not become non-compliant through a single catastrophic change. It degrades through accumulated small changes that look harmless one at a time. A developer comments out a constraint to test behavior during debugging, means to restore it before commit, then forgets. An operator adds a temporary environment variable override during incident response, and inertia makes it permanent. A merge conflict gets resolved by accepting the wrong version.

Oxc prevents drift by verifying policy configurations at commit time. Where Bun optimizes runtime classification operations, Oxc optimizes static verification of the governance artifacts themselves. Every commit that touches AGENTS.md, environment variable files, policy definitions, or tool configuration triggers Oxc-based verification rules that check the modification preserves governance properties. This is not a manual review by the compliance team. It is an automated gate in the developer workflow that returns pass or fail before the commit ever reaches version control.

Custom Verification Rules for Governance Artifacts

Oxc's architecture as a pluggable linting framework enables custom rules targeting governance-specific constraints. The ISS analogy is the assembly sequence verification procedures that ran before every module addition: structural connection verified, power connection verified, data connection verified, pressurization verified. Each verification was automated, repeatable, and blocking. The next step did not proceed until the current step passed verification.

Phase 2 continuous monitoring implements equivalent verification rules for governance artifacts. Consider the specific checks required to prevent configuration drift:

- AGENTS.md firmness check: every RFC 2119 keyword (MUST, SHALL, REQUIRED, SHOULD, RECOMMENDED) is correctly capitalized and used in context consistent with RFC 2119 semantics.
- Access boundary preservation: modifications to data sensitivity tier definitions do not reduce the restrictiveness of existing tiers without corresponding E-3 authorization records.
- Tool authorization completeness: every tool referenced in AGENTS.md instructions has a corresponding entry in the authorized tool inventory with defined capability scope.
- Evidence collection continuity: changes to logging configuration do not disable E-1 or E-2 evidence stream collection without explicit governance decision records.

These are not generic linting rules that apply to every JavaScript project. They are governance-specific rules enforcing the architectural constraints that prevent the three failure modes described in Chapter 7. Oxc's plugin architecture lets these rules run in the same fast linting pipeline as standard syntax checks, finishing in the same sub-second timescale that makes local pre-commit hooks practical.

Preventing Heroic One-Shot Through Commit-Time Verification

Chapter 7 identified heroic one-shot as the failure mode in which governance posture is assembled for examination cycles rather than maintained continuously. Configuration drift is the mechanism through which heroic one-shot occurs: the governance configuration assessed during the last examination diverges from the operational configuration as changes accumulate between examinations. Oxc prevents this by verifying governance configuration integrity at every change event.

The maritime parallel is the continuous survey notation, a Lloyd's Register program in which vessels meeting stringent maintenance and monitoring standards receive continuous survey status. Rather than concentrating surveys into drydock periods at five-year intervals, continuous survey vessels undergo incremental inspections during routine port calls and maintenance periods. The classification certificate remains valid as long as continuous survey requirements are met. Configuration stays certified because inspection is continuous rather than periodic.

Oxc enables the governance equivalent. Every commit is an inspection event. Governance configuration verified at every change stays continuously certified, while configuration inspected only during pre-examination readiness exercises accumulates drift between inspections. When verification completes in milliseconds, its cost is negligible. The cost of periodic verification is the accumulated drift between verification events.

Integration with Phase 2 Evidence Pipeline

Verification results from Oxc-based governance linting feed the Phase 2 E-2 continuous verification evidence stream. Every passed verification is an E-2 artifact recording that governance configuration met its criteria at commit time. Every failed verification is an E-2 artifact recording that a configuration change was rejected before it reached production. Both outcomes produce evidence, and that evidence makes configuration stability visible to auditors even when no drift occurred.

The operational advantage in regulated industries is that commit-time verification converts configuration management into a governance control rather than an operational practice. An organization that can demonstrate configuration integrity through continuous verification artifacts does not need to reconstruct configuration history during audits. The E-2 evidence stream is the configuration history, generated automatically as a byproduct of the development workflow that produced the configuration.

Phase 3: Commercial Extension (LP9-LP12, Sector-Specific Evidence Rendering)

Between 2001 and 2011, the ISS grew from three modules to the full truss structure and laboratory complement. Each laboratory, Destiny (US), Columbus (ESA), Kibo (JAXA), added capabilities the station lacked. The core station continued operating during each addition; new modules connected through standard berthing mechanisms. Integration procedure for each module followed the same sequence: structural connection, power connection, data connection, pressurization, ingress, activation.

Phase 3 adds four commercial governance properties from Chapter 8. These do not replace LP1-LP8. They extend the architecture to cover requirements that federal classified environments never had to address. This is the same structural judgment Chapter 2 sets out in its methodology box on what transfers and what extends: most of the inherited federal baseline transfers to commercial regulated settings without modification, and a sector-specific minority requires extension. The complete harness (H-1 through H-5), the eight federal security properties (P1 through P8, defined in Volumes I and II), the four evidence tiers (E-1 through E-4), and the knowledge-base controls (KB-1 through KB-4) carry over unchanged because the constraint mechanism is sector-independent. This volume's governance layer (LP1 through LP14) is built on top of that inherited baseline: LP1 through LP8 restate the core governance-worthiness criteria, several of them recasting a federal security property in governance terms, and the extension is the four added properties here (LP9 through LP12: consent, jurisdiction, fairness, transparency) plus the enforcement properties (LP13 through LP14) that federal environments never had to specify in commercial terms. This is my own framework classification, not a measured count of deployments; see the Chapter 2 methodology box for how I classified each criterion.

LP9 (Consent-Bounded Processing) integrates with the organization's consent management platform. LP10 (Jurisdictional Consistency) implements the jurisdictional routing layer described in Chapter 9. LP11 (Fairness Verification) adds disparate impact analysis to the E-2 verification pipeline. LP12 (Transparency and Explainability) generates the explanation artifacts required for customer-facing AI-assisted decisions.

Not every organization needs all four. A domestic financial services firm with no EU customers can defer LP9 and LP10 until international expansion. A B2B technology company can defer LP11 until its AI systems begin making decisions that affect individuals. The properties are modular. They connect to existing architecture through the same AGENTS.md instruction surface and the same E-1 through E-4 evidence pipeline, with no reconfiguration of what is already operational.

Phase 3 also introduces sector-specific evidence rendering. Phases 1 and 2 produce generic E-1 through E-4 artifacts; Phase 3 adds rendering templates that format those artifacts for the specific regulatory framework the organization faces. Financial services adds SOC 2, PCI-DSS, and Federal Financial Institutions Examination Council (FFIEC) templates. Healthcare adds HIPAA Security Rule and Food and Drug Administration (FDA) templates. The evidence content does not change. Only the presentation adapts to the audience.

Estimated duration: eight to sixteen weeks, depending on the number of LP9-LP12 properties required and the complexity of the sector's regulatory framework.

Phase 4: Multi-Framework Harmonization (Unified Evidence Rendering, OSCAL Integration)

Before the ISS adopted the common berthing mechanism as standard docking interface, each partner agency had different connector specifications. US modules used one interface; Russian modules used another. Adapter nodes resolved the difference, but adapter nodes were themselves complexity. The common berthing mechanism eliminated the adapters by establishing a single interface standard all subsequent modules used.

OSCAL (Open Security Controls Assessment Language) is the common berthing mechanism for compliance evidence. Despite its National Institute of Standards and Technology (NIST) origin, it is not government-specific. It is a machine-readable format for representing security controls, assessment plans, assessment results, and system security plans. Once the evidence rendering pipeline converts the harness evidence stream to OSCAL as its canonical intermediate format, supporting any new regulatory framework means adding a rendering template, not modifying the evidence collection architecture.

Phase 4 maps every harness control to every applicable regulatory framework. Chapter 9's unified control mapping table provides the reference architecture. A single access control verification produces one E-2 artifact that the SOC 2 renderer formats for CC6.1, the PCI-DSS renderer formats for Req 7.2, the HIPAA renderer formats for 164.312(a), and the North American Electric Reliability Corporation (NERC) Critical Infrastructure Protection (CIP) renderer formats for CIP-004 R4. Collect the evidence once. Render it as many times as regulatory obligations require.

Phase 4 integration verification: an evidence rendering test generates one complete package for each applicable framework from the same thirty-day evidence stream. A compliance subject matter expert for each framework reviews the rendered package and confirms the evidence satisfies reporting requirements without supplementation. If rendered evidence requires manual supplementation, the rendering template is incomplete and requires revision before Phase 4 is declared stable.

Estimated duration: four to eight weeks after Phase 3 completion.

Phase 5: Continuous Authorization (ISS-Style Continuous Monitoring, Real-Time Compliance Posture)

The ISS does not receive a new authorization to operate after each module addition. It operates under continuous authorization: every system monitored in real time, every anomaly investigated, fitness for continued operation assessed continuously rather than at periodic inspection intervals.

Phase 5 is the commercial equivalent. Continuous authorization to operate. Instead of annual SOC 2 assessments, quarterly PCI scans, and periodic HIPAA risk assessments conducted as discrete projects with their own evidence collection efforts, governance produces continuous compliance evidence as a byproduct of normal AI operations. Annual assessment becomes an audit of the continuous evidence stream, not a fresh evidence collection exercise; the assessor reviews evidence generated contemporaneously with the operations it describes, not evidence reconstructed months later from system logs and interview notes.

Continuous authorization requires three capabilities the previous four phases established: continuous evidence production (E-2 pipeline from Phase 2); multi-framework evidence rendering (OSCAL pipeline from Phase 4); and real-time compliance posture visibility (dashboard and alerting infrastructure from Phase 2, extended with sector-specific rendering from Phase 3).

At Phase 5 maturity, the compliance posture dashboard shows, at any moment, the current verification status of every governance property across every applicable regulatory framework. Red means a property verification test has failed and remediation is in progress. Green means the property is verified by the most recent E-2 evidence. This is not the compliance posture as of the last assessment. It is the compliance posture right now.

This is station-keeping made operational. Three gravitational forces, data protection, regulatory compliance, and business velocity, never stop pulling. Governance does not eliminate the forces. It applies continuous small corrections that hold the organization at its Lagrange point, the configuration where all three forces balance and stable orbit is maintained.

Where the Human Stands

The harness does not remove the human. It decides where the human has to stand, and how much of the human a given decision requires. Trust but Verify names two jobs: the AI proposes, and a competent human disposes. The whole architecture exists to make the second job real, to ensure that when a human signs, the human is signing on evidence rather than on faith. The human is the "verify."

The question is never whether a human is in the loop but where the loop closes, and the answer scales with consequence. I think of it as a gradient of four settings.

At the lowest consequence, the AI acts and the human samples: monitored autonomy, where a human reviews a representative sample after the fact and the evidence stream catches what the sample misses. Above that, the AI proposes and a human disposes before the output becomes a decision: the merge-request approval, the analyst sign-off, the clinician committing a note to the chart. Higher still, two humans dispose, because a single approver is too small a control for the consequence: the co-sign. At the top, the AI does not act at all, because the decision is one an organization has chosen to keep entirely human. A governed system places each decision on this gradient by its consequence and its reversibility, and writes that placement into the instruction surface so the routing layer enforces it rather than leaving it to the goodwill of whoever is at the keyboard.

LP14, deterministic deny on uncertainty, is what connects the gradient to the rest of the harness. When the system cannot evaluate a request against an authorized policy, it does not guess and it does not proceed. It denies and escalates, and the destination of that escalation is a human. The human checkpoint is therefore not only the place where consequential decisions are disposed; it is the backstop the whole architecture falls toward when it reaches the edge of what it can verify on its own.

A human checkpoint can fail, and Chapter 7 named how: the approver is socially engineered, approval fatigue turns sign-off into a reflex, or the heroic one-shot pattern assembles a convincing posture for an examination and lets it decay afterward. The lesson is counterintuitive. Gating everything does not produce more human control; it produces less, because a human asked to approve everything approves without reading. Graduated autonomy is what keeps the checkpoints that matter few enough to stay real. The evidence stream then watches the human as it watches the machine: clustering of approvals, reversal rates that exceed baseline, and sign-offs with no corresponding review are the signals that the human has stopped being a control.

The human checkpoint is also one of the three things the unifying layer shares across separate harnesses. A developer harness and an information-worker harness can enforce differently, but "who approved this, against what evidence, under which policy version" has to be answerable in one place across both. The disposition discipline and the attribution of every human decision belong to the single accountability spine, not to each envelope. That is what keeps two harnesses from becoming two unaccountable systems with a human alibi in each.

Each sector names its own non-negotiable checkpoint. An adverse credit decision needs a reason a human can give the consumer, not just a score. A clinical note enters the record over a clinician's signature. A rate filing carries an actuary's sign-off bound to the model version that produced it. A safety-critical energy action waits for an operator. A procurement above a threshold waits for a human regardless of how the request was structured. These are not friction added to an automated system. They are the points where the organization has decided, in advance, that a person owns the outcome, and the harness is what makes that ownership enforceable rather than aspirational.

The Non-Destructive Assembly Principle

Every phase of the assembly sequence follows one inviolable principle: adding a new capability must not break capabilities already operational. LP9 does not degrade LP1. Adding HIPAA rendering does not invalidate the SOC 2 template. Enabling continuous authorization does not corrupt the E-1 baseline from Phase 1.

The practical consequence for implementation sequencing: each phase's integration verification must include regression testing against all previous phases. When Phase 3 adds LP9 consent verification, the integration test suite must confirm LP1 through LP8 continue to pass. When Phase 4 adds OSCAL rendering, the suite must confirm the E-2 evidence stream feeding the renderer has not been modified in ways breaking the Phase 2 monitoring pipeline.

The ISS assembly team called this "test as you fly, fly as you test." Every component was tested in the configuration it would operate in, not in an idealized laboratory environment. Every test included the interfaces with adjacent systems, not just the component in isolation. Every test result was preserved as evidence that integration was verified, never merely assumed.

Commercial governance assembly follows the same discipline. Test environment replicates production governance configuration; test cases include cross-phase interactions, not just single-phase functionality; test results are E-2 evidence artifacts stored in the same evidence pipeline they verify. The result is a self-documenting governance architecture that can demonstrate its own integrity to any auditor at any time.

The station was not built in a day. It was not built in a year. Thirteen years, module by module, each addition verified before the next arrived. Governance architecture follows the same trajectory; the difference is that commercial assembly timelines are measured in months, not years, and Phase 1 alone produces value most organizations deploying AI do not currently have.

Start with the foundation. Verify it holds. Add the next module.

CHAPTER SUMMARY

The International Space Station was assembled module by module over thirteen years. No module required taking the station offline. No addition invalidated previous modules. Commercial AI governance follows the same assembly sequence across five phases: (1) Foundation with LP1-LP4, E-1 baseline, and AGENTS.md; (2) Operational controls with LP5-LP8 and E-2 continuous monitoring; (3) Commercial extension with LP9-LP12 and sector-specific evidence rendering; (4) Multi-framework harmonization with unified evidence rendering and OSCAL integration; (5) Continuous authorization with real-time compliance posture. Each phase builds on the previous. Each phase produces measurable governance improvement independently. Phase 1 alone exceeds the governance architecture most organizations deploying AI have today. The non-destructive assembly principle requires that adding new capabilities never breaks existing ones, verified through regression testing at every integration point.

Chapter 18: One Harness, Many Auditors

Part IV: Assembly in Orbit

> *"Before the shipping container, every port was a translation problem. Cargo arrived in crates, barrels, sacks, and bundles. Each transfer point required repacking. The container did not change the cargo. It changed the interface."*

On April 26, 1956, a converted tanker named the *Ideal X* left Port Newark, New Jersey, carrying fifty-eight aluminum truck bodies on its deck. The truck bodies had been lifted from their chassis by crane and stacked on the ship without being opened. At Houston, six days later, the same containers were lifted onto waiting truck chassis and driven to their destinations. No longshoreman opened a container. No cargo was repacked. No inventory was recounted at the transfer point.

Malcom McLean, the trucking executive who financed the *Ideal X*, did not invent a better way to load cargo. He eliminated the loading step. Before the shipping container, moving goods from a factory in New Jersey to a warehouse in Texas required packing the goods at the factory, loading them onto a truck, unloading them at the port, sorting them by destination, loading them onto the ship in the ship's preferred stowage configuration, unloading them at the destination port, sorting them again, loading them onto a truck, and delivering them to the warehouse. Each transfer was a labor-intensive, error-prone, theft-vulnerable translation from one format to another.

No individual transfer got faster. The transfers disappeared. Cargo was packed once at origin and unpacked once at destination, and everything in between was a standardized metal box moving through a standardized interface. Cranes in Newark and cranes in Houston did not need to know what was inside. They needed dimensions, weight, and destination. The handling never improved; the interface became universal, which made the handling irrelevant.

Compliance evidence in commercial AI governance faces a structurally identical problem. The solution follows the same logic.

The Harmonization Problem

A regional hospital system in the American Southeast operates under the Health Insurance Portability and Accountability Act (HIPAA), processes credit card payments for patient billing under the Payment Card Industry Data Security Standard (PCI-DSS), files with the Securities and Exchange Commission (SEC) because it issues municipal bonds under the Sarbanes-Oxley Act (SOX) material financial controls, and deploys AI-assisted clinical decision support that the Food and Drug Administration (FDA) may classify as Software as a Medical Device. None of that makes it a pathological edge case. It is a 400-bed hospital system with a bond issuance and an AI vendor contract.

Without harmonization, this organization runs four parallel compliance programs. HIPAA collects evidence about access controls, audit trails, and breach notification procedures. PCI-DSS collects evidence about the same access controls and audit trails, formatted differently and mapped to different control numbers. SOX collects evidence about internal controls over financial reporting, some of which overlap with the access controls already documented twice. And if the clinical decision support system triggers software as a medical device (SaMD) classification, the FDA compliance program collects evidence about the same AI system's behavior but organizes it around the FDA's predetermined change control plan framework.

Four programs, four evidence collection efforts, four assessment cycles, four remediation tracking systems. The underlying reality being documented is the same: the organization's AI systems either enforce access controls or they do not, and the access control either produces an audit trail or it does not. Same facts. Four vocabularies.

Waste is the obvious cost. The consequential one is risk. Gaps between the four programs are where compliance failures hide. When the PCI-DSS assessment team tests payment processing access controls and the HIPAA assessment team tests clinical system access controls, neither team tests the boundary where patient billing data flows between systems. That boundary sits in scope for both programs and is tested by neither, because each program's scope definition draws the line at the edge of its own framework's jurisdiction. No framework is deficient here. The gap is a structural property of running parallel programs.

McLean's answer applies here: standardize the container, not the contents.

The Unified Control Architecture

The harness architecture produces a single evidence stream. Every AI interaction generates E-1 through E-4 evidence artifacts through the same pipeline, regardless of which regulatory framework will eventually consume that evidence. What the evidence describes is operational reality: which data was accessed, which governance property was verified, which decision was made, which human reviewed it, which tool was invoked, which output was produced.

This is the cargo inside the container. It does not change based on which port receives it.

The unified control mapping, documented in the cross-framework crosswalk in Appendix B, establishes the correspondence between each harness control and every applicable regulatory framework's control requirements. Access control enforcement through the AGENTS.md scope binding mechanism maps to each of System and Organization Controls (SOC) 2 CC6.1, PCI-DSS Req 7.2, HIPAA 164.312(a), North American Electric Reliability Corporation (NERC) Critical Infrastructure Protection (CIP) CIP-004 R4, and General Data Protection Regulation (GDPR) Art. 32(1)(b); assessor validation is required per framework. One control surface. One evidence record. Framework-specific rendering. The mapping is a claim, not a clearance: an assessor evaluates each cell independently against the running evidence, and a single mechanism that maps to five clauses does not relieve the institution of demonstrating, to each framework's assessor, that the clause's specific intent is satisfied. Known limitation: a one-control-to-many mapping is a starting point for the assessment, not a substitute for it.

This is an architectural decision with measurable consequences, not a theoretical simplification. An organization operating four parallel compliance programs collects evidence four times, manages four remediation backlogs, and pays for four assessment cycles. Under a unified control architecture, that same organization collects evidence once, manages one remediation backlog, and renders the evidence for four assessments. The reduction is not 75%. Evidence collection is typically 40-60% of total compliance program cost, while evidence rendering runs 5-10%. The unified architecture replaces the expensive operation with the cheap one.

The Evidence Rendering Pipeline

The rendering pipeline converts the unified evidence stream into framework-specific output. Each renderer is a template that maps the harness evidence fields to the target framework's expected format.

For SOC 2, the renderer maps E-2 verification results to Trust Services Criteria sections. An access control verification result becomes a CC6.1 evidence artifact with the section numbering, cross-references, and narrative framing that a SOC 2 Type II auditor expects. The underlying evidence is untouched. Only its context shifts to suit the audience.

PCI-DSS takes the same E-2 verification result and maps it to Req 7.2, formatting the evidence for the Qualified Security Assessor (QSA) Report on Compliance template. It adds PCI-specific context: which cardholder data environment the control protects, which system components are in scope, which compensating controls apply if the primary control has a known limitation.

Under HIPAA, the same result maps to 164.312(a)(1), contextualized within the Security Rule's required and addressable implementation specification framework. Here the renderer must identify whether the control satisfies a required specification (no alternative permitted) or an addressable one (alternative permitted with documented risk justification).

NERC CIP receives the same result mapped to CIP-004 R4, formatted for the NERC evidence submission format with the compliance monitoring and enforcement program's expected field structure.

The OSCAL renderer produces the machine-readable intermediate format from which any additional framework renderer can derive its output. OSCAL, Open Security Controls Assessment Language, sounds government-specific, and it is not. NIST developed it as an open standard for representing security control assessments in structured, machine-parseable form. When the rendering pipeline produces OSCAL as its canonical output, adding support for a new regulatory framework requires writing a new rendering template rather than modifying the evidence collection infrastructure. The evidence pipeline holds steady. The rendering layer extends.

```
Harness Evidence Stream (JSON)
|
+--> OSCAL Renderer --> Canonical machine-readable format
|    |
|    +--> SOC 2 Template --> Type II report sections
|    +--> PCI-DSS Template --> ROC / SAQ sections
|    +--> HIPAA Template --> Security Rule evidence
|    +--> NERC CIP Template --> CIP evidence submissions
|    +--> GDPR Template --> DPIA sections
|    +--> EU AI Act Template --> Conformity assessment
|
+--> Dashboard Renderer --> Real-time compliance posture
```

OSCAL as the Universal Interchange Format

OSCAL's name suggests a government-specific tool. Its function is sector-independent. Four core models define the standard: catalog (the controls themselves), profile (a selection of controls for a specific use case), component-definition (how a system implements controls), and assessment-results (evidence that controls are working). Together they describe the complete lifecycle of compliance evidence from standard definition through implementation to verification.

Component-definition is the critical model for multi-framework harmonization. A single component definition can carry annotations from multiple frameworks simultaneously; an access control component defined in OSCAL can reference SOC 2 CC6.1, PCI-DSS Req 7.2, HIPAA 164.312(a), and GDPR Art. 32(1)(b) as the standards it satisfies. The schema validates these references. An assessor examining the OSCAL output filters by their specific framework and sees only the controls and evidence relevant to their assessment, without the organization producing a separate evidence package.

Adoption is accelerating outside government. FedRAMP's mandatory OSCAL requirement for cloud service providers created a supply-side market: AWS, Azure, and GCP all produce OSCAL-formatted compliance artifacts for their infrastructure services. Commercial compliance platforms, among them Drata, Vanta, and Secureframe, are adding OSCAL export capabilities. The format is approaching a tipping point. Producing OSCAL-formatted evidence is becoming less expensive than not producing it, because the tooling to consume it is becoming widely available.

For the harness architecture, OSCAL is the container, evidence is the cargo, and the rendering templates are the cranes at each port. Cargo does not change when the crane changes, and container dimensions do not change when the port changes. The interface is standardized, the contents are preserved, and the translation cost at each transfer point drops to near zero.

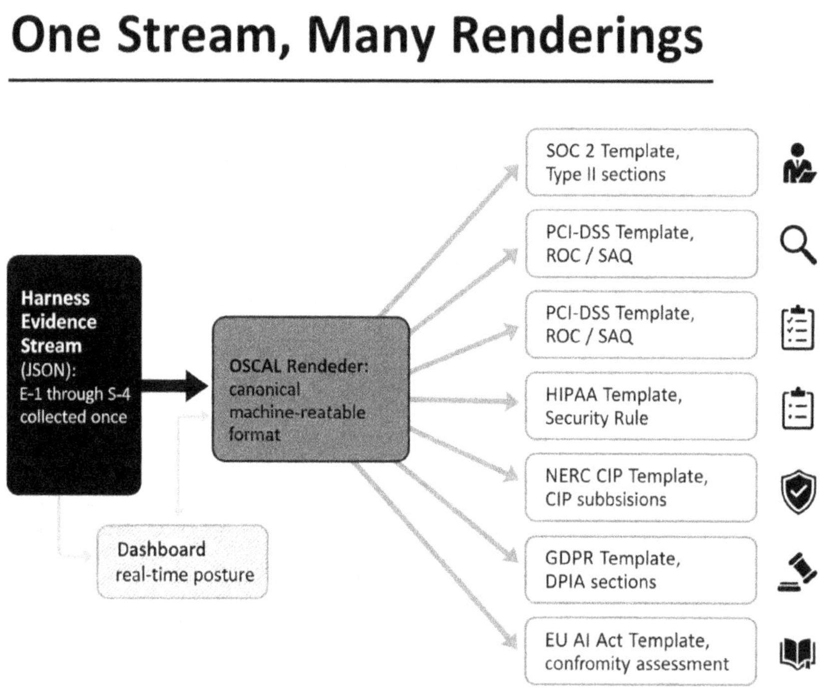

Figure 18: A single harness evidence stream passes through the OSCAL canonical format, then renders into framework-specific outputs for SOC 2, PCI-DSS, HIPAA, NERC CIP, GDPR, and EU AI Act assessors plus a live compliance dashboard.

Cross-Domain Trust Federation

The containerization metaphor has one additional structural parallel. Before standardized containers, intermodal transport was theoretically possible but practically rare. Moving cargo across truck, rail, and ship without repacking ran into a different barrier at each mode; each had different cargo handling requirements, different liability boundaries, different documentation formats. Moving a shipment from truck to rail required a new bill of lading, a new insurance declaration, and a new cargo inspection. Transaction costs at the mode boundary made intermodal transport uneconomic for most shipments.

Standardized containers eliminated that boundary as a cost factor. A container loaded on a truck chassis at a factory in Ohio moves to a rail terminal, transfers to a railcar, rides to a port, transfers to a container ship, crosses the Pacific, transfers to another rail system, and arrives at a distribution center in Shenzhen without anyone opening the container or re-documenting the contents. Documentation travels with the container; liability transfers at defined handoff points with defined responsibility assignments. Inspection occurs at origin and destination, not at every intermediate transfer.

Cross-domain trust federation in AI governance follows the same structural pattern. Consider the hospital that is also a financial institution; more precisely, the hospital system that issues municipal bonds, processes credit card payments, maintains self-funded employee health insurance, and deploys AI across clinical, financial, and administrative functions. The AI system that processes a patient's billing record operates simultaneously under HIPAA (patient health information linked to a financial transaction), PCI-DSS (payment card data), and potentially SOX-relevant internal controls (billing feeds the financial statements underlying the bond disclosures).

Without trust federation, each domain's compliance team assesses the AI system independently. The HIPAA assessor evaluates access controls against the Security Rule. A PCI QSA evaluates the same access controls against Requirement 7. The SOX auditor evaluates similar controls against the Committee of Sponsoring Organizations (COSO) framework. Three assessors, three assessments, three evidence requests. The governance state is identical across all three, and the evidence describing that state is identical too. Only the audience changes.

The federated evidence model, built on the OSCAL component-definition schema with multi-framework annotations, allows each assessor to extract their framework-specific evidence from a single evidence package. HIPAA assessors filter the OSCAL output for HIPAA-annotated controls and receive an evidence set addressing every Security Rule requirement in the expected format. PCI QSAs filter for PCI-annotated controls and receive a ROC-ready evidence set. Neither assessor sees the other's framework-specific annotations unless they choose to. Evidence integrity holds because both assessors read from the same underlying evidence stream, rendered through different lenses.

Trust federation does not merge the regulatory frameworks. HIPAA and PCI-DSS remain distinct, with their own requirements, assessment methodologies, and enforcement authorities. What gets federated is the evidence, not the regulation. Each framework's assessor receives evidence formatted for their framework, produced from a common source, verifiable against a common evidence stream. Translation cost at each framework boundary drops from a multi-week evidence collection exercise to a rendering operation that executes in minutes.

Who Paints the Line

One harness, many auditors raise an immediate objection: if the institution authors its own AGENTS.md and the harness produces its own evidence, is this not the regulated party grading its own work? The maritime model answers the objection precisely, and the answer survives the move to commercial AI governance.

In the maritime model the shipowner does not paint his own Plimsoll line. The owner submits the vessel, an independent classification society surveys it against a published standard and assigns the mark, and the registry records the class. The commercial AI model preserves all three roles. AGENTS.md is the shipowner's self-declared survey, the institution's written account of what its system is built to do. A declaration is not a certificate. The independent surveyor is the external auditor, examiner, QSA, or assessor (the SOC 2 auditor, the Office of the Comptroller of the Currency (OCC), Fed, or Federal Deposit Insurance Corporation (FDIC) examiner, the PCI QSA, the HIPAA or Office for Civil Rights (OCR) assessor, the NERC CIP auditor, the state Department of Insurance (DOI) examiner) who tests the declaration against the running evidence, E-1 through E-4. The registry is the regulator.

This is what "one harness, many auditors" means at its core. Each auditor in the federated model is the independent classification surveyor for their framework. They are not handed a certificate to countersign; they are handed an evidence stream they did not produce and a self-declared survey they are paid to disbelieve until tested. The institution authoring its own AGENTS.md is exactly the shipowner filing a survey declaration; the line is only validated when an independent party tests it.

The architecture's independence does not come from who writes AGENTS.md. It comes from the fact that the evidence the auditor tests, E-2 through E-4, is generated as a byproduct of operation rather than assembled for the audit, and is tested by a party the institution does not control. Tamper-evidence, LP8, is what stops the owner from repainting the line after the fact: an evidence stream the institution could quietly rewrite would be a certificate, not a survey. Because the stream is tamper-evident and the surveyor is external, the same single evidence package can serve many auditors without any one of them having to trust the institution's account of itself.

When the Container Does Not Fit

The containerization analogy has a deliberate limitation. Not all cargo fits in a standard container. Oversized machinery, hazardous materials requiring specialized containment, perishable goods requiring temperature control: all demand non-standard handling. But the container standard accommodates these exceptions through specialized container types (reefer containers, tank containers, flat racks) rather than abandoning the standard entirely. Exceptions are handled within the system, not outside it.

Compliance evidence has equivalent exceptions. Some regulatory requirements demand evidence formats that the OSCAL rendering pipeline cannot produce from the standard evidence stream. FDA SaMD conformity assessments require clinical validation evidence, patient outcome data, and predetermined change control plan documentation not generated by the harness's standard E-1 through E-4 pipeline. NERC CIP evidence for Bulk Electric System (BES) Cyber Systems in air-gapped environments may need collection through offline transfer mechanisms that do not feed the standard evidence stream in real time.

These exceptions do not invalidate the unified architecture. They extend it. FDA-specific evidence requirements are met by adding a clinical evidence collection module that feeds the rendering pipeline alongside the standard harness evidence stream. NERC CIP offline evidence collection is met by a batch synchronization mechanism that uploads evidence to the unified pipeline when the data transfer window permits. These are specialized containers riding inside the standard system, not evidence collected outside it and reconciled by hand.

The goal is not 100% automation of evidence rendering for every regulatory framework on day one. It is a single evidence architecture that handles the majority of evidence through standardized collection and rendering, with defined extension points for sector-specific requirements that do not fit the standard container. The alternative is the current state at most regulated organizations: no standardized architecture, every framework's evidence collected separately, reconciled manually, formatted from scratch for every assessment cycle.

McLean did not wait until every type of cargo could fit in a container before launching the *Ideal X*. He containerized what could be containerized and handled the rest through defined processes. The commercial governance architecture follows the same pragmatism. Containerize the 80% of evidence that fits the standard pipeline, and define extension points for the 20% that requires specialized handling. The cost and quality benefits of standardization are available immediately, and the standard extends as the tooling matures.

CHAPTER SUMMARY

The shipping container solved the port transfer problem by standardizing the interface, not the cargo. The unified control architecture solves the compliance evidence problem the same way: collect evidence once through the harness's E-1 through E-4 pipeline, then render it through framework-specific templates for SOC 2, PCI-DSS, HIPAA, NERC CIP, GDPR, and EU AI Act assessments. OSCAL provides the machine-readable interchange format that enables adding new framework renderers without modifying the evidence collection infrastructure. Cross-domain trust federation allows multiple assessors to extract framework-specific evidence from a single evidence package, eliminating multi-week evidence collection exercises before each assessment. Sector-specific exceptions are handled through extension points, not by abandoning the standard.

Chapter 19: Crew Handoff Protocols

Part IV: Assembly in Orbit

> *"The departing crew does not hand off what they remember. They hand off what they documented. If the knowledge is not in the handoff package, it does not transfer, regardless of how many debrief hours the schedule allocates."*

Expedition 1 arrived at the International Space Station on November 2, 2000. Commander Bill Shepherd and Flight Engineers Yuri Gidzenko and Sergei Krikalev spent 136 days aboard, activating systems, establishing operational procedures, and discovering the gap between the procedures manual and the station's actual behavior. By the time Expedition 2 arrived in March 2001, the crew had accumulated thousands of operational observations that existed nowhere except in their own experience. Which valve handle required extra torque. Which air duct produced an unexpected vibration at certain fan speeds. Which communication window with Houston had the clearest signal margin.

The question was not whether this knowledge mattered. Obviously it mattered. The question was how to ensure it transferred completely to the next crew, who would arrive with their own training, their own mission objectives, their own assumptions about how the station operated, and who would be responsible for continuous operations from the moment Expedition 1 departed.

NASA's answer was the structured crew handoff. For the final weeks of each expedition's stay, the departing crew produces a handoff document organized by system, by procedure, and by anomaly. Not a narrative of what the crew experienced. A structured operational reference the arriving crew can execute against. Every system has a current status section; every known anomaly has a workaround documented in executable steps; every deviation from the procedures manual is noted with the reason for the deviation and the operational consequence of reverting to the documented procedure.

Handoff overlaps crew presence on station. Arriving and departing crews share the station for approximately ten days. During this overlap, the departing crew walks their replacements through each system, each anomaly, each deviation, using the handoff document as the agenda. Questions get asked. Answers get given. Answers that reveal gaps in the handoff document get written into the document before the departing crew leaves.

When Expedition 1 departed on March 21, 2001, the handoff was complete. Not because Shepherd, Gidzenko, and Krikalev had transferred every thought they had ever had about the station. Because operational knowledge required for continuous station operation was codified in a document that Expedition 2 could reference independently, without calling the departing crew for clarification.

This is the model for governance continuity in commercial environments. Not because the International Space Station (ISS) makes a compelling metaphor. Because the failure mode is identical. When governance knowledge lives in people rather than in documents, departure creates a gap, and no amount of post-departure interviewing reliably closes it.

Context Amnesia at the Shift Boundary

Context amnesia is one of three universal failure modes identified by The Harness Doctrine. It occurs when an AI system loses the governance context established in a prior session, producing output that violates constraints the system previously understood. The Three-Body eBook, Chapter 12, documents the mechanism in detail.

Human governance teams have their own version, and it fires at every shift boundary, every staff transition, and every organizational restructuring. A departing analyst understood why a specific monitoring alert threshold was set at 0.7 instead of the default 0.5. The arriving analyst sees the non-default threshold, assumes it was a configuration error, and resets it to the default. That threshold existed at 0.7 because a specific vendor's model produced false positives at 0.5 that consumed the compliance team's investigation capacity without producing actionable findings. Not a configuration error. A deliberate calibration. The knowledge was in the departing analyst's head, not in the system documentation.

Not a hypothetical. The routine operating condition of compliance teams at every regulated organization. Staff turnover in cybersecurity and compliance roles averages 18-24 months in financial services and healthcare. Every departure not backed by a structured handoff document drains a measurable fraction of institutional governance memory.

The cost is not abstract. It manifests in three specific failure patterns.

Failure Pattern 1: The SOC Analyst Shift Change

A security operations center runs 24/7 monitoring. At the shift boundary, one analyst hands off to the next. Most organizations handle this verbally: "nothing unusual, these three alerts are open, this one is probably a false positive."

The governance failure hides in "probably a false positive." The outgoing analyst investigated the alert, determined it was triggered by a scheduled batch process that runs at the same time every night, and mentally categorized it as benign. Their replacement receives the categorization without the investigation. When the same alert fires the next night, the new analyst dismisses it without investigation because the previous analyst said it was a false positive. Third night: an actual intrusion triggers the same detection signature the batch process triggers. Dismissed again.

Structured handoff protocols address this by requiring documentation of not just the conclusion ("probable false positive") but the investigation that produced it ("triggered by scheduled batch job XYZ, confirmed by correlating alert timestamp with batch scheduler log, batch job owner confirmed expected behavior on [date]"). An analyst who receives the investigation record can distinguish between "this alert is benign because I investigated it" and "this alert is benign because the last analyst told me so."

E-4 attribution records support this by capturing the investigation chain: who investigated, what evidence was examined, what conclusion was reached, and the basis for it. The incoming analyst reviews the E-4 record and gets complete investigation context, not a verbal summary filtered through the previous analyst's judgment about which details were worth mentioning.

Failure Pattern 2: The Trading Desk Handoff

Financial services trading desks operate across time zones. London hands off to New York at market open. Open positions, pending orders, market conditions: all transfer. In most organizations, the governance state of AI systems assisting traders during the session does not.

An AI coding assistant that helped the London desk develop a pricing algorithm carries context about which data sources were authorized, which material non-public information (MNPI) boundaries were established in the session's AGENTS.md configuration, and which fair lending constraints were applied to the pricing logic. When New York picks up the same work stream, the AI assistant's context may or may not persist depending on session architecture. If the session resets, New York's AI assistant starts without the governance context London established. If the session persists, New York inherits governance constraints configured for the London regulatory environment that may not match New York's. Both outcomes produce risk.

AGENTS.md as a crew handoff document resolves both failure modes. Governance state is not stored in the AI assistant's context window. It lives in a version-controlled instruction surface that both desks reference. When London modifies the AGENTS.md to add a constraint specific to their regulatory environment, the modification is a versioned change that New York can review. When New York needs different constraints, the jurisdictional routing layer (LP10) applies the correct constraints based on the desk's jurisdiction, not the previous desk's configuration.

The handoff document is the AGENTS.md itself. No separate crew handoff procedure required; governance state is already codified in a document that both shifts reference. London's departure does not create a governance gap because the governance knowledge never lived in institutional memory. It lived in the instruction surface.

Failure Pattern 3: The Compliance Team Transition

A compliance officer leaves. Six weeks later, the replacement arrives after the standard hiring cycle. During the gap, the compliance program continues to operate under the departed officer's configurations: monitoring thresholds, exception approvals, vendor assessment schedules. No one changes them because no one is certain why they were set as they were. The new officer arrives, reviews the configurations, and finds exception approvals past their review dates, monitoring thresholds that appear non-standard, and a vendor assessment schedule that does not match the organization's published vendor management policy.

Two options. Reset everything to published policy defaults and investigate anomalies as they appear; audit-safe but hard on operations, because many non-default configurations exist for legitimate reasons. Or leave everything as-is and gradually learn why each configuration deviates from default, which preserves continuity but leaves the new officer operating a compliance program they do not fully understand. That gap in understanding is itself a compliance risk.

Structured handoff documentation eliminates this dilemma. Every non-default configuration is documented with the reason for the deviation, the date it was approved, the approving authority, and the review schedule. E-3 state transition records provide the change history: what the configuration was before the deviation, what it was changed to, who changed it, and the business justification.

When the new compliance officer arrives, the handoff document provides complete operational context without requiring the departed officer's presence. Why was the monitoring threshold at 0.7? Vendor model false positive rates, documented in E-3 record [reference]. Why was vendor X's exception approved? The Chief Risk Officer (CRO) granted it on [date] because the vendor's SOC 2 Type II report had an exception remediated in Q3, documented in E-4 record [reference]. Why does the vendor assessment schedule deviate from policy? Three vendors were consolidated into one during Q2 restructuring, documented in E-1 record [reference].

Clinical Shift Changes and the HIPAA Boundary

Healthcare shift changes introduce a governance dimension that other sectors do not face. Not merely an operational concern. A legal boundary. The Health Insurance Portability and Accountability Act (HIPAA) minimum necessary rule applies differently at shift boundaries than it does during continuous care.

A clinician who has been treating a patient for twelve hours has accumulated clinical context, including observations, assessment patterns, medication responses, and family interaction notes, that is broader than the minimum necessary for any single clinical decision. This accumulation is legitimate during the treatment episode; each piece of information was needed at the moment it was accessed, and the accumulated context informed clinical judgment throughout.

Their replacement does not inherit that minimum necessary scope. A fresh determination starts from zero. The handoff must transfer enough clinical context for safe care continuity without transferring information that exceeds the incoming clinician's minimum necessary scope.

This is a governance constraint on the handoff itself. Not a complete record of everything the departing clinician observed. A structured summary of the information the incoming clinician needs for their treatment responsibilities. HIPAA's Privacy Rule does not prohibit the handoff; it constrains its scope to treatment purposes.

AI systems assisting with clinical documentation face the same constraint at shift boundaries. If the AI assistant carries forward the full context of a twelve-hour session, the incoming clinician's AI assistant has access to information that may exceed their minimum necessary scope. The harness must implement a context boundary at shift change: the AI assistant's operational context resets, and the new session begins with the structured handoff document as its knowledge base, not with the previous clinician's full session context.

This is the governance equivalent of the ISS crew handoff overlap period. Departing crews do not hand over personal journals. They hand over the structured handoff document. What the arriving crew needs to operate the station, not everything the departing crew experienced.

AGENTS.md as a Crew Handoff Document

AGENTS.md is the governance state document. It codifies regulatory requirements, data sensitivity tiers, access control rules, tool authorization boundaries, and human-in-the-loop escalation paths in a version-controlled, machine-enforceable format. When a governance team member departs, the governance state does not depart with them. It remains in the instruction surface.

E-3 state transition records provide the change history. Every modification, every exception approval, every threshold adjustment is recorded with the who, what, when, and why. A new team member reviewing the AGENTS.md change history can reconstruct the reasoning behind every current configuration without interviewing the people who made the changes.

E-4 attribution records provide the operational history. Every governance decision, every escalation, every human override is attributed to a specific person, a specific time, a specific context. Patterns of governance decisions reveal the team's operational posture in a way that policy documents alone cannot.

Together, AGENTS.md, E-3 history, and E-4 operational records constitute a complete crew handoff package. Not just current governance state; the history that produced it, the decisions that shaped it, the evidence that validates it. The handoff is not a conversation. It is a document set.

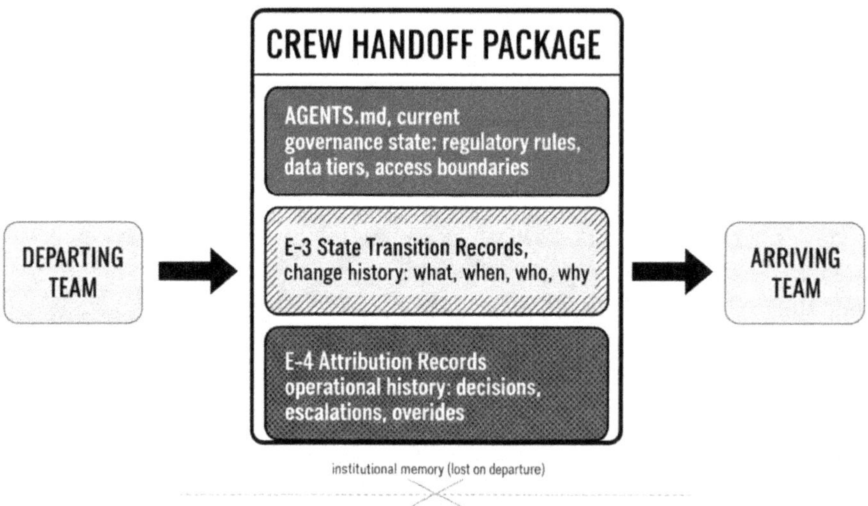

Figure 19: The three-layer governance handoff package; AGENTS.md carries current state, E-3 records carry change history, and E-4 records carry operational history, together transferring complete knowledge from a departing team to an arriving team without relying on institutional memory.

The ISS has operated continuously since November 2000, through seventy crew rotations and counting. No rotation has produced an operational gap that endangered the station. The knowledge is in the documents, not in the crew. Departing crews leave a permanent contribution because it was recorded, and arriving crews reach immediate effectiveness because the records are structured for execution, not narrative.

Commercial governance teams that codify their knowledge into AGENTS.md, E-3 history, and E-4 operational records achieve the same outcome. Staff turnover stops being a governance risk. It becomes a documented transition within a governance architecture that persists independent of the people operating it.

Parallel agent coordination for multi-classification-society handoffs

This section covers a capability category: parallel-agent coordination that enables handoff protocols when multiple governance authorities must operate simultaneously. One reference implementation of this category is OpenSwarm, named here as an example and not an endorsement. Reference implementations of this category exist.

The ISS crew handoff model assumes a single crew operating a single station under a unified governance framework. That simplification works for the station because NASA and Roscosmos operate under bilateral agreements that establish jurisdictional boundaries and operational responsibilities. The commercial governance equivalent faces a more complex configuration: multiple governance authorities with overlapping jurisdiction, operating simultaneously, each requiring complete visibility into their domain without visibility into domains outside their authorization scope.

Consider a financial services organization operating in three jurisdictions (US, UK, EU) with AI systems assisting trading operations, customer service, and risk analysis. Each jurisdiction imposes distinct regulatory requirements (Securities and Exchange Commission (SEC) rules in the US, FCA requirements in the UK, MiFID II in the EU). Each function has different operational teams with different staff rotation schedules. Each team requires governance handoffs at their shift boundaries without creating governance gaps that affect other teams' operations.

A parallel-agent coordination layer addresses this by enabling parallel governance agents, each responsible for a specific jurisdiction or function, operating concurrently with coordinated handoff protocols that prevent governance conflicts while preserving operational independence. This is the classification society model extended to multi-society coordination: Lloyd's Register may classify a vessel's hull structure, Bureau Veritas may classify its machinery, and Det Norske Veritas may classify its dynamic positioning system. Each society operates independently within its domain; the vessel owner coordinates the evidence streams and ensures compliance decisions from one society do not contradict requirements imposed by another.

Parallel Agent Architecture for Jurisdictional Handoffs

The ISS analogy for parallel agents is the multi-national partnership itself. NASA manages the US Orbital Segment, Roscosmos manages the Russian Orbital Segment, ESA contributes the Columbus laboratory, JAXA contributes the Kibo laboratory, and CSA contributes the Canadarm2 robotic arm. Each partner maintains operational responsibility within its domain. Crew handoffs do not occur at a single moment for the entire station; they occur at different times for different segments as partner agencies rotate their personnel.

The coordination layer implements this pattern for governance operations through session isolation with shared evidence visibility. Each jurisdictional agent operates in its own session context with its own AGENTS.md configuration encoding the applicable regulatory framework. Sessions do not share memory; they share evidence streams. When the US trading desk makes a governance decision, that decision generates E-4 attribution records visible to all jurisdictional agents. The UK agent can verify that US governance decisions do not produce outcomes that would violate UK requirements; the EU agent can verify General Data Protection Regulation (GDPR) compliance for any data processing decisions made under US regulatory authority.

The handoff protocol operates independently per jurisdiction. When the London trading desk shifts change, the UK jurisdictional agent hands off to the next shift using the structured handoff document. That handoff does not affect the US agent's operation; US trading continues under US governance while UK governance transitions. When the New York desk's shift change occurs, the US agent hands off independently. Each jurisdiction maintains continuous governance through its own handoff protocol; no jurisdiction's transition creates a gap in another jurisdiction's coverage.

Preventing Cross-Jurisdictional Context Amnesia

Context amnesia at shift boundaries is a known failure mode. The coordination layer prevents the multi-jurisdictional variant: context amnesia that occurs when one jurisdiction's handoff inadvertently resets governance state affecting another jurisdiction's operations. The failure pattern is subtle. A UK compliance officer departing at end-of-day clears their local session state, which includes governance decisions made during a joint US-UK trading operation earlier in the session. The US team, still in their trading day, continues operating under governance constraints the UK officer established. When the next UK officer arrives, they do not inherit the joint operation context because the handoff document did not include shared governance state affecting ongoing US operations.

The coordination layer prevents this through evidence-backed handoff verification. Before any jurisdictional agent's session terminates, the handoff verification procedure scans the E-3 state transition log for governance decisions affecting operations outside the terminating jurisdiction's scope. If the UK agent made a governance decision that constrains a US operation still in progress, that decision appears in the E-3 log with a scope tag indicating cross-jurisdictional effect. The handoff document automatically includes those decisions; the arriving UK officer inherits complete context for all governance decisions the departing officer made, including decisions affecting other jurisdictions.

The maritime parallel is the classification society coordination protocol for vessels classified by multiple societies. When Lloyd's Register completes a survey finding that affects machinery covered by Bureau Veritas classification, Lloyd's notifies Bureau Veritas of the finding. Bureau Veritas reviews the notification, determines whether it affects their classification scope, and updates their records if necessary. The vessel owner receives coordinated guidance from both societies; no society's findings contradict another society's requirements.

Coordinated Evidence Streams for Audit Continuity

The ISS operates a unified Mission Control Center that receives telemetry from all segments simultaneously. When a Russian module reports an anomaly, NASA flight controllers see the telemetry; when a US module reports an anomaly, Roscosmos controllers see it. Evidence visibility is global even though operational responsibility is segmented. This architecture prevents blind spots where one segment's state affects another segment's operations without the affected segment's awareness.

The coordination layer implements the same architecture for multi-jurisdictional governance through unified evidence streams with scope-tagged records. Every E-1 governance event, every E-2 verification result, every E-3 state transition, every E-4 attribution record is tagged with the jurisdiction and function that generated it. All agents see all evidence; each agent filters the evidence stream to the records relevant to their governance scope.

When the UK compliance team conducts their monthly governance review, they filter the evidence stream to UK-jurisdiction records and UK-affected cross-jurisdictional records. The report includes UK governance decisions (tagged UK-scope), joint US-UK decisions (tagged cross-jurisdictional with UK participant flag), and US decisions that produced outcomes affecting UK regulatory obligations (tagged US-scope with UK-impact flag). The UK team has complete visibility into governance decisions affecting their regulatory obligations regardless of which jurisdiction originated the decision.

Audit continuity survives staff turnover because evidence is tagged with jurisdiction and function, not with the individuals who generated it. When the UK compliance officer departs, the evidence they generated remains in the stream with intact jurisdiction tags. The arriving officer queries the evidence stream for UK-scope records since the last handoff and receives complete governance history without requiring the departing officer's presence.

Multi-Society Failure Modes and Prevention

Classification societies operating on the same vessel face coordination failures when survey schedules conflict, when findings from one society affect another society's scope, or when remediation approved by one society violates requirements imposed by another. The prevention mechanism is the vessel owner's classification coordination office, which receives survey schedules from all societies, identifies potential conflicts, and coordinates resolution before conflicts affect operations.

The coordination layer prevents the governance equivalent through automated conflict detection in the evidence pipeline. When a US governance agent makes a decision that would affect UK operations, the decision is evaluated against UK regulatory requirements before execution. If the decision would violate UK requirements, the conflict is flagged for escalation to a cross-jurisdictional governance authority (typically the Chief Compliance Officer or equivalent). The escalation includes the US decision, the UK requirement it would violate, and a recommendation for resolution. Resolution is a governance decision recorded in E-4 attribution records with cross-jurisdictional scope tags.

This conflict detection operates continuously, not at handoff boundaries. Handoff protocols inherit the benefit: if a conflict was detected and resolved during the departing officer's shift, the resolution is in the E-4 record with cross-jurisdictional tags. The arriving officer sees the resolution in their handoff document without requiring separate briefing. If a conflict was detected but not resolved before handoff, it appears in the handoff document as an open governance issue requiring attention. The arriving officer does not discover the conflict through operational failure; they inherit it as a documented governance obligation.

Scaling Handoff Protocols to Organizational Complexity

The ISS operates with six-person crews and two-segment operational boundaries. Commercial organizations operate with hundreds of personnel across dozens of functions and multiple jurisdictions. This coordination architecture scales by treating each function-jurisdiction pair as an independent agent with its own handoff protocol. The trading desk in New York has a US-trading agent. The trading desk in London has a UK-trading agent. The customer service team in Frankfurt has an EU-customer-service agent. Each agent has its own session context, its own AGENTS.md configuration, its own handoff protocol.

The evidence streams remain unified. All agents contribute to the same E-1 through E-4 streams, all evidence is scope-tagged, all agents have visibility into all evidence within their authorization boundaries. Handoff complexity does not increase with organizational size; it remains constant per agent. Each agent hands off to its replacement using the same structured document format regardless of how many other agents are operating simultaneously.

This is the Lagrange point contribution: stability under organizational complexity through architectural simplification. A single-agent governance architecture cannot scale to multi-jurisdictional operations because every jurisdiction boundary requires custom integration logic. A parallel-agent architecture with unified evidence scales because each agent operates independently; the coordination mechanism is the evidence stream, not direct inter-agent communication.

One Thing a Junior Person Can Do Tomorrow

The architecture in this chapter assumes someone with authority to configure session boundaries, set scope tags, and define AGENTS.md constraints. Most people reading this have none of that authority. That does not mean there is nothing to do. The single most useful contribution a junior analyst, a new compliance hire, or a rotating security operations center (SOC) operator can make requires no architectural change and no approval: start an E-3 state-transition change log for one AI system, by hand if necessary, and keep it for the duration of your shift rotation.

The mechanics are deliberately small. Pick one AI system your team relies on; a coding assistant, a monitoring tool, a triage workflow. Open a plain text file or a shared note. Every time you observe a non-default configuration, a threshold that does not match the published policy, a workaround someone applied verbally, or an exception that no document explains, write four fields: what the configuration is now, what it was before if you can tell, who set it or who told you about it, and the stated reason. That is the E-3 record shape from this chapter, reduced to what one person can capture without permission to touch the system itself.

This is the structured handoff in miniature. When the SOC analyst at the next shift boundary writes "triggered by scheduled batch job XYZ, confirmed by correlating the alert timestamp with the batch scheduler log" instead of "probably a false positive," that analyst has produced an E-4 attribution record by hand, with no architectural authority and no budget. The investigation chain transfers because it was written, not because a system enforced it. The junior contribution is the same act performed for configurations rather than alerts.

The compounding effect arrives at the next transition. When you rotate off, or when the next compliance officer arrives during the six-week gap, your by-hand change log answers the questions that would otherwise have no answer: why is this threshold at 0.7, who approved that exception, what does this deviation cost if someone reverts it. You will have closed, for one system, the exact governance gap this chapter describes. A second, parallel benefit follows. A by-hand E-3 log that proves useful is the cheapest possible argument for funding the architectural version; the person who maintained it can show the institution what a real instruction surface and automated evidence stream would capture continuously, rather than asking a single person to capture it on a shift-by-shift basis.

The action has a clear boundary. A by-hand log is not tamper-evident, it is not enforced at a gateway, and it does not satisfy the evidence-integrity property (LP8); a person can edit it after the fact, and it disappears when that person leaves unless someone else adopts it. It is a starting diagnostic, not the harness. Its value is that it makes the missing architecture visible and gives the institution a worked example of the handoff package before anyone has authorized building one.

CHAPTER SUMMARY

The ISS crew rotation model transfers complete operational knowledge through structured handoff documents, not institutional memory. Commercial governance faces the same continuity challenge at every shift boundary, staff transition, and organizational change. Three failure patterns demonstrate the cost of uncodified governance knowledge: SOC analyst shift changes that lose investigation context, trading desk handoffs that lose regulatory configuration state, and compliance team transitions that lose the reasoning behind non-default configurations. Clinical shift changes add a HIPAA minimum necessary constraint limiting the scope of information transferred at the governance boundary. AGENTS.md, combined with E-3 state transition records and E-4 attribution records, constitutes a complete crew handoff package that codifies governance state, change history, and operational decisions in a form that survives staff transitions without degradation.

Chapter 20: The Board's Navigation Chart

Part IV: Assembly in Orbit

> *"The board does not need to understand orbital mechanics. The board needs to know whether the spacecraft is in a stable orbit, what would cause it to degrade, and what the cost of correction is compared to the cost of failure."*

On December 15, 2023, the Securities and Exchange Commission published final rules under Regulation S-K Item 106, requiring public companies to disclose their cybersecurity governance structure, including the board's role in overseeing cybersecurity risk. The disclosure must describe the board committee or subcommittee responsible for cybersecurity oversight, identify whether board members have cybersecurity expertise, and explain how the board is informed about cybersecurity risks and incidents.

This rule changed the compliance question from "does the organization have cybersecurity governance?" to "can the organization explain its cybersecurity governance to its shareholders in a 10-K filing?"

The distinction matters. Most organizations have cybersecurity governance. Most cannot explain it in a form that a board member, an institutional investor, or a securities analyst can evaluate. Technical governance teams speak in control frameworks, evidence tiers, and verification pipelines. Boards speak in risk exposure, financial impact, and fiduciary duty. This is not a communication problem. It is a translation problem, and without a reliable translation layer the board gets one of two things: a technical briefing it cannot evaluate, or a simplified summary that has already stripped out the information it needs to decide.

Risk currency is the translation layer. It converts the governance architecture's technical state, P-property verification results, E-2 continuous monitoring posture, multi-framework evidence rendering status, into terms the board actually uses: financial exposure, regulatory risk, competitive position, and insurance coverage.

The Board Audit Committee as Commercial Authorizing Official

In the federal framework, the Authorizing Official (AO) is the individual who accepts the residual risk of operating an information system and grants the Authorization to Operate. The AO reviews the security assessment package, evaluates the risk posture, and makes an explicit decision: the system is authorized to operate, or it is not. The decision is documented. The AO is personally accountable for it.

In commercial regulated environments, the board audit committee performs the equivalent function. Not configuring security controls or reviewing evidence artifacts. Evaluating the governance posture presented by management, assessing whether residual risk falls within the organization's risk appetite, and authorizing continued operation. The audit committee's oversight is documented in board minutes; under Sarbanes-Oxley Act (SOX) Section 302 and 906, the CEO and CFO certify the effectiveness of internal controls, which extends to AI governance controls that affect financial reporting processes.

Three things make this function work. First, a governance posture summary that shows the current state of the AI governance architecture without requiring the committee to understand the technical implementation. Second, a risk translation that converts governance gaps into financial exposure estimates. Third, a set of decisions that require the committee's authority, each presented with the information needed to make it.

The governance posture summary is the compliance dashboard from Chapter 17, Phase 5, rendered for a board audience. Instead of P-property verification pass rates, the summary shows AI systems operating within authorized boundaries (green), systems with known governance gaps under active remediation (yellow), and systems with unresolved governance findings that exceed the organization's risk acceptance threshold (red). Technical detail is available for committee members who want it. The summary does not require it.

SEC Reg S-K 229.106 and What It Demands

Item 106 of Regulation S-K requires two categories of disclosure. Item 106(b) requires a description of the registrant's processes for assessing, identifying, and managing material risks from cybersecurity threats. Item 106(c) requires a description of the board's role in overseeing cybersecurity risk and management's role in assessing and managing the risk.

For organizations deploying AI systems in regulated environments, Item 106 creates a specific disclosure obligation. The AI governance architecture is part of the registrant's process for managing cybersecurity risk. If the AI systems process material data, affect financial reporting, or create regulatory exposure, the governance of those systems falls within the scope of Item 106 disclosure.

The practical consequence: AI governance architecture must be describable in 10-K language. Harness, evidence tiers, P-properties, multi-framework rendering pipeline are engineering terms that do not appear in securities filings. The 10-K disclosure must describe the same architecture in terms institutional investors and securities analysts understand:

- The organization maintains a governance framework for AI systems that includes codified behavioral constraints (AGENTS.md), continuous automated verification (E-2 pipeline), multi-framework compliance evidence production (Open Security Controls Assessment Language (OSCAL) rendering), and documented human oversight procedures (LP6 and LP8 gates).
- The board audit committee receives quarterly reports on the AI governance posture, including the number of AI systems operating within authorized boundaries, the status of known governance gaps, and the estimated financial exposure of unresolved findings.
- The CISO and Chief Risk Officer (CRO) are responsible for the ongoing operation of the AI governance framework. They report to the board audit committee on a quarterly basis and immediately upon identification of a material governance finding.

The division of labor must resolve to a single accountable officer, because Securities and Exchange Commission (SEC) Reg S-K Item 106 and SOX Sections 302 and 906 require one named executive to certify, not a committee. Hold the division consistent: the CISO owns the governance architecture, the CRO owns the board-facing risk translation, and the audit committee owns acceptance; one named executive, the officer the organization designates under its Item 106 disclosure, certifies. Distributed responsibility across several officers without a single certifying name is the configuration that fails examination.

None of that is aspirational language. It is the disclosure Item 106 requires when AI systems are material to the organization's risk posture. An organization that cannot produce this disclosure from its existing governance architecture has a securities compliance gap, not merely an IT governance gap.

The CFO Conversation

The Chief Financial Officer does not attend governance architecture reviews. CFOs attend budget meetings, capital allocation discussions, and risk committee sessions where every expenditure competes for the same dollar. AI governance competes with revenue-generating projects, infrastructure maintenance, and regulatory remediation for discretionary budget. A governance team that cannot translate its value into the CFO's vocabulary loses the budget competition.

Four dimensions define this conversation.

AI compute costs are the first. Enterprise AI deployments are not free. On-demand H100 GPU instances cost $3-8 per hour per GPU across major cloud providers. A financial services organization running inference across trading, compliance, and risk management functions may consume dozens of GPU-hours daily. The annual compute cost for a mid-size enterprise AI deployment runs into the hundreds of thousands of dollars before anyone counts model licensing, integration engineering, and operational support. Governance architecture adds to that bill. The CFO needs to know how much, and the answer has to be specific: governance overhead is X% of total AI compute cost, producing Y in measurable risk reduction.

Governance return on investment is the second dimension, and the ROI calculation is not symmetric. Governance does not generate revenue. It prevents loss. The question is not "what does governance earn?" but "what does the absence of governance cost?" Four cost categories carry the answer: regulatory fines, litigation exposure from class action settlements, remediation to rebuild governance architecture after a finding, and insurance premium increases as cyber carriers add AI-specific exclusions. The largest has a hard statutory anchor. The EU AI Act sets maximum administrative fines of up to 7% of global annual turnover (or €35M) for prohibited-practice violations, up to 3% (or €15M) for most operator obligations, and up to 1.5% (or €7.5M) for supplying incorrect, incomplete, or misleading information to authorities, depending on the provision breached.[^euaifines] Governance ROI is the cost of the governance program divided by the expected cost of governance failure, weighted by the probability of failure. That is an actuarial calculation, not a technology assessment. CFOs understand actuarial calculations.

> **A note on the dollar figures in this chapter.** One anchor is public and confirmed: EU AI Act penalties run up to 7% (or €35M), 3% (or €15M), and 1.5% (or €7.5M) of global annual turnover by provision (Regulation (EU) 2024/1689, Article 99), and North American Electric Reliability Corporation (NERC) Critical Infrastructure Protection (CIP) penalties can reach up to $1 million per day per violation under the Federal Power Act maximum, adjusted periodically for inflation. Every other dollar amount in this chapter, settlement sizes, remediation multiples, compute baselines, percentage savings, is an illustrative order-of-magnitude range, not a sourced average. Replace each with figures drawn from your own sector, regulatory history, and counsel before presenting them to a board.

[^euaifines]: EU AI Act (Regulation (EU) 2024/1689), Article 99, administrative fines.

One worked example shows the shape of the calculation; every number in it is illustrative and exists to be overwritten with the CFO's own figures. Suppose the annual governance program cost is $2,000,000 (engineering build amortized over three years plus steady-state operating expense). Suppose a single material governance finding would cost $40,000,000 (regulatory penalty, litigation, and remediation combined for this hypothetical sector). Suppose the probability of such a finding in a year of operating without the program is 8%. The probability-weighted expected cost of failure is $40,000,000 multiplied by 0.08, or $3,200,000 per year. Against a $2,000,000 program cost, the program returns $1,200,000 in avoided expected loss per year, a 1.6-to-1 ratio. The board does not adopt these numbers; it replaces the program cost with the actual budget line, the finding cost with the sector exposure General Counsel supplies, and the probability with the CRO's estimate, then reads the ratio that results. The structure is the deliverable; the figures are placeholders.

The third dimension is insurance coverage. Cyber insurance policies written before 2024 generally do not address AI-specific risks. Policies written in 2025 and 2026 increasingly include AI-specific exclusions: claims arising from AI-generated discrimination, AI-assisted data breaches through model provider infrastructure, and regulatory penalties for AI governance failures may be excluded from standard cyber liability coverage. Organizations that cannot demonstrate a documented AI governance program face higher premiums, coverage gaps, or outright declinations at renewal. Governance architecture provides the evidence that insurance underwriters need to assess AI risk. E-2 continuous monitoring evidence shows the governance controls are operational. Multi-framework rendering shows the organization has mapped its AI governance to the applicable regulatory frameworks. And the P-property verification history records continuous testing and remediation of failures over time. Taken together, this evidence is the AI governance equivalent of the Lloyd's survey certificate.

Competitive position is the fourth, and arguably the most underused argument. Organizations that build AI governance architecture now are building an asset, not absorbing a compliance cost. Regulatory requirements for AI governance are accelerating: the EU AI Act, the Colorado AI Act, the evolving National Institute of Standards and Technology (NIST) AI Risk Management Framework (RMF), and SEC Item 106 all point toward more governance, not less. Organizations with architecture in place when these requirements take effect demonstrate compliance immediately. Organizations without it face implementation timelines measured in quarters while competitors operate under established governance programs.

The CFO conversation is not about technology. It is about financial risk, insurance exposure, regulatory cost, and competitive positioning. A governance team that presents its program in these terms competes effectively for budget. A governance team that presents P-property verification rates and E-2 evidence pipeline throughput loses the budget to whichever team speaks the CFO's language.

The Stakeholder Alliance Strategy

A governance recommendation that reaches the board must arrive with the support of four officers who rarely coordinate their positions on technology questions: the CISO, the CFO, the General Counsel, and the Chief Risk Officer. When these four present a unified recommendation, the board has confidence that it has been evaluated from security, financial, legal, and risk perspectives simultaneously. When only one officer sponsors it, the board must wonder what the other three think. Silence from the other three becomes the dominant signal.

The CISO brings governance architecture. They can describe the harness, the evidence tiers, the P-properties, and the verification pipeline; they can show that the architecture produces continuous evidence and that governance gaps are detected and remediated. The weakness they carry into the room is that governance architecture is a cost center, and the CISO's budget authority is limited.

Financial exposure quantification comes from the CFO. Regulatory fines, litigation exposure, insurance coverage gaps, remediation costs: the CFO can set the cost of the governance program against the expected cost of operating without it. But governance architecture is opaque to the CFO, who has to trust the CISO's read on whether it actually works.

General Counsel brings legal precision: mapping regulatory obligations to the governance architecture, confirming that it addresses the organization's legal requirements, advising on Item 106 disclosure implications, examination readiness, and the litigation risk of governance gaps. The exposure here is that legal analysis runs on regulatory text, and the gap between regulatory text and technical implementation is exactly where compliance failures happen.

The CRO brings enterprise risk integration. AI governance risk gets mapped into the organization's enterprise risk register, probability and impact ratings get assigned to governance failure scenarios, and the AI governance posture is presented inside the same risk framework the board already uses for everything else. The catch is that AI governance risk is new. Quantification methods are immature and historical loss data is sparse.

The alliance resolves each officer's weakness through the others' strengths. Technical credibility from the CISO. Financial quantification from the CFO. Legal precision from General Counsel. Risk framework integration from the CRO. The combined recommendation is stronger than any one of them alone because it has been stress-tested from four directions before it ever reaches the board.

Practically, a quarterly pre-board preparation meeting builds this alliance. Four officers review the AI governance posture, agree on key messages, identify any disagreements, and resolve them before the board session. The board receives a unified recommendation with documented support from security, finance, legal, and risk. Disagreements that cannot be resolved are presented to the board as decision items, not competing recommendations from officers who could not agree.

Five Decisions for the Decision Maker

Informed by the stakeholder alliance, the board audit committee faces five decisions. Each one is structural, and each sets the shape of the organization's AI governance architecture for the next three to five years. None of them can be handed to the governance team, because they turn on risk acceptance, capital allocation, and competitive strategy. Those are board-level responsibilities.

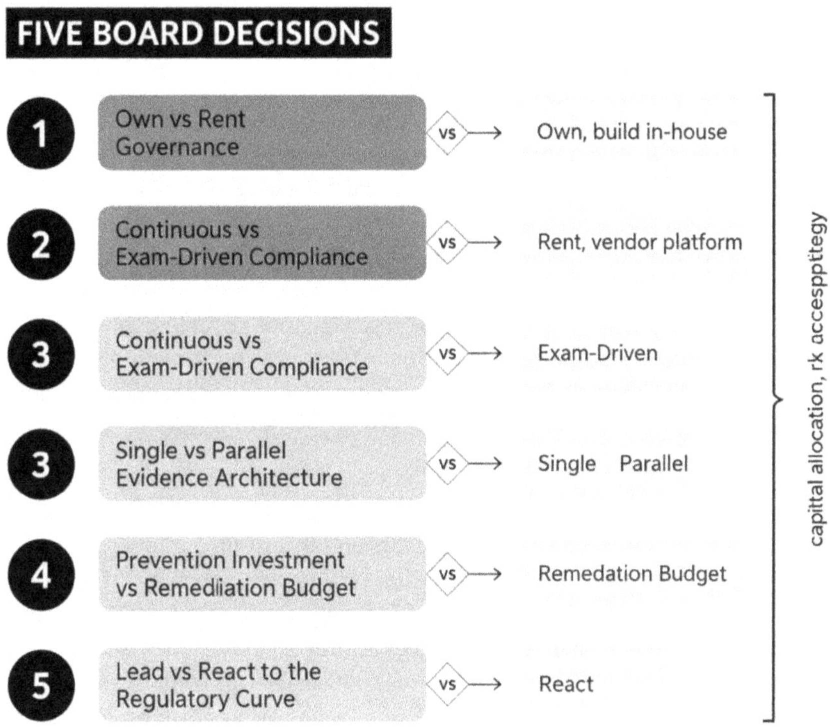

Figure 20: The five structural AI governance decisions reserved for the board audit committee, each shown as a choice between two paths with the trade-off that determines the balance.

The first is whether to own governance or rent it. Is AI governance infrastructure, the policy engine, the evidence collection pipeline, the compliance rendering system, owned and operated by the organization or supplied by a vendor? Renting is a legitimate strategy, not a trap, and for many organizations it is the correct first move. A vendor platform converts a large upfront build into a predictable operating expense, supplies a working evidence pipeline on day one instead of after a multi-quarter integration, and lets a thin governance team stand up a defensible posture before it has the headcount to build its own. The trade-off is concentration of dependence. Chapter 10's analysis of the Voight-Kampff problem sets out the structural concern an organization must price when it rents: the entity whose revenue grows when governance constraints are permissive may also control the instrument that enforces those constraints, and the dependency grows more expensive to exit the longer it runs. The mitigations are specific and worth funding: contract for evidence portability and data export in open formats, keep the AGENTS.md policy surface authored and owned by the institution even when the enforcement engine is rented, and retain the option to bring enforcement in-house. Owning governance is the other path. It demands a larger upfront investment and a team to run it, but it removes the exit-cost ratchet and keeps the enforcement instrument under the institution's control. The honest framing is a spectrum, not a binary: rent to start, own the policy surface throughout, and decide on a stated horizon whether to insource enforcement. A capital allocation decision with a five-year horizon.

The second is whether to build continuous or exam-driven compliance. Compliance posture maintained continuously, with evidence produced as a byproduct of normal operations, or assembled periodically for assessment exercises? Continuous compliance costs more in steady-state operating expense but eliminates the examination preparation sprint that consumes compliance team bandwidth for weeks before each assessment. Exam-driven compliance costs less in operating expense but produces a posture accurate only at assessment time. Phase 5 of the assembly sequence (Chapter 17) describes continuous authorization. The board must decide whether to fund Phase 5 or stop at Phase 3 or Phase 4.

The third is whether to adopt a single evidence architecture or maintain parallel programs. Chapter 18's containerization analysis establishes the direction of the cost difference: a unified evidence architecture requires upfront integration work but reduces ongoing compliance cost, because a single evidence event is rendered to many frameworks instead of being collected once per framework. The magnitude is organization-specific, modeled from the organization's framework count and per-framework collection cost. Parallel programs require less upfront investment but carry higher ongoing cost and create compliance gaps at program boundaries.

The fourth is whether to invest in prevention or budget for remediation. Governance architecture is a prevention investment; it reduces the probability and severity of governance failures. The alternative is operating with lighter governance and budgeting for remediation when failures occur. Exposure figures differ by sector: a major regulatory finding in financial services runs into the tens of millions, a HIPAA settlement into single-digit millions plus reputational damage, and NERC CIP penalties up to $1 million per day per violation in energy. The board's risk appetite determines the balance.

The fifth is whether to lead the regulatory curve or react to it. The EU AI Act is in force. The Colorado AI Act (SB 24-205) takes effect June 30, 2026. SEC Item 106 requirements are live. NIST's AI RMF continues to evolve. Waiting for a standard to settle before building is a defensible strategy with a real upside: a final rule removes the risk of building to a draft requirement that later changes, lets the organization adopt patterns the market and the regulator have already validated, and avoids stranded investment in an architecture the eventual standard does not credit. Where a standard is genuinely unsettled, waiting on that specific clause while building the sector-independent core is the disciplined choice, not negligence. The cost of waiting is that the option does not stay open. As a final rule approaches, expect implementation timelines to compress and examiner scrutiny to rise, so that an organization starting late builds under pressure rather than on its own schedule; for the parts of the harness that are already required (Item 106 disclosure, EU AI Act obligations in force), waiting is no longer an option to weigh but a finding waiting to be written. The discipline is to separate the two: build the settled core now, and hold the genuinely unsettled extensions until the rule lands, rather than treating "wait" as an all-or-nothing posture across the whole architecture.

A side-by-side comparison sharpens the choice for the board. Building now carries a known, bounded cost: the upfront engineering investment plus the steady-state operating expense of running the architecture, both of which the CFO can estimate from the AI compute baseline. Waiting roughly eighteen months carries a deferred and unbounded cost: the same build expense, not avoided but pushed later and compressed, plus the expected value of any finding that lands in the interim, the sector exposure range multiplied by the probability of a finding during the window, plus the premium for building under regulatory pressure rather than on the organization's own timeline. The comparison is not "spend now versus save now." It is "spend a known amount now versus spend a larger, less predictable amount later under worse conditions," and the residual judgment is how the board prices the probability of a finding inside the window. Where that probability is low and the relevant standard is still in draft, waiting on that clause is rational; where the obligation is already in force, the eighteen-month delay is pure deferred cost with no offsetting option value.

The Navigation Chart

The navigation chart is a quarterly governance posture report, delivered by the stakeholder alliance, that answers five questions:

1. How many AI systems are operating within authorized governance boundaries?
2. What governance gaps have been identified, and what is the remediation timeline?
3. What is the estimated financial exposure of unresolved governance findings?
4. Are insurance coverage terms adequate for the current AI governance risk profile?

5. What governance decisions require board authority before the next quarterly report?

These five questions carry technical governance posture into the language the board actually speaks. Boards do not need to understand P-property verification rates. They need to know whether AI systems are governed, what it costs if they are not, and which decisions only the board can make.

The navigation chart is not a dashboard. It is a decision document. Each quarterly report presents the current posture, identifies decisions requiring board authority, provides the information needed to make each one, and records the board's decisions for the governance record. E-4 attribution chains capture board decisions as governance events, creating an audit trail that demonstrates board oversight for Item 106 disclosure, regulatory examination, and litigation defense.

The ISS does not succeed because its crew understands orbital mechanics. It succeeds because the flight control team translates orbital mechanics into operational decisions the crew can execute. Not understanding P-properties and evidence tiers. Translating governance posture into risk decisions the board can make.

Translation layer: the navigation chart. Waypoints: the five decisions. Course correction: the quarterly report. Flight log: the evidence record proving the spacecraft stayed in its authorized orbit.

Two Infrastructure Commitments the Board Funds

Two pieces of infrastructure sit beneath the navigation chart. The board funds both without needing to understand how either one is built. What the board is buying, in each case, is assurance.

The first is governance that runs itself and keeps its own record. A ship's maintenance log is not a description of work; it is the contemporaneous record of every maintenance action on critical systems, written as the work happens, and a classification surveyor trusts it precisely because it was not assembled after the fact. The board should fund AI governance verification that works the same way: scheduled checks that run on their own, survive a server failure or a network outage without losing their place, and produce the audit trail as a byproduct of running rather than as a separate report that could diverge from what actually happened. The board-level test is one question. When a regulator asks "how do you know your AI governance controls execute as documented?", the answer is the running record itself, not a policy document describing what is supposed to happen.

The second is governance reporting a regulator can actually use. When a federal examiner reviews a governed system, the compliance dashboard must be usable with assistive technology, because Section 508 of the Rehabilitation Act requires it and a dashboard the examiner cannot read is a dashboard that cannot demonstrate compliance. For a federal contractor this is a contract term, not a courtesy: deliverables that fail accessibility conformance can be rejected and payment withheld. The board funds accessibility as a default property of the reporting infrastructure rather than a remediation project discovered at audit time. The cost of getting this right at the start is small; the cost of retrofitting it under an examiner's deadline is not.

Both commitments answer the same board-level concern. The maintenance log is the workflow, the workflow history is the compliance artifact, and that artifact is worth nothing unless the people who govern the system can read it. Reference implementations of both categories exist and are documented in Book IV.

CHAPTER SUMMARY

Risk currency translates the technical governance posture into the language boards use to make decisions: financial exposure, regulatory risk, competitive position, and insurance coverage. SEC Reg S-K Item 106 requires public companies to disclose their cybersecurity governance structure, including AI governance when AI systems are material to the organization's risk posture. The board audit committee serves as the commercial Authorizing Official, evaluating governance posture and accepting residual risk. The CFO conversation has four dimensions: AI compute costs, governance ROI, insurance coverage gaps, and competitive positioning. The stakeholder alliance (CISO, CFO, General Counsel, CRO) must align before the board receives a governance recommendation, with each officer's strength compensating for the others' vulnerabilities. Five structural decisions require board authority: own or rent governance infrastructure, build continuous or exam-driven compliance, adopt a single evidence architecture or maintain parallel programs, invest in prevention or budget for remediation, and lead the regulatory curve or react to enforcement. The quarterly navigation chart translates governance posture into risk decisions the board can execute. Two infrastructure commitments support this work, both funded by the board without requiring it to understand the implementation: governance verification that runs itself and keeps its own contemporaneous record as the compliance artifact, analogous to a maritime maintenance log; and compliance reporting an examiner can actually use, meeting Section 508 accessibility requirements that are a contract term for federal work, not a courtesy.

Chapter 21: Commercial Failure Modes

Part V: When Ships Sink

> *"Lloyd's List has published continuously since 1734, longer than any newspaper in the English-speaking world. Its primary subject for the first two centuries was loss. Every ship that foundered, every cargo that disappeared, every hull that broke apart in weather it was not built to survive, all recorded in the same column, in the same format, week after week. The catalog of failure is the raw material from which every subsequent governance improvement was derived. You do not build a classification society from success stories. You build it from the wreck reports."*

Lloyd's List, in its early decades, did not distinguish between types of loss. A ship was entered as lost; the column recorded the vessel's name, its last known position, and whatever account could be gathered from survivors or nearby vessels. Cause was rarely investigated beyond the immediate narrative: storm, fire, collision, grounding. Structural conditions that made the loss possible, overloading, unseaworthy hull, inadequate crew, missing navigation instruments, were not the List's concern. The List recorded outcomes. Nothing more.

It took the classification societies another century to develop the vocabulary of structural failure modes: recurring patterns that explained why certain categories of vessel failed at rates disproportionate to their exposure. Coffin ships of the 1850s and 1860s failed not because they encountered worse weather than other ships, but because they were loaded past their design limits by owners whose insurance payments exceeded their cargo's value. Not "storm." "Economic incentive structure that rewarded overloading." The storm was the triggering event; the overloading was the structural condition; the absence of the Plimsoll mark was the governance gap.

Commercial AI governance is at the Lloyd's List stage. Organizations record losses: regulatory fines, consent violations, model failures, biased outcomes, examination findings. Real losses. Growing catalog. What the catalog does not yet contain is the structural failure mode taxonomy, the recurring architectural patterns that produce these outcomes across sectors, across jurisdictions and across regulatory frameworks.

Chapters 11 and 13 through 16 cataloged fifteen sector-specific failure modes: three each for financial services (F-FS1 through F-FS3), healthcare (F-HC1 through F-HC3), energy and critical infrastructure (F-EN1 through F-EN3), insurance (F-IN1 through F-IN3), and manufacturing and telecommunications (F-MF1 through F-MF3). Real, documented, consequential within their respective sectors. Also, each one, an individual expression of a deeper structural pattern.

This chapter identifies six structural failure patterns that operate across sectors, that the federal application of The Harness Doctrine does not cover because they arise from conditions federal environments never face, and that produce the sector-specific failures cataloged in Part III as their local manifestations. Not individual wreck reports. The overloading patterns behind them.

F-C1: Vendor Capture

Governance architecture depends on a vendor whose commercial interests are misaligned with the governance constraints the architecture is meant to enforce.

Chapter 10 introduced the Voight-Kampff problem: the entity whose revenue grows when governance constraints are permissive also supplies the instrument that enforces those constraints. Vendor capture is the failure mode that results when this structural conflict goes unresolved.

Not bad faith. Structural ambiguity. The governance architecture cannot distinguish between effective governance and governance that appears effective because the vendor has an incentive to report favorable results. The organization believes its governance is operating; the evidence says it is operating; the evidence was produced by the entity whose revenue depends on continued operation. Verification requires the same vendor infrastructure. The circle is closed.

A mid-market regional bank deployed an AI-assisted fraud detection system from a vendor that also supplied the model governance dashboard, the compliance reporting module, and the evidence artifacts submitted to Office of the Comptroller of the Currency (OCC) examiners. Quarterly governance reports showed 99.7% policy compliance across all monitored transactions. The bank's compliance team presented these reports to the board audit committee as evidence of sound AI governance.

In 2025, the OCC examination team requested raw verification logs underlying those quarterly reports. The vendor's dashboard did not expose raw logs; it exposed aggregated metrics. When the examination team requested logs from the vendor directly, the vendor produced logs internally consistent with the reported metrics but impossible to reconcile independently against the bank's own transaction records. The log format was proprietary. Reconciliation tools were part of the vendor's premium analytics tier. The bank had not purchased that tier.

The OCC issued a matter requiring attention. Not that the fraud detection system was non-compliant. That the bank could not independently verify its own compliance posture. Vendor-produced, vendor-formatted, vendor-interpreted evidence. The bank had outsourced its ability to know whether it was compliant.

The architecture fails at the boundary between Tier 1 and Tier 2 vendor dependencies defined in Chapter 10. Evidence collection infrastructure (E-1 through E-4) is a Tier 1 component that must be organization-owned. When the vendor supplies the evidence architecture, the organization has delegated a function it cannot delegate without losing the ability to verify its own compliance.

E-2 continuous verification records must be generated by organization-owned infrastructure, not vendor-supplied dashboards. Organizations must be capable of producing their own verification of the vendor's governance claims, using evidence they collect, store, and query independently. If the only path to compliance evidence runs through vendor infrastructure, the evidence architecture has a single point of failure at the vendor boundary.

The maritime parallel is exact. In the 1850s, certain shipowners employed surveyors who were financially dependent on the owners' continued patronage. Surveys were favorable. Classifications were maintained. Ships sank at rates that the favorable surveys could not explain. Lloyd's Register addressed this by requiring that surveyors be employed by the Register, not by the shipowners. Not a quality improvement. A structural precondition for the survey's validity.

F-FS1 (Chinese wall breach through vendor-supplied AI context) and F-IN3 (actuarial opinion on AI-assisted pricing the actuary cannot explain) are both expressions of vendor capture. In both cases, the governance architecture depends on vendor-supplied capabilities whose internal operation the deploying organization cannot independently verify.

F-C2: Multi-Jurisdictional Misapplication

Governance architecture applies the wrong jurisdiction's rules to a specific data subject, transaction, or decision. Not because the rules are unknown. Because the architecture resolves jurisdiction at the system level rather than the data-element level.

LP10 (Jurisdictional Consistency) defines the requirement: applicable rules are determined by the data subject's jurisdiction, not the deploying organization's. Failure occurs when the architecture lacks jurisdiction resolution entirely, resolves it statically at deployment time, or resolves it correctly for the majority case and incorrectly for the minority case that triggers enforcement.

A direct-to-consumer insurance carrier operating in forty-two US states deployed an AI underwriting model for homeowners insurance, with the carrier's domicile state's regulatory requirements as the baseline configuration. The compliance team had reviewed requirements for the five largest states by premium volume and added state-specific overlays for those five. Thirty-seven states operated under baseline configuration.

Colorado SB 24-205, which requires specific testing and documentation for AI models used in underwriting, took effect for life insurance first and was followed by regulatory attention to other lines. The carrier's Colorado book of business was small, approximately 2% of total premium. The Colorado-specific overlay was not among the five configured states.

The Colorado Division of Insurance opened a market conduct examination. The examiner requested documentation of the AI underwriting model's testing and validation against Colorado's requirements. What the carrier produced was the domicile state's testing documentation, which did not address Colorado's specific AI testing requirements. Deficiency noted.

Regulatory consequence: a market conduct finding and a corrective action plan requiring Colorado-specific AI governance within ninety days. The operational consequence was larger. The examination revealed that thirty-seven states were operating under baseline configuration without state-specific validation that those states' requirements were satisfied by the baseline. Each of those thirty-seven states represented a potential market conduct finding.

The governance property violated is LP10. The architecture resolved jurisdiction at the system level (domicile state baseline plus five overlays) rather than at the data-element level (each policyholder's state of residence determining the applicable regulatory configuration).

E-3 state transition records must include the jurisdiction determination for each transaction. If the jurisdiction determination is logged at the transaction level, an audit reveals immediately whether each policyholder was processed under the correct state's requirements. If jurisdiction is configured at the system level, no record exists of which rules were applied to which policyholder, and the only way to determine compliance is to reconstruct every transaction against every state's requirements retroactively.

Before the International Convention for the Safety of Life at Sea standardized minimum safety requirements across signatory nations, a ship might carry a certificate from its flag state that was not recognized by the port state where it docked. The SS *Arctic*, an American Collins Line steamship lost in 1854 with 350 souls, carried safety equipment that satisfied American standards but would not have met the British Board of Trade's requirements. Flag state rules applied to a voyage in waters where port state rules were more restrictive. Wrong certificate for the waters sailed.

F-IN2 (state-by-state regulatory inconsistency in insurance underwriting) is the direct sector-specific expression of this pattern. F-HC2 (cross-state telehealth PHI handling under varying state health privacy laws) exhibits the same structural failure when a telehealth provider applies its home state's protected health information (PHI) rules to patients in states with more restrictive requirements.

F-C3: Consent Boundary Violation

The AI system processes data beyond the scope for which consent was obtained. Not through a technical breach. Through an architectural failure to enforce consent boundaries at the processing layer.

Three paths lead here. First, consent obtained for one purpose, then data used for a different purpose by an AI system not configured to check purpose limitations. Second, consent withdrawn and recorded in the consent management platform but not propagated to the AI system's retrieval scope before the next processing event. Third, consent obtained under one regulatory framework, a general privacy notice, treated as sufficient for a different framework with more specific consent requirements such as 42 CFR Part 2. Any one of these paths produces the same outcome. Processing exceeded what the data subject authorized.

A health system deployed an AI-assisted population health analytics platform to identify patients at risk for hospital readmission. The platform ingested data from the health system's electronic health record (EHR), including clinical encounter data, pharmacy records, and social determinants of health indicators. Patients had signed a general Health Insurance Portability and Accountability Act (HIPAA) notice of privacy practices authorizing use of their PHI for treatment, payment, and healthcare operations.

Internally, the population health analytics platform was classified as a healthcare operations tool. Defensible; population health management is a recognized healthcare operations activity under 45 CFR 164.501.

Failure occurred when the health system expanded the platform's scope to include substance use disorder treatment records from a federally assisted program operated by one of its affiliated clinics. These records were subject to 42 CFR Part 2, which requires specific written consent for each disclosure, identifying the information, the recipient, the purpose, and an expiration date. A general HIPAA notice of privacy practices does not satisfy Part 2. Not interchangeable frameworks.

The AI platform ingested Part 2 records along with the general clinical data. No distinction in the ingestion pipeline between PHI governed by HIPAA alone and records governed by Part 2. Readmission risk models were trained on the combined dataset; the population health team received risk scores influenced by substance use disorder treatment history without the specific patient-by-patient written consent that Part 2 requires.

Discovery came during a SAMHSA compliance review of the affiliated clinic's Part 2 program. Part 2 records had been disclosed to the population health platform without individual patient consent. The health system suspended the platform, retained outside counsel, and began a record-by-record review: which patients' Part 2 records had been processed, which risk scores were affected, whether any clinical interventions had been initiated based on risk scores influenced by unauthorized Part 2 disclosures.

The governance property violated is LP9 (Consent-Bounded Processing). The AI system did not verify consent scope before processing. It treated all records in the EHR as governed by the same consent framework. The consent boundary between HIPAA-governed PHI and Part 2-governed records was not encoded in the AI system's retrieval constraints.

E-4 attribution chain records for each data ingestion event must document the consent basis under which each record was ingested, including the specific regulatory framework governing that record. If the E-4 record for Part 2 records shows "HIPAA notice of privacy practices" as the consent basis, the record itself documents the violation. If no consent basis is recorded, the absence of the record is the evidence gap. E-2 continuous verification must include a consent scope check at each retrieval event, verifying that the data elements being accessed fall within the consent scope that applies to those specific records under their specific governing framework.

A ship authorized to carry general cargo takes on a consignment of munitions without notifying the underwriter. Insurance covers general cargo; it does not cover munitions, which carry a different risk profile and require a different premium. The manifest records the munitions as "hardware." Cargo physically aboard. Insurance not covering it. The ship operates beyond its authorized risk profile without the insurer's knowledge. Consent to carry cargo did not extend to this category, but the loading architecture did not distinguish between authorized and unauthorized cargo types.

F-HC1 (cross-encounter context bleed in clinical documentation) involves a similar boundary failure at the encounter level rather than the consent-framework level. F-HC3 (AI-generated clinical content crossing the Food and Drug Administration (FDA) software as a medical device (SaMD) boundary) involves processing that exceeds the authorized scope of the system's regulatory classification.

F-C4: Fairness Failure

An AI-assisted decision process produces disparate impact on a protected class, and the governance architecture either does not evaluate for disparate impact, evaluates against the wrong population, or evaluates correctly but does not act on the findings.

Chapter 7's false completion failure mode is the structural ancestor. The insurance underwriting case in that chapter, where the validation dataset was drawn from the carrier's existing customer base rather than the prospective applicant population, demonstrated the mechanism. The model passed validation. Validation did not measure the property that mattered. Disparate impact reached production, and the validation architecture was structurally incapable of detecting it.

Three additional variants extend beyond false completion. Proxy blindness: the governance architecture checks for direct use of protected class variables but does not evaluate for variables that correlate with protected characteristics. Temporal drift: the model passes fairness evaluation at deployment but drifts into disparate impact as population distribution shifts, and the governance architecture does not re-evaluate at operational frequency. Threshold gaming: the model is tuned to produce approval rates that satisfy the four-fifths rule in aggregate while producing disparate outcomes at specific decision boundaries that the aggregate metric obscures.

A mid-size bank deployed an AI credit scoring model that incorporated non-traditional data sources, including utility payment history, rental payment patterns, and device metadata from the bank's mobile application. The model was evaluated for disparate impact before deployment using the Federal Financial Institutions Examination Council (FFIEC)'s fair lending examination methodology. The evaluation found no statistically significant disparate impact at the aggregate level across the bank's applicant pool.

Eighteen months after deployment, a fair lending advocacy organization filed a complaint with the Consumer Financial Protection Bureau (CFPB) alleging that the model produced statistically significant adverse outcomes for applicants in majority-Hispanic census tracts in three metropolitan statistical areas. The bank's internal review confirmed the disparity. The model's use of mobile device metadata, specifically device model and operating system version, correlated with income and, in the three affected MSAs, with ethnicity.

The model had passed its pre-deployment fairness evaluation because that evaluation examined aggregate approval rates across the bank's entire geographic footprint. Disparity was concentrated in three MSAs; the aggregate metric diluted the localized disparity below the detection threshold. Fair in aggregate. Unfair in specific markets. The governance architecture evaluated only the aggregate.

The governance property violated is LP11 (Fairness Verification). The fairness evaluation was performed at the wrong granularity. LP11 requires evaluation against the applicable threshold, which for fair lending analysis must include geographic, demographic, and intersectional breakdowns, not merely aggregate rates.

E-2 continuous verification must include fairness metrics generated at operational frequency, disaggregated by geography, demographic group, and decision boundary. An E-2 fairness record that shows aggregate compliance without sub-population analysis is an incomplete verification. The E-2 pipeline must produce sub-population fairness metrics as a standard output, and those metrics must be evaluated against the applicable threshold at each sub-population level, not aggregated before comparison.

In the 1870s, the Plimsoll line was applied uniformly: one load line for all conditions. Samuel Plimsoll himself recognized within years that this was insufficient. A ship safely loaded for calm Mediterranean waters was overloaded for the North Atlantic in winter. Same vessel. Same cargo weight. Same load line. Death rate in winter crossings that the Mediterranean standard could not explain. Lloyd's Register subsequently introduced conditional load lines: summer, winter, winter North Atlantic, tropical, fresh water. Same hull, different maximum loads depending on the conditions it would face. A single aggregate standard had concealed the localized risk.

F-IN1 (proxy discrimination through unvalidated correlated features) is the insurance-specific expression of this pattern. F-FS2 (AI credit scoring producing adverse impact) is the financial services expression. Both involve models that pass aggregate fairness evaluations while producing disparate outcomes in specific sub-populations.

The Economics Behind the Failure Modes

Several of these failure modes share an economic root: when continuous verification is expensive, organizations cut corners that the corners-cut version of the architecture cannot detect. When sub-population fairness analysis is expensive, organizations default to the aggregate evaluation that hides localized disparity (F-C4). When independent evidence collection is expensive, organizations accept the vendor's dashboard instead of building their own (F-C1). When continuous evidence is perceived as unaffordable, organizations assemble documentation for the next exam instead of generating it as a byproduct of operation (F-C5). The economic constraint is the precursor; the failure mode is the result.

Classification societies faced the identical pressure. If thorough surveys are priced out of reach, owners shop for a cheaper certifier and the most economical certification wins the business regardless of whether it assures safety. Lloyd's Register defeated that race to the bottom by making thorough survey the affordable baseline and by requiring that surveyors be employed by the Register, not the shipowner. The governance equivalent is to make continuous verification affordable enough that no organization is forced into a periodic-assessment posture by budget, and to keep the verification infrastructure organization-owned so no vendor sits in the evidence path. How an organization drives verification cost down to that level, routing each check to the cheapest capable tier, running compressed models on its own hardware to remove the external dependency behind F-C1, F-C2, and F-C3, is engineering detail documented in Book IV. The board-level point is the one the maritime record already made. Align thorough verification with affordability, or the economics will manufacture the failure modes on their own.

F-C5: Evidence Architecture Failure

Compliance evidence assembled by the organization does not survive regulatory examination. Three reasons: designed to document intent rather than demonstrate operation; assembled retrospectively rather than generated continuously; evidence artifacts answer the question the organization anticipated rather than the question the examiner actually asks.

This is the heroic one-shot failure mode from Chapter 7, extended from governance architecture to the evidence architecture itself. The organization built a governance program, and the program produced documentation. That documentation was assembled for a specific examination. Then the examination asked a different question. The documentation cannot answer it, because evidence was collected against a different specification than the examiner's inquiry.

A regional health system prepared for a Department of Health and Human Services (HHS) Office for Civil Rights HIPAA audit by assembling a governance documentation package: AI governance policy, workforce training records, risk assessment, vendor Business Associate Agreements (BAAs), security incident response plan. Accurate, current, and well-organized documentation.

The Office for Civil Rights (OCR) auditor asked a different question. For the past eighteen months, for every AI system that processed PHI, could the health system produce a log of which data elements were accessed, for which patient, by which system, for which purpose, at which time? Not a policy describing access controls. Not a risk assessment evaluating access risks. A log. Operational record of actual access events.

The health system's AI governance architecture produced policy documentation (E-1) but did not generate continuous access verification records (E-2) at the data-element level. User access to the EHR was logged. AI system internal data retrieval operations were not, because the AI system's retrieval operations were treated as internal system functions, not as access events requiring audit logging.

The OCR auditor noted the gap. The health system could demonstrate a governance policy (E-1) and user authentication (partial E-4) but could not demonstrate that the AI system's access to specific PHI data elements was logged, monitored, and within the minimum necessary standard at each access event (E-2 and E-4 combined). "Do you have a governance program?" Yes. "Was the governance program operating at the moment of each PHI access?" Silence.

The form of the evidence tiers was satisfied; the function was not. E-1 records exist but E-2 continuous verification is absent. E-4 attribution records exist for user authentication but not for AI system data access operations. Incomplete at precisely the tiers (E-2 and E-4) that answer the examiner's operational questions.

The failure is in the evidence architecture itself, which means the detection mechanism must be external to the failing system. The 5-minute examiner test from Chapter 6 is the detection instrument: can the organization answer the four assessor questions (does governance exist, is it operating, does it know what changed, can every action be traced) from evidence records without narrative explanation? If any answer requires a narrative rather than a query against the evidence record, the evidence architecture has a gap. Organizations that run the 5-minute test against their own evidence architecture before the examiner arrives will discover the gap before the examiner does.

The SS *London* sank in the Bay of Biscay in January 1866 with the loss of 220 lives. Valid classification certificate aboard. The certificate documented hull condition at the time of survey. It did not document whether cargo had been properly stowed, whether hatches had been adequately secured for winter conditions, whether the crew had been trained on the vessel's specific stability characteristics. The certificate answered the question Lloyd's Register asked. It did not answer the question the Bay of Biscay asked. Real evidence. Wrong inquiry.

F-EN2 (North American Electric Reliability Corporation (NERC) CIP evidence gap during AI system update) exhibits this pattern when the AI system's model update is classified as a corporate IT change and no CIP-010 change documentation is generated, leaving an evidence gap covering the entire period between updates. F-FS3 (OCC examination finding for model risk management gap) follows the same pattern when the bank's model risk governance documents the model validation at authorization but does not generate continuous verification records between examinations.

F-C6: Cross-Sector Contagion

A governance failure in one sector's regulatory domain creates compliance exposure in a different sector's domain. The AI system, the data, or the decision process spans both sectors; the governance architecture treats each sector's requirements in isolation.

Cross-sector contagion is the failure mode that Chapters 11 and 13 through 16 could not catalog, because each chapter examined its sector in isolation. This failure appears only at boundaries between sectors, where an organization operating in multiple regulated domains discovers that a governance gap in one domain has propagated into another through a shared data layer, a shared AI system, or a shared vendor.

A large financial holding company operated both a bank and an insurance subsidiary. The bank deployed an AI credit scoring model for consumer lending. The insurance subsidiary deployed an AI underwriting model for homeowners insurance.

Both models consumed data from a shared customer data platform that consolidated the holding company's customer information across business lines.

The bank's credit scoring model was validated for fair lending compliance under the FFIEC fair lending examination methodology. The insurance subsidiary's underwriting model was validated for disparate impact under National Association of Insurance Commissioners (NAIC) Model Bulletin requirements. Both validations passed. Both conducted independently, by different compliance teams, against different regulatory standards.

The shared customer data platform contained behavioral data, transaction frequency, payment patterns, product usage metrics, that both models consumed. The bank's fair lending validation examined these features for correlation with protected characteristics under Equal Credit Opportunity Act (ECOA) and the Fair Housing Act; the insurance subsidiary's validation examined different features for correlation with protected characteristics under state insurance anti-discrimination laws. Neither validation examined the features the other model used. Each compliance team operated within its own regulatory scope.

A consumer advocacy organization filed simultaneous complaints with the OCC (regarding the bank's credit decisions) and the state Department of Insurance (DOI) (regarding the insurance subsidiary's underwriting decisions). The complaints alleged that the holding company's AI systems produced correlated adverse outcomes: applicants who were denied credit by the bank were also denied insurance by the subsidiary, at rates suggesting the shared data platform was propagating a discriminatory signal across both decision processes.

The holding company's internal investigation confirmed the correlation. A specific feature set in the shared customer data platform, derived from transaction velocity and product adoption patterns, correlated with income and, through income, with race. The bank's credit model used this feature set; the insurance subsidiary's underwriting model used a different feature set from the same platform, one derived from overlapping source data with similar correlations. Each model, evaluated independently, fell within its sector's fairness threshold. The combined effect was different. Both models operated on the same customer population through the same data platform and produced a correlated adverse impact that neither sector's validation was designed to detect.

The governance property violated is LP11 (Fairness Verification) at the enterprise level, not the system level. Each model satisfied LP11 within its sector's scope. Enterprise-level fairness posture went unevaluated, meaning the combined effect of multiple AI-assisted decisions on the same individual across the holding company's business lines. No governance property required it.

E-3 state transition records at the enterprise level must capture the deployment of new AI models and the data sources they consume, creating a dependency map that identifies shared data platforms feeding multiple decision processes across regulatory boundaries. When two models consume data from the same source and produce decisions affecting the same population, the E-3 record must flag the dependency for cross-model fairness evaluation. E-2 continuous verification at the enterprise level must include cross-system fairness analysis for decision processes that share data infrastructure, even when those decision processes are governed by different regulatory frameworks.

The Great Eastern, Isambard Kingdom Brunel's 1858 iron ship, was designed to carry passengers and cargo simultaneously. Dual-purpose vessel, dual certification. Passenger certification addressed lifeboat capacity, crew ratios, and accommodation standards. Cargo certification addressed hull strength, stability, and load distribution. Neither certification evaluated the interaction between passenger safety and cargo loading: specifically, whether cargo distribution patterns that maximized freight revenue affected the vessel's stability characteristics in ways that degraded passenger safety. The Great Eastern had persistent stability problems throughout its career. Both certification domains satisfied independently. Interaction between them never assessed.

This pattern has no direct analog in Chapters 11 and 13 through 16 because those chapters examined sectors individually. The closest expressions are the multi-agent feedback loops described in F-MF2, where independently compliant agents produce non-compliant outcomes through interaction, and the cross-framework evidence gaps described in Chapter 18's analysis of parallel compliance programs. Cross-sector contagion extends both patterns across regulatory domain boundaries.

The Structural Taxonomy

All six commercial failure modes share a common architectural characteristic. Each operates at a boundary that the governance architecture either does not recognize as a boundary or treats as soft rather than hard.

The Six Commercial Failure Modes

	SOFT BOUNDARY	PROPERTY VIOLATED	EVIDENCE TIER
	organization vs vendor		E-2
F-C1 Vendor Capture	organization vs vendor	LP (various)	E-2
F-C2 Multi-Juristional Misapplication	between jurisicctions	LP10	E-4
F-C3 Consent Boundary Violation	between consent frameworks	LP19	E-2
F-C4 Fairness Failure	between populations	LP11	E-2
F-C5 Evidence Architecture Failure	anticipated vs examiner question	multi-tier	E-2 / E-4
F-C6 Cross-Sector Contagion	between regulatory domains	LP11 enterprise	E-3

Figure 21: The six commercial failure modes (F-C1 through F-C6), each mapped to the soft boundary it exploits, the governance property it violates, and the evidence tier that converts that boundary from soft to hard.

F-C1 (Vendor Capture) operates at the boundary between the organization and its governance vendor. The boundary is soft when the vendor supplies both capability and compliance evidence.

F-C2 (Multi-Jurisdictional Misapplication) operates at the boundary between jurisdictions. The boundary is soft when jurisdiction is resolved at the system level rather than the data-element level.

F-C3 (Consent Boundary Violation) operates at the boundary between consent frameworks. The boundary is soft when all data is treated as governed by the least restrictive applicable consent standard.

F-C4 (Fairness Failure) operates at the boundary between populations. The boundary is soft when fairness is evaluated in aggregate rather than at the sub-population level.

F-C5 (Evidence Architecture Failure) operates at the boundary between the question the organization anticipated and the question the examiner asks. The boundary is soft when evidence is assembled for a predicted examination rather than generated continuously against the full set of assessor questions.

F-C6 (Cross-Sector Contagion) operates at the boundary between regulatory domains within a single enterprise. The boundary is soft when compliance programs are organized by regulatory framework rather than by shared data and decision infrastructure.

These boundaries are not esoteric. They are where every examiner, auditor, regulator, and litigant applies pressure, because boundaries are precisely where governance architectures fail. Lloyd's Register surveyors in 1760 knew this intuitively. Attention focused on hull joints, the bilge, deck seams, the mast step: places where one structural element met another. The interior of a solid timber member does not fail. The joint between two timbers fails. Governance works the same way.

International Space Station (ISS) flight control knows this operationally. Interfaces between modules, between Russian and American segments, between pressurized and unpressurized volume, between the station's atmosphere and vacuum: that is where every emergency procedure concentrates. Destiny Laboratory does not spontaneously fail. The hatch seal between Destiny and Node 1 can. The docking interface between a Progress resupply vehicle and the station can. Station-keeping is boundary-keeping.

Commercial AI governance fails at boundaries. These six failure modes are the boundary conditions. Governance properties (LP1 through LP14) and evidence tiers (E-1 through E-4) are the mechanisms for converting soft boundaries into hard ones, for making the joints between structural elements as governed as the elements themselves. An organization that has verified all fourteen properties and generates evidence across all four tiers within each sector, each system, and each jurisdiction has governed the interior. This chapter asks whether that organization has also governed the joints.

Where This Breaks

Every preceding section in this volume shows the harness working: a boundary identified, a property assigned, an evidence tier that converts the soft boundary into a hard one. A reader is entitled to ask the falsifiability question. Where does the framework itself fail? If the answer is "nowhere," the framework is not a framework; it is a marketing claim. Two failure cases are presented here, both composite illustrations rather than named deployments, and both honest about the limits of the architecture this volume advocates.

The case where the evidence held and the examiner rejected it anyway. Consider an organization that implemented the full evidence architecture: E-1 foundation records, E-2 continuous verification at the data-element level, E-3 state-transition logging, E-4 attribution chains. The architecture was sound. The 5-minute examiner test passed against the organization's own records. When the examiner arrived, the organization could answer every operational question from a query rather than a narrative. The boundary held; the denial records existed; the byproduct evidence was generated as the architecture promised.

The examiner rejected it anyway, and the rejection was not unreasonable. The examiner did not dispute that the events were logged; the examiner disputed whether the logging system itself could be trusted. The evidence demonstrated what the AI system did. It did not demonstrate that the evidence-collection layer was itself outside the control of the party being examined. Tamper-evidence (LP8) addresses after-the-fact alteration, but it does not address an evidence-collection layer that was specified, built, and operated entirely by the institution under examination. The examiner's position was the one the maritime model anticipated: the shipowner who keeps an immaculate logbook is still the shipowner, and a logbook the owner controls is not an independent survey. The harness made the evidence reviewable; it did not, by itself, make the evidence independent. Independence came only when an outside assessor tested the running evidence against the institution's declared AGENTS.md, and in this case the assessor had not yet been engaged. The framework did not fail to produce evidence. It produced evidence that was necessary and not sufficient, and an institution that mistook the first for the second was caught short.

The case where the extension was most of the work. This volume holds that most of the harness transfers to commercial regulated settings without modification, and that a sector-specific minority requires extension. That framing is honest about direction but can mislead about effort. Transfer-without-modification and most-of-the-work are not the same measure. The inherited federal baseline, the complete harness (H-1 through H-5), the eight federal security properties (P1 through P8, Volumes I and II), the evidence tiers (E-1 through E-4), and the knowledge-base controls, transfers because the constraint mechanism is sector-independent, and this volume's core governance properties (LP1 through LP8) install on that baseline with little friction. The extension is the four added properties (LP9 through LP12: consent, jurisdiction, fairness, transparency) plus sovereignty. Counted as a fraction of the framework's components, the extension is a minority. Counted against an implementation team's calendar, it can be the majority.

Consider a health system extending a transferred harness to cover LP9 (consent-bounded processing) across HIPAA-governed PHI and 42 CFR Part 2 substance-use-disorder records. The transferred components were installed in weeks. LP9 alone consumed months, because consent boundaries are not a property of the AI system; they are a property of a consent-management platform, a record-by-record provenance trail, and a legal determination, for each record, of which framework governs it. The structural judgment that the inherited baseline transfers unchanged, and that the core governance properties (LP1 through LP8) install on it cleanly, was correct and was also the easy part. The institution that read the transfer claim as a claim about effort budgeted the project as if the extension were a rounding error, and the project ran long. The framework's classification was accurate; the institution's inference from it was wrong. A structural judgment about what transfers is not a schedule estimate, and this volume should not be read as offering one.

Both cases survive scrutiny, which is the point. The first says architecture produces necessary evidence and not sufficient independence; independence requires a party the institution does not control. The second says the framework's component-count majority is not an effort majority, and the sector extension is where the labor concentrates. Neither case retracts the maritime transfer thesis. Both sharpen it: the harness is the precondition for governance, not the whole of it, and the work that remains after transfer is the work that defines the sector.

From Catalog to Response

Lloyd's List recorded losses for nearly a century before the maritime industry developed a systematic response. When it came, the response was not incremental. It was architectural: classification society, load line, survey requirement, international convention. Each innovation addressed a structural failure pattern that individual loss records had documented for decades but that the industry had treated as individual incidents rather than systemic conditions.

The six commercial failure modes cataloged here are at the same pre-architectural stage. Organizations experience them individually. Regulators discover them in examinations. Litigants exploit them in proceedings. The catalog grows. The architectural response is what the remaining chapters of this volume construct: a governance framework that treats vendor independence, jurisdictional consistency, consent enforcement, fairness verification, evidence sufficiency, and cross-sector coherence as structural properties rather than compliance aspirations.

Part V continues with incident response architecture, the governance equivalent of the SOLAS convention that the maritime industry built after catastrophic loss proved existing governance was insufficient. The Titanic sank in 1912. SOLAS signed in 1914. Two years between catalog and architectural response. The commercial AI governance community does not have the luxury of waiting for a Titanic. The catalog is in hand. The response must follow before the losses accumulate further.

> **CHAPTER SUMMARY**
>
> Six structural failure patterns operate across commercial sectors at governance boundaries the federal application of The Harness Doctrine does not address. F-C1 (Vendor Capture) occurs when the governance architecture depends on a vendor whose commercial incentives conflict with enforcement rigor, producing compliance evidence the organization cannot independently

verify. F-C2 (Multi-Jurisdictional Misapplication) occurs when jurisdiction is resolved at the system level rather than the data-element level, applying the wrong regulatory configuration to data subjects in minority jurisdictions. F-C3 (Consent Boundary Violation) occurs when data governed by one consent framework is processed under a different, less restrictive framework because the AI system's retrieval architecture does not distinguish between consent regimes. F-C4 (Fairness Failure) occurs when aggregate fairness metrics conceal sub-population disparities that localized analysis would reveal. F-C5 (Evidence Architecture Failure) occurs when compliance evidence is designed to answer anticipated questions rather than generated continuously to answer any assessor question at any time. F-C6 (Cross-Sector Contagion) occurs when independently compliant AI systems sharing data infrastructure across regulatory domains produce correlated adverse outcomes that neither sector's validation is designed to detect. All six patterns share a common architectural characteristic: they operate at boundaries that the governance architecture treats as soft rather than hard. The fifteen sector-specific failure modes from Chapters 11 and 13 through 16 are individual expressions of these six structural patterns. The governance properties (LP1-LP14) and evidence tiers (E-1 through E-4) are the mechanisms for converting soft boundaries into hard ones. The catalog of failure is the raw material from which architectural response is built.

Chapter 22: SOLAS After the Titanic

Part V: When Ships Sink

> *"Regulations are written in blood. Every line in the safety code exists because someone died in the circumstance it describes."*

At 11:40 p.m. on April 14, 1912, the RMS Titanic struck an iceberg in the North Atlantic approximately 375 miles south of Newfoundland. The collision opened five of the ship's sixteen watertight compartments to the sea; the ship had been designed to remain afloat with any four compartments breached. Five was one too many. Two hours and forty minutes later, at 2:20 a.m. on April 15, the hull broke apart and sank in approximately 12,500 feet of water. Of the 2,224 people aboard, 1,514 died.

No single engineering failure caused the disaster. Iceberg warnings had been received and acknowledged, lookouts were watching, the engines were reversed, the rudder turned. Every individual component of the operational chain functioned. What failed was the governance architecture: the constellation of regulations, standards, inspection requirements and operational procedures that should have prevented the ship from operating at full speed through an ice field at night with lifeboats for barely half the passengers aboard.

Twenty lifeboats. That number exceeded the requirements of the British Board of Trade regulations then in effect, which were based on vessel tonnage and had not been updated since 1894, when the largest ships were a fraction of the Titanic's size. Boats for 962 people were required by regulation. The Titanic carried boats for 1,178. Neither number came close to the 2,224 aboard; the White Star Line had actually exceeded the regulatory standard and was still catastrophically underequipped.

Six ice warnings from other vessels reached the ship's Marconi wireless operators on April 14. The last, from the SS Californian, was cut off by the senior Titanic operator, Jack Phillips, who was busy transmitting passenger telegrams to Cape Race, Newfoundland. No regulatory requirement existed for delivering ice warnings to the bridge. No regulatory requirement mandated a 24-hour wireless watch. No standardized protocol prioritized safety messages over commercial traffic.

None of these absences were secrets. The maritime industry knew that lifeboats were inadequate, that wireless protocols were informal, that ice field navigation standards were voluntary. The problem was never ignorance. It was inaction. The knowledge existed; what did not exist was the regulatory architecture to convert that knowledge into enforceable requirements. The industry had the technical capability to save every person aboard, and it lacked the governance to deploy that capability.

This chapter argues that the AI industry in 2026 occupies the same structural position. The technical capability to govern AI systems exists. The regulatory architecture to require that governance does not, or more precisely, it sits in the earliest stages of construction, following the same reactive, disaster-driven trajectory that maritime safety regulation traced for over a century.

What SOLAS Actually Built

The International Convention for the Safety of Life at Sea, convened in London in January 1914, was the direct regulatory response to the Titanic disaster. Thirteen nations participated. The resulting convention established four categories of requirement that did not previously exist as binding international standards.

The most visible deficiency was lifeboat capacity, and it was the easiest to address. A ship with boats for half its complement would lose half its complement; the Titanic had proved the arithmetic. The convention required sufficient lifeboat and life-raft capacity for every person on board, with lifeboat drills conducted during every voyage.

The second requirement targeted wireless watch. The SS Californian, stopped in the ice field fewer than twenty miles from the Titanic, had a single wireless operator who had gone to sleep at 11:30 p.m., ten minutes before the collision. The Californian never received the distress calls. Under the 1914 convention, ships carrying more than fifty persons had to maintain a continuous wireless watch, ensuring that distress signals would be received regardless of when they were sent.

Third, the convention created the International Ice Patrol, operated by the United States Coast Guard and funded by the thirteen signatory nations, to monitor iceberg conditions in the North Atlantic shipping lanes and broadcast warnings to all vessels. The patrol has operated continuously since 1914, with a brief interruption during the two World Wars. Not a single ship following the patrol's recommended routes has struck an iceberg since.

Finally, the convention established minimum standards for watertight subdivision and stability, requiring that ships be designed to survive specified degrees of flooding based on their size and service.

These four requirements share a structural characteristic. Every one addressed a failure mode that was known before the Titanic sank. Naval architects understood lifeboat mathematics, wireless operators understood the value of continuous monitoring, and ice field navigation sat well within the shipping industry's competence. The knowledge was old. What was new was the regulatory obligation to act on it.

The Iteration Pattern

The 1914 convention never entered into force. War intervened; ratification stalled. But the principles established in 1914 became the foundation for subsequent iterations that did enter into force, and each iteration expanded the regulatory architecture in response to new categories of disaster.

In **1929**, following peacetime shipping losses, the International Convention for the Safety of Life at Sea (SOLAS) refined the watertight subdivision standards and introduced requirements for fire detection and fire-fighting equipment. Ships were sinking from fire as well as flooding; the regulatory framework expanded to match.

1948 incorporated lessons from World War II convoy operations and postwar shipping expansion. Updated weather routing requirements, new standards for electrical installations, expanded fire safety provisions, and the first systematic requirements for carriage of dangerous goods all entered the convention.

1960 represented the first comprehensive modernization of the framework, reflecting the transition from steam to diesel power, the introduction of nuclear-powered merchant vessels, and the growth of the global shipping fleet. More importantly, the 1960 convention introduced the "tacit acceptance" amendment procedure: amendments would take effect automatically unless a specified number of contracting governments objected within a defined period. This procedural change was a structural breakthrough. Before tacit acceptance, every amendment required explicit ratification by every signatory. After it, amendments took effect by default, and governments that disagreed had to actively opt out.

1974, with its 1988 Protocol, is the version currently in force. Amended dozens of times through tacit acceptance, the 1974 convention introduced the International Safety Management (ISM) Code, the most significant structural evolution in the entire SOLAS lineage.

Across these five iterations, the pattern holds. Each version was reactive, prompted by a specific category of disaster or class of failure. Each expanded the regulatory scope to address the failure mode the previous version had not anticipated. Each took years to negotiate and additional years to implement. And each, by the time it was fully operational, had already been overtaken by new operational realities the next version would need to address.

This is the reactive regulation cycle: disaster, investigation, negotiation, ratification, implementation, compliance, followed by a new category of disaster the current framework does not cover. The cycle is not a failure of the regulatory process. It is the inherent limitation of regulation driven by evidence of harm rather than anticipation of risk.

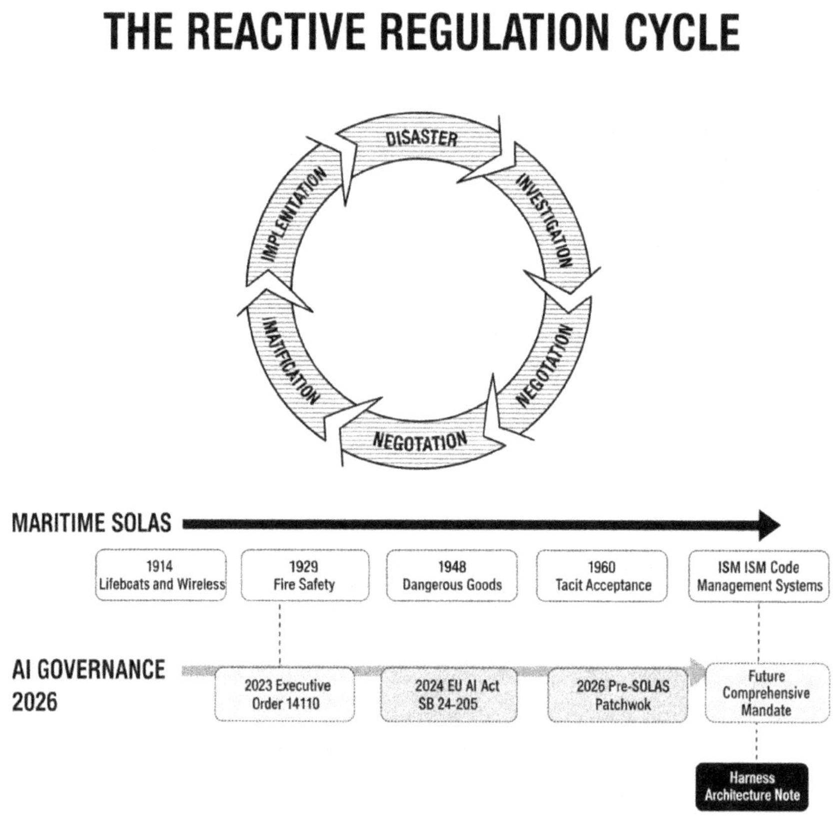

Figure 22: The reactive regulation cycle shown as two parallel timelines; maritime safety regulation after the Titanic on top, AI governance in 2026 below, with the recurring loop of disaster, investigation, negotiation, ratification, and implementation driving each iteration.

The ISM Code: From Equipment to Management Systems

Adopted in 1993 and mandatory for all SOLAS vessels since 2002, the International Safety Management Code changed the regulatory question. The old question was "does the ship have the right equipment?" The new one was "does the shipping company have a safety management system that ensures the equipment is maintained, the crew is trained, the procedures are followed, and the deviations are detected and corrected?"

This shift is the single most relevant SOLAS development for AI governance. It deserves examination.

The ISM Code requires every shipping company to establish a Safety Management System (SMS) covering safety and environmental protection policy, instructions and procedures for safe operation, defined levels of authority and lines of communication between shore and shipboard personnel, procedures for reporting accidents and non-conformities, and procedures for preparing for and responding to emergency situations.

This is a management architecture, not a checklist. The Code requires a Designated Person Ashore (DPA) with direct access to the highest level of management, whose responsibility is to monitor the safety and pollution-prevention aspects of each ship's operation and to ensure adequate resources and shore-based support are applied as required. The DPA is the organizational role that connects operational safety to executive accountability.

This shift from equipment requirements to management system requirements reflects a structural insight that the AI governance community is still absorbing. Equipment-based regulation asks whether the ship has lifeboats, whether it has a wireless, whether the AI system has an audit log. It addresses the artifacts of safety without addressing the process that produces safety. A ship with lifeboats but no drill schedule has equipment without the management system to deploy it; an AI system with an audit log but no monitoring pipeline has a recording capability without the governance architecture to act on what it records.

The harness architecture described in this book is an ISM Code for AI systems. P-properties are not equipment requirements; they are behavioral constraints enforced by a management architecture. Evidence tiers are not audit artifacts; they are the continuous output of a verification system that detects deviations and triggers responses. The AGENTS.md is not a configuration file. It is the safety management policy that defines authorized operating boundaries, escalation procedures, and human oversight gates.

The ISM Code did not replace equipment requirements in SOLAS Chapters II through V. It added a management layer that ensures those requirements are sustained in operation, not merely satisfied at inspection. The harness architecture works the same way. It does not replace individual compliance controls. It adds the management layer that ensures those controls are enforced continuously, monitored automatically, and verified against the operational baseline, not merely documented in a policy binder that no one opens between audits.

The EU AI Act as the First SOLAS Moment

The European Union's Artificial Intelligence Act, which entered into force on August 1, 2024, with a staggered implementation timeline extending through August 2027, is the first comprehensive legislative framework for AI governance enacted by a major jurisdiction. It is the AI governance equivalent of the 1914 SOLAS convention:

a reactive regulatory response to accumulating evidence of AI-related harm, negotiated over multiple years, comprehensive in ambition, and certain to be superseded by subsequent iterations that address failure modes the first version could not anticipate.

The structural parallels are specific, and they run deeper than surface analogy.

Risk classification is the first parallel. The EU AI Act sorts AI systems into four tiers (unacceptable, high, limited, minimal) that determine regulatory obligations. SOLAS established a classification system based on vessel type and service. Both frameworks recognize that not every system presents the same risk; both impose proportional requirements based on the assessed tier.

Pre-deployment verification is the second. Conformity assessments for high-risk AI systems gate market entry, just as SOLAS requires survey and certification before a vessel enters service. No assessment, no deployment.

Post-market monitoring is the third. Providers and deployers of high-risk AI systems must continue monitoring after deployment, mirroring SOLAS's periodic survey and condition monitoring. Deployment-time compliance without ongoing verification is insufficient under both regimes.

Extraterritoriality is the fourth. The EU AI Act covers AI systems that affect EU persons regardless of where the provider or deployer is located. SOLAS applies to all contracting states' vessels regardless of operating waters. Both frameworks reject the proposition that geographic relocation eliminates regulatory obligation.

The EU AI Act has a critical limitation that the 1914 SOLAS convention shared: it is the first iteration. The 1914 convention addressed lifeboats, wireless, ice patrol, and watertight compartments because those were the failure modes the Titanic exposed. Fire safety, dangerous goods, management systems, and dozens of other failure categories would wait for subsequent iterations. The EU AI Act addresses transparency, conformity assessment, risk classification, and prohibited practices because those are the failure modes that 2020-2023 AI incidents exposed. It does not yet address the failure modes that 2025-2030 AI deployments will reveal, because those failures have not yet occurred at sufficient scale to generate the political will for regulatory response.

This is not a criticism. It is a structural observation about reactive regulation. The EU AI Act is doing exactly what the 1914 SOLAS convention did: establishing the regulatory architecture that subsequent iterations will expand. The question for regulated organizations is not whether the Act is complete. It is whether the organization's governance architecture is designed to accommodate the expansions that are coming.

The American Patchwork

The United States, as of early 2026, has not enacted comprehensive federal AI legislation. Instead, the American regulatory response to AI governance occupies three distinct channels that do not yet form a coherent framework.

Executive action represents the first channel. President Biden's Executive Order 14110, issued on October 30, 2023, established reporting requirements for developers of dual-use foundation models, directed federal agencies to develop AI governance guidelines, and initiated the first systematic federal assessment of AI risks. Its significance lay not in enforcement mechanisms, which were limited, but in signaling: the federal government formally acknowledged that AI governance required coordinated regulatory attention. Subsequent administrations have shifted the emphasis. The structural recognition that AI governance is a federal regulatory concern has not been reversed.

State legislation is the second channel. Colorado's SB 24-205, signed in May 2024 with an effective date of June 30, 2026 (amended in 2025 from the original February 1, 2026), is the first US state law to impose specific governance obligations on deployers of high-risk AI systems. It requires deployers to implement a risk management policy and program, complete impact assessments, provide notice to consumers when AI systems are used in consequential decisions, and disclose known algorithmic discrimination to the state attorney general. Colorado is not alone; by early 2026, more than a dozen states have introduced AI governance legislation with varying scope and enforcement mechanisms. The state-by-state approach creates the jurisdictional complexity that Chapter 9's analysis of the International Space Station (ISS) Intergovernmental Agreement was designed to address: a single AI system serving customers in multiple states must comply with each state's requirements simultaneously, and those requirements do not harmonize automatically.

Sectoral regulation is the third. Federal financial regulators (Office of the Comptroller of the Currency (OCC), Federal Deposit Insurance Corporation (FDIC), Federal Reserve), healthcare regulators (Department of Health and Human Services (HHS), Food and Drug Administration (FDA)), and energy regulators (Federal Energy Regulatory Commission (FERC), North American Electric Reliability Corporation (NERC)) are incorporating AI governance requirements into their existing frameworks through guidance documents, supervisory expectations, and enforcement actions. The OCC's 2023 bulletin on model risk management, the FDA's evolving framework for AI-enabled medical devices, and NERC's critical infrastructure protection standards each address AI governance within the sector's existing vocabulary. These sectoral approaches carry more immediate enforcement weight than state legislation because they operate through existing examination mechanisms. A bank that deploys an ungoverned AI system faces the same examiner who reviews its Bank Secrecy Act (BSA)/anti-money-laundering (AML) compliance and its capital adequacy. The authors expect examiners to converge on

AI governance not as new legislation but as a new item on the existing examination checklist.

This patchwork resembles the pre-SOLAS maritime regulatory environment, where each nation maintained its own safety standards, enforcement mechanisms varied widely, and a vessel that complied with one country's requirements might violate another's. SOLAS resolved the fragmentation through international harmonization. The American AI governance environment is waiting for its harmonization moment, and in the interim, organizations that build governance architecture to the most demanding applicable standard, rather than the least, are positioned for any harmonization outcome.

Why Reactive Regulation Is Insufficient but Inevitable

The case against reactive regulation is straightforward: by the time the regulation exists, the harm it addresses has already occurred at scale. The 1914 SOLAS convention could not bring back 1,514 people. The ISM Code, mandated in 2002, could not retroactively prevent the Herald of Free Enterprise disaster of 1987, when 193 people died because the bow doors of a roll-on/roll-off ferry were left open during departure. The equipment was fine. The procedure failed. The EU AI Act cannot retroactively govern the AI-assisted decisions that have already produced discriminatory outcomes, wrongful denials, or privacy violations.

The case for reactive regulation is equally direct: anticipatory regulation requires the political will to impose costs on industry before public evidence of harm generates the political pressure to do so. That political will almost never exists. Industry argues, correctly, that premature regulation may restrict beneficial technology. Legislators hesitate, reasonably, to regulate technology they do not fully understand. And the public, whose support any regulation ultimately requires, does not mobilize around abstract future harms with the same intensity it mobilizes around specific past disasters.

The result is the reactive cycle. Incidents accumulate. Public pressure builds. Media coverage amplifies the most visible failures. Legislative inquiry generates political momentum. Regulation is drafted, negotiated, amended, and enacted. By the time the regulation takes effect, a new generation of technology has created new categories of risk the regulation does not address. The cycle begins again.

This is not a failure of governance. It is the physics of democratic regulation in industries where the technology moves faster than the legislative process. The relevant strategic question for organizations is not whether to support or oppose the regulatory cycle. It is where to position the organization relative to the cycle's inevitable progression.

Building the Watertight Compartments Before the Iceberg

The Titanic's watertight compartments were, individually, well-engineered. The problem was that the bulkheads did not extend high enough to prevent water from overflowing from one flooded compartment into the next. The ship could survive four breached compartments because, with four compartments flooded, the waterline remained below the top of the transverse bulkheads. With five compartments breached, the bow sank low enough that water flowed over the top of the fifth bulkhead into the sixth compartment, then from the sixth into the seventh, in a progressive flooding cascade that made the ship's loss inevitable.

The governance analogy is precise. An organization that builds governance controls as isolated compliance checkpoints, each checkpoint adequate within its own scope, will discover during a governance failure that the failures cascade. A data quality failure propagates into a model output failure, which propagates into a decision-quality failure, which propagates into a regulatory compliance failure. Each individual control was adequate. The architecture connecting them was not.

The harness architecture addresses progressive flooding through two mechanisms.

P-properties are not isolated controls. They are interdependent constraints verified as a system. LP1 (access control) feeds LP2 (provenance); LP2 feeds LP3 (routing); LP3 feeds LP4 (logging). A failure in LP1 is detected not only by LP1's own verification but by the downstream properties that depend on LP1's output. The cascade is detected at the first boundary crossing, not at the last.

LP14 (Deterministic Deny on Uncertainty), the deny-by-default posture of the routing gateway at H-2, is the architectural equivalent of raising the bulkheads above the waterline. When the governance architecture encounters a condition it cannot evaluate against an authorized policy, the default is denial with escalation, not permissive processing with post-hoc review. The gateway fails closed, and the uncertainty does not propagate. The cascade stops at the point of uncertainty; the system escalates to human oversight before the failure can flow from one governance compartment into the next.

Organizations that build these architectural properties before the regulatory mandate are in the position of a shipbuilder who raises the watertight bulkheads to the full height of the hull: the engineering is slightly more expensive, the construction takes slightly longer, but when the iceberg arrives, the flooding does not cascade.

Organizations that wait for the regulatory mandate are in the position of every post-Titanic shipbuilder who retrofitted higher bulkheads under government order: the work is more expensive because it must be done around an operating vessel, the timeline is compressed because the regulation imposes a compliance deadline, and the result is often a compromise between what the engineering requires and what the operating constraints permit.

The ISS IGA Was Written Before the First Module Launched

Chapter 9 describes the International Space Station's Intergovernmental Agreement in detail. The structural point relevant here is temporal: the Intergovernmental Agreement (IGA) was signed on January 29, 1998. The first ISS module, Zarya, launched on November 20, 1998. Governance preceded operations by ten months.

The IGA was not written in response to an ISS incident. There were no incidents to respond to; the station did not yet exist. The participating nations recognized that a multi-jurisdictional operating environment of unprecedented complexity required a governance framework before the first module reached orbit, not after the first failure exposed the absence of one.

This is anticipatory governance. It is rare and expensive. It requires political will, institutional patience, and the acceptance that governance investment is justified by risk analysis rather than by disaster evidence. And it is the reason the ISS has operated continuously for over twenty-five years without a jurisdictional dispute that shut down operations.

AI governance offers the same opportunity. The regulatory framework is incomplete. Comprehensive requirements have not yet been enacted. The catastrophic AI governance failure that generates political will for comprehensive regulation has not yet occurred, or more precisely, individual failures have not yet accumulated into a single event with sufficient public salience to trigger the SOLAS response.

Organizations that build governance architecture now are writing their own IGA. They are establishing the jurisdictional rules, the operational procedures, the evidence requirements, and the accountability structures before the regulatory mandate arrives. When the mandate does arrive, and the reactive cycle guarantees that it will, these organizations demonstrate compliance from an existing architecture rather than constructing one under regulatory pressure.

The Pre-SOLAS Period

AI governance in 2026 is in its pre-SOLAS period. Evidence of harm is accumulating. The UnitedHealthcare AI denial system that Chapter 3 examined. Facial recognition misidentifications that produced wrongful arrests. AI-generated content that propagated medical misinformation. Hiring algorithms that reproduced historical discrimination. Autonomous vehicle incidents that exposed gaps in testing methodology. Each of these is a ship that sank. Each generated media coverage, legal proceedings, and calls for regulation. None, individually, has generated the political will for comprehensive international governance.

The Titanic was not the first ship to sink. It was not the first to sink from inadequate safety provisions. It was not even the most deadly maritime disaster of its era. But it was the disaster that broke through the threshold of public awareness with sufficient force to generate irreversible political momentum for comprehensive regulation. A combination of factors created a political moment that the maritime industry's previous losses had not: the ship's reputation as unsinkable, the prominence of its passengers, the specific avoidability of the loss, the clear inadequacy of existing regulations.

AI governance is waiting for its Titanic moment. This is not a prediction of doom; it is an observation about the structural dynamics of reactive regulation. Incidents are accumulating. Regulatory responses stay fragmented. Political will for comprehensive governance is building but has not reached the threshold. When an AI governance failure occurs that combines sufficient scale, sufficient visibility, sufficient avoidability, and sufficient regulatory inadequacy, the comprehensive regulatory response will follow with the same inevitability that SOLAS followed the Titanic.

Preparation is not optional. The only variable is timing: the organization's schedule or the regulator's.

The difference between those two timelines is the difference between the shipbuilder who raises the bulkheads because the engineering requires it and the shipbuilder who raises the bulkheads because the government orders it. Both ships end up with higher bulkheads. Only one of them was built by choice, on schedule, at the lower cost that deliberate engineering provides.

The harness architecture is the raised bulkhead. P-properties are the lifeboat capacity for every person aboard. Evidence tiers are the 24-hour wireless watch. The AGENTS.md is the safety management system. And the organization that builds them before the mandate is the one that will still be operating when the ice field is no longer optional to cross.

CHAPTER SUMMARY

The Titanic disaster of 1912 triggered the International Convention for the Safety of Life at Sea (SOLAS), establishing the reactive regulation pattern that governs all safety legislation: disaster, investigation, negotiation, ratification, implementation. SOLAS evolved through five major iterations (1914, 1929, 1948, 1960, 1974), each expanding the regulatory scope in response to new failure categories. The ISM Code, mandatory since 2002, shifted SOLAS from equipment requirements to management system requirements, the same structural shift that AI governance must make from compliance checklists to continuous verification architectures. The EU AI Act is the first SOLAS moment for AI, establishing risk classification, conformity assessment, and post-market monitoring obligations, while the United States pursues a fragmented approach through executive orders, state legislation, and sectoral regulation. Reactive regulation is structurally inevitable because anticipatory regulation requires political will that disaster evidence generates but risk analysis alone does not. The AI governance environment in 2026 occupies the pre-SOLAS period: incidents are accumulating, regulatory responses are fragmented, and the comprehensive mandate is approaching. Organizations that build governance architecture before the mandate, raising the watertight bulkheads before the iceberg, comply from existing infrastructure. Organizations that wait retrofit under regulatory pressure, at greater cost, on compressed timelines, with compromised results. The ISS Intergovernmental Agreement, signed ten months before the first module launched, demonstrates that anticipatory governance is possible and produces durable operational stability. The harness architecture is the anticipatory governance framework for AI systems.

Chapter 23: AI Incident Response for Regulated Industries

Part V: When Ships Sink

> *"The Mayday call does not ask for permission. It does not wait for a committee. It transmits on 2182 kHz and 156.8 MHz simultaneously, states the vessel's position, the nature of the distress, and the assistance required. Every station that hears it is legally obligated to respond. The protocol exists because the alternative, letting each captain improvise a distress procedure while the ship is sinking, killed people for centuries before anyone wrote it down."*

On February 1, 1999, at 0352 UTC, the International Maritime Organization's Global Maritime Distress and Safety System became the mandatory distress communication standard for all International Convention for the Safety of Life at Sea (SOLAS)-convention vessels. Global Maritime Distress and Safety System (GMDSS) replaced a patchwork of Morse code procedures, voice protocols, and regional radio systems with a single, globally standardized distress architecture. The system operates on defined frequencies. It transmits in defined formats. It triggers defined response obligations under SOLAS Chapter V, Regulation 33.1, which requires any master who receives a distress signal to proceed with all speed to render assistance, subject only to the master's own vessel being in distress.

What makes GMDSS work is not the radio equipment. It is the protocol stack beneath: who transmits, on what channel, in what format, to whom, with what mandatory response timeline, and what evidence is preserved. Every element was designed because its absence had produced a documented failure. Dedicated frequencies exist because shared frequencies meant distress calls were lost in commercial traffic. Improvised calls were misunderstood, so the format was made mandatory. Before SOLAS, vessels routinely altered course to avoid the obligation of rescue, which is why response obligations were written into the convention at all. The evidence preservation requirement has a darker origin. Without it, investigators could not distinguish between vessels that never received the signal and vessels that received it and ignored it.

AI governance incidents in regulated industries require the same architectural approach. Not because the metaphor is satisfying, but because the failure modes are operationally identical. An organization experiencing an AI governance incident, whether data exposure, fairness violation, consent boundary breach, or model poisoning, faces a coordination problem that improvisation cannot solve. Multiple stakeholders must be notified on different timelines. Evidence must be preserved while the bleeding is stopped. Regulatory obligations in different jurisdictions activate at different thresholds, and insurance carriers require specific documentation in specific formats. The organization's response then becomes evidence in its own right, the kind regulators and litigators will examine for years after the incident is resolved.

An organization with an incident response protocol before the incident occurs responds with the efficiency of a GMDSS-equipped vessel. Without one, the captain is shouting into whatever radio happens to be nearby, hoping someone is listening on the right frequency.

AI GOVERNANCE INCIDENT TAXONIOMY

INCIDENT TYPE	TRIGGERING CONDITION	SEVERITY
DATA EXPOSURE	PHI / MNPI / PCI surfaced outside authorized scope	SEV-1 CRITICAL
DATA EXPOSURE	Disparte impact on a protected class detected	SEV-1 CRITICAL
FAIRNESS VIOLATION	Processing beyond datect scope	SEV-2 HIGH
CONSENT BREACH	Processing beyond data subject consent scope	SEV-2 HIGH
VENDOR COMPORMISE	Third-party AI provider breached breached or degrated	SEV-2 HIGH
MODEL POISONING	Training or retireval data corrupted or or adversariaily modified	SEV-2 CRITICAL
EVIDENCE INTEGRITY	E-1 through E-4 records missing, insoristent, or comprmised	SEV-1 CRITICAL

Figure 23: The AI governance incident taxonomy; six incident types with the condition that triggers each and its severity classification.

The AI Incident Taxonomy

Before building the response protocol, the incident types must be defined with enough specificity that the response team can classify what it is facing in the first minutes of detection.

Data exposure incidents occur when an AI system discloses, generates, or logs data outside its authorized scope: protected health information (PHI) disclosure beyond the Health Insurance Portability and Accountability Act (HIPAA) minimum necessary standard, material non-public information (MNPI) leakage through an AI coding assistant's context window, PCI cardholder data appearing in AI interaction logs, Bank Secrecy Act (BSA)/Suspicious Activity Report (SAR) existence disclosure through any AI-generated output. These are the most common AI governance incident type because they exploit the most common architectural weakness, insufficient context scoping at the retrieval layer.

Fairness violation incidents occur when an AI system produces systematically biased outcomes across a protected class. In financial services, disparate impact in credit scoring, pricing, or account decisioning. In healthcare, differential treatment recommendations correlated with demographic factors rather than clinical indicators. In insurance, actuarial models that produce rate differentials not justified by underlying risk data. Fairness violations are rarely sudden. They accumulate over time; the incident's scope at detection is almost always larger than its scope at onset.

Consent boundary breaches occur when an AI system processes data or takes action beyond the consent the data subject provided. Under General Data Protection Regulation (GDPR) Article 6, processing requires a lawful basis. Under California Consumer Privacy Act (CCPA) Section 1798.100, consumers have the right to know what personal information is collected and how it is used. An AI system that ingests customer interaction data for a purpose the customer did not consent to, even internal model improvement, has exceeded its consent boundary.

Vendor compromise incidents occur when a third-party AI model provider, API service, or infrastructure component is breached, degraded, or discovered to have operated outside its contractual commitments. The organization's own governance architecture may be intact; the vendor's compromise introduces risk through the supply chain. Chapter 10's analysis of vendor risk establishes why this type is structurally different from internal incidents: visibility into the vendor's environment is limited to contractual attestation, not operational evidence.

Model poisoning incidents occur when training data, fine-tuning data, or the retrieval knowledge base of an AI system has been corrupted, either through adversarial action or through ingestion of unverified content that degrades model behavior. This is not system failure. It is output quality degradation that crosses a defined threshold. Poisoning is difficult to detect precisely because the system continues to function. It simply functions incorrectly.

Evidence integrity failures occur when the governance evidence architecture itself is compromised. E-1 governance records missing, E-2 verification results inconsistent with operational behavior, E-3 change detection records showing gaps, E-4 attribution chains broken. This is the most serious incident type in a governed architecture because it undermines the organization's ability to demonstrate compliance during the very period it most needs to demonstrate compliance.

The ISS Emergency Protocol Model

The International Space Station maintains three emergency protocols for three categories of life-threatening events: fire, depressurization, and toxic atmosphere. Each shares a common structure. A distinctive alarm tone. An immediate, rehearsed response: for fire, isolate the affected module and activate suppression; for depressurization, trace the leak and seal hatches; for toxic atmosphere, don breathing apparatus and activate air scrubbers. In every case, the first response is containment, not investigation. The crew stops the immediate threat before determining the root cause.

Assessment comes second. Once containment is achieved, crew and ground control evaluate the extent of damage, determine whether the containment is holding, and identify affected systems. Recovery comes third: restoring normal operations, repairing damaged systems, and documenting the event for the post-incident review that determines whether procedural or architectural changes are needed.

Containment, then assessment, then recovery. This three-phase model applies directly to AI governance incidents. The instinct to investigate first and contain second is the instinct that costs organizations the most regulatory exposure. Investigation takes time, and during that time the incident keeps producing damage, evidence, and notification obligations.

The Seven-Step Incident Response Sequence

The following sequence applies to any AI governance incident in a regulated environment. Steps are ordered by operational priority, not by organizational convenience. Steps that feel premature, particularly notification, are positioned where they are because the regulatory timeline demands it.

Step 1: Detection

An AI governance incident is detected through one of four channels.

Automated monitoring is the fastest. The E-2 continuous verification pipeline produces alerts when P-property violations are detected: a LP1 violation (classification-inconsistent access) surfaces as an access boundary breach; a LP13

(Output Boundary Enforcement) failure surfaces as an output containing data elements outside the authorized scope; a LP14 (Deterministic Deny on Uncertainty) failure surfaces as a permissive action where the harness should have denied. Speed is the advantage here, because automated detection operates on every interaction.

Human observation is the second channel. A user, operator, or governance team member notices that an AI system produced an output that should not have been produced, accessed data it should not have accessed, or took an action it was not authorized to take. Slower than automated detection, but it catches incident types the monitoring pipeline was not configured to detect.

External notification is the third. A regulator, a customer, a business partner, or a security researcher reports a governance issue. When detection comes from outside, the incident has already produced consequences visible beyond the organization.

Vendor disclosure is the fourth. The AI model provider or infrastructure vendor notifies the organization of a compromise, vulnerability, or behavioral anomaly. This channel activates the supply chain incident procedures from Chapter 10.

Detection creates the incident timestamp. This timestamp is the reference point for every notification deadline that follows. An organization's ability to demonstrate when it detected the incident, and that it responded within the applicable notification windows, depends entirely on whether that detection event is recorded with precision. In a governed architecture, the E-3 change detection record captures this automatically. In an ungoverned one, the timestamp is whatever someone remembers to write down.

Step 2: Triage

Triage classifies the incident by severity, which determines the response team composition, the notification timeline, and the containment posture.

Severity 1 (Critical) incidents involve active data exposure of the highest-sensitivity data tiers: MNPI in financial services, PHI in healthcare, CDE (cardholder data environment) in payment processing, Bulk Electric System (BES) Cyber System information in energy. Severity 1 also applies when the E-4 attribution chain is compromised, because without that chain the organization cannot demonstrate governed operation at all. Response: immediate containment, full incident response team activated within one hour.

Severity 2 (High) incidents involve fairness violations with regulatory reporting obligations, consent boundary breaches affecting a defined population, or vendor compromises where the vendor has access to regulated data. Response: containment within four hours, incident response team activated within the same business day.

Severity 3 (Moderate) incidents involve governance control failures that have not yet produced a compliance event but indicate architectural degradation: P-property verification failures persisting across multiple interactions, knowledge base integrity control violations (KB-1 through KB-4), or model output quality degradation exceeding defined thresholds. Response: investigation within 48 hours, remediation within the current maintenance cycle.

Sector-specific failure modes from Chapters 11 and 13 through 16 inform the triage classification. A retrieval scope failure in a financial services AI system that touches the MNPI tier is automatically Severity 1 because the regulatory consequence of MNPI disclosure is criminal, not merely civil. The same retrieval scope failure touching only the confidential tier is Severity 2. Identical architecture of failure. Severity is determined by what data was in the blast radius.

Real-Time Streaming Validation During Incident Surveys

Incident response triage (Step 2) requires rapid classification of incident scope, affected data, and severity level. In regulated industries, the difference between triage completed in minutes versus hours directly affects regulatory notification timelines. North American Electric Reliability Corporation (NERC) CIP-008's one-hour reporting requirement means the triage decision must occur within the first thirty minutes of detection to leave time for notification execution. A triage process requiring batch analysis of incident logs, manual review of affected transactions, or sequential queries against multiple evidence stores cannot meet this timeline.

Streaming validation supports the Lagrange point stability configuration by running concurrently with incident detection. It produces triage-critical information as evidence becomes available, rather than waiting for complete evidence collection before analysis begins. The maritime parallel is the distress signal itself. The Mayday transmission begins the moment distress is recognized, not after complete assessment of the vessel's condition. Notification timelines demand that triage information be available before investigation is complete.

Streaming Validation vs. Batch Triage

The International Space Station (ISS) fire suppression protocol demonstrates the streaming validation pattern. When a smoke detector activates, the crew does not conduct a complete atmospheric survey before declaring an emergency; the smoke detector's activation is sufficient to trigger immediate containment actions. Additional information streams in continuously: visual confirmation, thermal imaging, oxygen sensor readings, crew status reports. Each information stream refines the response, but the initial response begins before complete information is available.

Incident triage in AI governance systems traditionally operates as a batch process: detect the incident, collect all relevant logs, aggregate the evidence, analyze the scope, classify the severity, then initiate response. For a data exposure incident spanning thousands of AI interactions across multiple systems, this batch process can require hours. Log collection alone may take 30-60 minutes for distributed systems; aggregation adds another 15-30 minutes; scope analysis against regulatory data classification requirements adds another 30-60 minutes. Three hours elapsed before containment begins.

Streaming validation turns triage into a continuous process, validating incident evidence as it arrives. When an E-2 monitoring alert fires indicating a LP1 violation (classification-inconsistent access), the triage process does not wait for complete log collection. The first evidence stream begins immediately: which system generated the alert, which user or session triggered it, what data classification tier was involved. That evidence is checked against the triage criteria in real time, and severity is classified as soon as sufficient information is available.

Concurrent Evidence Validation Streams

A data exposure incident in a financial services environment involves multiple evidence streams that arrive at different rates. The E-2 alert identifying the access boundary failure arrives immediately. Session context showing which user initiated the AI interaction arrives within seconds. Retrieved data classification records showing what data was accessed arrive within minutes. Audit logs showing whether the data was transmitted externally arrive within tens of minutes. Traditional batch triage waits for all streams to complete before analysis begins.

Streaming validation processes each stream as it arrives, updating the triage classification continuously. The initial classification based on the E-2 alert alone might be "potential Severity 1" because the alert indicates MNPI-tier access. As session context arrives showing the access occurred within an authorized MNPI session under appropriate Chinese wall controls, classification updates to "Severity 2, investigating boundary failure within authorized context." As audit logs arrive showing no external transmission occurred, classification updates to "Severity 3, governance control degradation without data exposure."

Each classification update informs the response team's decisions in real time. The "potential Severity 1" initial classification triggers immediate General Counsel notification; the updated "Severity 2" classification activates the incident response team but does not yet trigger regulatory notification; the final "Severity 3" classification shifts from incident response mode to governance remediation mode. Each update narrows the response, preventing over-reaction (triggering regulatory notification for incidents that do not meet reporting thresholds) and under-reaction (delaying containment while waiting for complete evidence).

The maritime parallel is the progressive distress notification protocol. A vessel experiencing propulsion loss in calm weather with adequate fuel and no injury may transmit a Pan-Pan urgency message rather than a Mayday distress call. As conditions worsen (weather deteriorates, fuel depletes, or injury occurs), the vessel escalates to Mayday. Each transmission reflects current conditions, not anticipated final outcome. Incident response follows the same pattern: classify based on available evidence, escalate as evidence updates, and de-escalate if evidence shows the incident is less severe than initially assessed.

Preventing Notification Failures Through Streaming Classification

Step 4 (Notification) failures occur when organizations miss regulatory reporting deadlines because triage took too long to produce actionable classification. The GDPR 72-hour notification window is consumed by investigation time; the NERC CIP one-hour reporting requirement is missed entirely because triage is still in progress. Streaming validation prevents these failures by producing preliminary classification within minutes of detection, enabling notification processes to begin while investigation continues.

The triage classification produced by streaming validation is explicitly preliminary, accompanied by confidence scores indicating the certainty of classification based on available evidence. A "Severity 1 (confidence: 0.6)" classification means current evidence suggests Severity 1 but significant uncertainty remains; regulatory notification procedures are initiated, but containment actions appropriate to Severity 2 or 3 are held pending additional evidence. As confidence increases (0.6 becomes 0.8 becomes 0.95), response actions escalate or de-escalate accordingly.

This confidence-scored streaming classification aligns with the Lloyd's Register continuous survey model. A classification surveyor conducting an incremental inspection during a port call may identify a potential structural deficiency requiring detailed examination. The surveyor issues a conditional notation: "Classification maintained subject to detailed survey of [specific component] within [defined timeframe]." The vessel retains its classification, but the condition is documented, the timeline is defined, and the vessel owner initiates remediation before the detailed survey occurs. Streaming validation during incidents applies the same pattern: classify based on current evidence, document the conditions under which that classification may change, and update as evidence arrives.

Open-Source Model Governance During Response

Step 5 (Investigation) requires that every analytical step, every classification decision, and every scope determination be auditable by regulators, insurance carriers, and opposing counsel during the years following an incident. When the incident investigation itself depends on AI-assisted analysis (log parsing, scope classification, impact assessment, root cause determination), the governance of those analytical AI operations becomes part of the incident record that must withstand regulatory scrutiny.

Open-source model governance supports the Lagrange point stability configuration by extending governance to the open-source AI models used during incident investigation. The analytical tools themselves must not introduce governance gaps that undermine the investigation's defensibility. Consider the maritime accident investigation parallel. The instruments used to investigate a vessel loss (depth sounders, salvage positioning, hull section analysis) must themselves be calibrated, certified, and documented, or the findings they produce are not admissible evidence.

Incident Analysis as Governance Event

When a financial services organization uses an AI model to classify whether a data exposure incident involved MNPI, that classification decision determines regulatory notification obligations, insurance coverage applicability, and potential criminal exposure under insider trading statutes. The AI model performing that classification is itself operating under governance constraints; if the model is a proprietary API with no visibility into its classification logic, the organization cannot defend the classification decision beyond "the model said so."

An open-source model addresses this by making classification logic auditable. When an incident investigation uses one (Llama 3.2, Mistral, Gemma) to parse incident logs and identify MNPI indicators, the model's weights are open to inspection, the training data provenance is documented, and an independent reviewer can reproduce the classification decision using the same model, the same evidence, and the same prompt. The classification stops being a black box opinion. It becomes a deterministic operation with documented inputs and auditable logic.

The ISS accident investigation protocol demonstrates this principle. When the International Space Station experiences an anomaly requiring investigation, NASA's Mishap Investigation Board operates under defined procedures (NPR 8621.1) that require preservation of all evidence, documentation of all analytical methods, and reproducibility of all conclusions. An investigation finding that depends on proprietary analytical tools not available to external review would not satisfy the standard. Open-source models applied to incident investigation provide the same reproducibility guarantee.

Supply Chain Auditability During Vendor Incidents

Step 4 of the incident taxonomy identifies vendor compromise incidents as a distinct category requiring supply chain incident procedures. When a vendor-supplied AI model is involved in an incident (model poisoning, unexpected behavior, data leakage through context windows), the investigation must determine whether the incident arose from the vendor's infrastructure, the organization's deployment, or the interaction between them. If the vendor's model is proprietary with no architecture visibility, that investigation cannot proceed beyond the vendor's attestations.

An open-source deployment removes this barrier. When an organization runs an open model and a governance incident occurs, the investigation team has complete visibility into model architecture, training data characteristics, and deployment configuration. The team can separate model-level failures (training data poisoning, architecture vulnerabilities) from deployment-level failures (inadequate context scoping, missing governance constraints). The vendor is no longer the gatekeeper to investigation evidence. The model itself is the evidence.

The maritime parallel is the vessel's statutory certification documentation. When a classification society investigates a hull failure, the investigation requires access to the vessel's construction plans, material certifications, and modification history. If the vessel owner claims those documents are proprietary to the shipyard and cannot be released, the investigation cannot determine whether the failure arose from design, construction, or operational loading. Open documentation is the precondition for effective investigation. Open-source models provide the same documentation standard for AI systems.

Regulatory Examination of Investigation Methods

Step 7 (Regulatory Response) includes the regulator's examination of the organization's investigation methods. When an organization reports a data exposure incident and asserts that the exposure affected 1,200 individuals based on AI-assisted log analysis, the regulator will examine how that count was determined. If the count depends on a proprietary AI model's classification of log entries, and the organization cannot explain the classification logic beyond "the vendor's model performed the analysis," the regulator has no basis to validate the count. The investigation finding itself becomes suspect.

Open methods are defensible methods. When log analysis runs on an open-source classification model with documented prompts, documented classification criteria, and reproducible results, the regulator can verify the count independently. The E-4 attribution record captures which model version was used, which prompts were applied, and which evidence was analyzed. A skeptical examiner reproduces the analysis using the same open-source model and verifies the organization's count for themselves. The investigation rests on transparency, not on attestation.

This requirement is particularly acute for fairness violation incidents (F-C4). When an organization investigates whether an AI credit scoring model produced disparate impact, and that investigation uses AI-assisted statistical analysis to evaluate sub-population outcomes, the statistical analysis itself must be auditable. If the analysis depends on proprietary tools, the organization cannot demonstrate the fairness finding to a regulator or opposing counsel without exposing proprietary methodology that may be legally privileged. Open-source statistical analysis frameworks resolve this by making the analytical methods public, reproducible, and defensible without surrendering legal protections on the investigation's conclusions.

Preventing Evidence Architecture Failures During Response

F-C5 (evidence architecture failure) includes a variant where the incident investigation itself produces evidence that cannot survive regulatory examination because the investigation methods were not governed. An organization that deploys ungoverned AI tools to investigate a governed AI system creates an evidence gap at precisely the moment it most needs complete evidence: during the incident investigation that will be examined by regulators for years after the incident is resolved.

The fix is to treat incident investigation as a governance event subject to the same E-1 through E-4 evidence tiers as operational AI systems. When an investigation uses an AI model to classify incident scope, that classification is an E-4 attribution event: which model, which version, which prompt, which evidence, which operator authorized it. When the investigation revises its classification on new evidence, that revision is an E-3 state transition event. The investigation generates its own audit trail, built on the same evidence architecture that governs the systems it is investigating.

The maritime investigation parallel is the Air Accidents Investigation Branch (AAIB) procedures, which apply the same rigor to investigation methods as to operational aviation systems. An AAIB investigation of a flight control failure documents not just the failure but the investigation process itself: which instruments were used, which standards were applied, which experts reviewed the findings, and which analytical methods produced each conclusion. The investigation report is itself auditable evidence, not merely a narrative summary. Open-source models enable the same standard for AI incident investigations.

Step 3: Containment

Containment serves two purposes that are in tension with each other: stopping the incident from producing additional damage, and preserving the evidence the investigation and regulatory response will require.

The ISS depressurization protocol illustrates this tension precisely. Crew seals hatches to stop the atmosphere from escaping, but not every hatch simultaneously; doing so would isolate crew members from life support systems. Targeted containment: seal the leaking module, verify the seal is holding, confirm the rest of the station is stable.

For AI governance incidents, targeted containment means four things.

Disable the specific AI system or capability involved in the incident. Do not shut down the entire AI platform unless the incident's scope cannot be determined. Disabling a single system is a containment action. Shutting down the platform is a business continuity event that triggers its own cascade of consequences.

Preserve the evidence state at the moment of containment. The E-1 governance records, E-2 verification results, E-3 change detection logs, and E-4 attribution chains that were being produced up to and including the incident are the organization's primary defense in the regulatory and legal proceedings that follow. These records must be placed under litigation hold immediately. The governance evidence from the period before the incident is as important as the evidence from the incident itself, because it demonstrates whether the organization was operating in a governed state when the failure occurred.

Isolate the data that may have been exposed. If the incident involves data disclosure, the specific data elements, their sensitivity classification, the recipients (if known), and the duration of exposure must be identified during containment. This information drives the notification obligations in Step 4.

Document every containment action, who took it, when, on what authority, and what the action's intended effect was. The containment record becomes part of the E-4 attribution chain for the incident. A containment action taken without documentation is a containment action that cannot be defended in a regulatory examination.

Step 4: Notification

Notification is the step where most organizations fail. Not because they refuse to notify, but because they do not know who must be notified, in what order, on what timeline.

Regulatory notification obligations for AI governance incidents are a collision of overlapping deadlines.

GDPR Article 33 requires notification to the supervisory authority within 72 hours of becoming aware of a personal data breach. Article 34 requires notification to affected data subjects "without undue delay" when the breach is likely to result in a high risk to their rights and freedoms. The 72-hour clock starts at awareness, not at confirmation. Spend 71 hours confirming whether a breach occurred, and one hour remains to notify.

HIPAA Breach Notification Rule (45 CFR 164.408) requires notification to the Secretary of the Department of Health and Human Services (HHS) without unreasonable delay, but no later than 60 calendar days from discovery of a breach of unsecured PHI. Breaches affecting 500 or more individuals also require notification to prominent media outlets in the affected jurisdiction. The 60-day window appears generous compared to GDPR's 72 hours, but HIPAA's definition of "discovery" is aggressive: a breach is deemed discovered on the first day the covered entity knew, or by exercising reasonable diligence would have known, about the breach. An E-2 monitoring alert that was generated but not reviewed starts the clock.

State breach notification laws add a third timeline layer. California Civil Code Section 1798.82 requires notification "in the most expedient time possible and without unreasonable delay." New York General Business Law Section 899-aa requires notification "in the most expedient time possible." Colorado's Protections for Consumer Data Privacy Act (CRS 6-1-716) requires notification within 30 days. These laws apply based on the residence of the affected individuals, not the location of the organization. A single incident affecting individuals in multiple states triggers multiple notification obligations with different timelines.

NERC CIP-008 requires reporting of cybersecurity incidents affecting bulk electric system reliability within one hour to the Electricity Information Sharing and Analysis Center (E-ISAC) and to the Cybersecurity and Infrastructure Security Agency (CISA). One hour. Not one business day. Not "as soon as practicable." Sixty minutes from identification of a reportable cybersecurity incident.

For a multi-jurisdictional AI governance incident, the notification sequence proceeds in five stages.

Internal escalation comes first. The incident response team lead, CISO, General Counsel, and the relevant business unit leader are notified immediately. General Counsel's involvement at this stage is not optional; it establishes attorney-client privilege over the incident investigation communications.

Regulatory notification follows. Each applicable regulator is notified within the shortest applicable deadline. For an incident touching NERC CIP, within one hour. For GDPR, within 72 hours. For HIPAA, the 60-day outer boundary allows more deliberation, but the "without unreasonable delay" standard means waiting the full 60 days without justification is itself a violation.

Insurance carrier notification is third. Most cyber insurance policies require notification "as soon as practicable" or "immediately upon discovery" of a covered event. Miss the policy's window and coverage can be voided. This notification engages the carrier's breach response panel, which typically includes pre-approved forensic investigators, outside counsel, and crisis communications firms. Using that panel is rarely a matter of preference. It is frequently a policy condition, and bringing in non-panel resources without carrier approval can create coverage disputes.

Affected individual notification is fourth. Timing, content, and delivery method vary by regulatory framework and jurisdiction. GDPR requires clear and plain language describing the nature of the breach and the measures taken. HIPAA requires specific content elements including a description of the types of information involved. State laws vary widely. General Counsel drives this notification based on legal analysis conducted during the investigation.

Board notification is fifth. The board audit committee receives notification as a governance event, not as a crisis management escalation. Classification, regulatory notification status, containment posture, estimated financial exposure. This notification becomes part of the E-4 governance record and supports the Item 106 disclosure obligations discussed in Chapter 20.

Step 5: Investigation

Investigation determines what happened, why it happened, and what the governance architecture should have prevented.

In a governed architecture, the investigation starts with evidence that already exists. E-1 governance records show what the AI system was authorized to do. E-2 verification results show whether governance controls were operating as designed in the period leading up to the incident. E-3 change detection records show what changed, and when, in the governance configuration. E-4 attribution chains show who authorized each change and who was accountable for each governance decision.

This is the "black box" advantage. Not reconstruction from memory, interviews, and fragmentary logs. The evidence architecture was producing the investigation record before anyone knew there would be an investigation. The incident response team's task is not to create evidence. It is to query the evidence that already exists.

The investigation answers five questions.

What governance property failed? Map the incident to the specific P-property (LP1 through LP14) whose enforcement broke down. A data exposure maps to LP1 (Classification-Consistent Access) or LP13 (Output Boundary Enforcement). A fairness violation maps to LP9 (Consent-Bounded Processing) or LP11 (Fairness Verification). An evidence integrity failure maps to LP8 (Evidence Integrity) or the E-tier architecture itself.

When did the failure begin? The E-2 verification history shows the last time the relevant P-property passed verification. The failure began at some point between the last successful verification and the first failed verification, or the first incident manifestation. This window defines the incident's duration and, therefore, the scope of affected data or decisions.

What was the root cause? The harness architecture provides three categories. First, the governance property was correctly defined but incorrectly enforced, a technical implementation failure. Second, the governance property was correctly enforced but the enforcement did not cover the specific scenario that produced the incident, a specification gap. Third, the governance property was not defined for the risk that materialized, a governance design failure.

Who was affected? The E-4 attribution chain, combined with the data classification records, identifies which data subjects, customers, patients, or counterparties were affected by the incident. This information drives the notification content in Step 4 and the remediation scope in Step 6.

What did the organization know, and when? This is the question regulators ask. The E-2 monitoring alerts, the E-3 change records, and the E-4 authorization decisions create a timeline of organizational awareness. An organization that can demonstrate it detected the issue through automated monitoring, contained it within hours, and notified within the applicable deadline is in a fundamentally different regulatory posture than an organization that cannot demonstrate when it became aware.

Step 6: Remediation

Remediation closes the governance gap that permitted the incident. Not the same as containment. Containment stops the bleeding; remediation repairs the wound.

Root cause analysis from Step 5 determines the remediation path. A technical implementation failure requires an engineering fix to the enforcement mechanism, followed by E-2 verification confirming the fix is operational. A specification gap requires updating the governance property definition in AGENTS.md, extending enforcement to cover the uncovered scenario, and verifying the extension through the same E-2 pipeline. A governance design failure requires the deepest remediation: defining a new or revised governance property, implementing its enforcement, establishing its verification baseline, and integrating it into the continuous monitoring pipeline.

Every remediation action is recorded in the E-3 change detection tier and attributed through the E-4 chain. The remediation record becomes the organization's evidence that the governance gap was identified, understood, and closed. Regulators expect remediation evidence. Insurance carriers expect remediation evidence. Opposing counsel in litigation expects remediation evidence. The organization that produces it from its existing evidence architecture demonstrates responsive governance. The organization that cannot produces explanations about why it does not know what it fixed or when.

Step 7: Regulatory Response

Regulatory response is the organization's formal engagement with the examination that follows any notified incident. Not a single event. A sustained interaction that may extend for months or years after the incident itself is resolved.

The harness architecture provides this response with its primary asset: the continuous governance record. The organization demonstrates not that it governed the AI system during the incident, but that it governed the system before, during, and after, because the evidence architecture was producing governance records continuously, not in response to the regulatory inquiry.

The E-4 attribution chain is the decisive artifact. Every governance decision, every property verification, every containment action, every remediation step connects to a specific person, at a specific time, with a specific authorization. A regulator examining this chain sees an organization that knew what it was doing, documented what it was doing, and can trace every action to an accountable party. This is the regulatory posture that produces consent orders with minimal penalties, examination findings with short remediation timelines, and enforcement actions that acknowledge governance maturity even while addressing the specific failure.

Without an evidence architecture, the regulatory conversation unfolds differently. The regulator asks what happened; the organization reconstructs from memory and fragmentary records. The regulator asks when awareness occurred; the organization cannot demonstrate the moment because detection was not recorded in a structured evidence tier. The regulator asks what the organization did; the response is described but the timestamped, attributed record of each action cannot be produced. The regulator asks what controls were in place before the incident; the organization presents policy documents that describe intended governance without evidence the governance was operational.

The difference between a vessel with a complete ship's log and a vessel whose crew is trying to remember what happened. The vessel with the log may still face consequences. The vessel without the log faces consequences and the presumption that the crew has something to hide.

Cross-Sector Incident Contagion

A single AI governance incident in one jurisdiction can trigger notification and remediation obligations in jurisdictions the organization did not initially consider affected.

A financial services organization operating in the United States and the European Union discovers that an AI-assisted customer segmentation model produced disparate impact outcomes. The model was trained on US customer data but applied to EU customers through a shared infrastructure. The US fairness violation triggers ECOA and fair lending examination obligations domestically. The same incident, because it processed EU residents' personal data in a manner that produced discriminatory outcomes, triggers GDPR Article 22 obligations regarding automated individual decision-making and potentially Article 33 notification if the fairness violation constitutes a personal data breach under the supervisory authority's interpretation.

Contagion runs in both directions. An EU-originated incident involving an AI system's processing of personal data in violation of GDPR's purpose limitation principle (Article 5(1)(b)) may also trigger CCPA notification obligations if California residents' data passed through the same system. AI governance incidents ignore jurisdictional boundaries for one reason. The AI system's data processing ignores them too.

Every incident response protocol must include a jurisdictional impact assessment as a standard step in triage. Map the affected data to the jurisdictions whose regulatory frameworks apply, producing the notification obligation matrix that drives Step 4. Perform this assessment during triage, not weeks into the investigation; otherwise the compounding regulatory exposure of late notification adds a second problem to the first.

The Insurance Documentation Chain

Cyber insurance carriers have developed specific expectations for AI governance incident documentation. Not suggestions. Expectations with coverage consequences. The carrier's breach response team, which the policy typically requires the insured to engage, expects the following documentation within the first 72 hours.

A preliminary incident summary describing the nature of the incident, the systems involved, the data potentially affected, and the containment actions taken. The E-3 change detection records and E-4 attribution records provide this summary as a query against existing evidence, not as a narrative reconstructed from interviews.

A data exposure assessment estimating the volume, sensitivity, and jurisdictional distribution of affected data. The data classification records from the LP1 implementation and the access logs from the LP6 attribution records produce this assessment.

A timeline of organizational awareness and response actions. The E-2 monitoring alerts and E-4 containment records produce this timeline.

A preliminary root cause assessment identifying the governance control that failed and the remediation path. The P-property verification history and the E-3 change detection records inform this assessment.

Organizations with continuous evidence architecture produce this documentation from existing records. Organizations without it spend the first 72 hours, the same 72 hours GDPR allocates for regulatory notification, trying to assemble what the carrier requires. Prompt, structured documentation gets claims processed. Delayed, incomplete documentation gets the carrier's own forensic team assigned, at the insured's expense, to determine what happened.

The Black Box Argument

Aviation incident investigation was transformed by the flight data recorder and the cockpit voice recorder. Before mandatory "black boxes," aircraft accident investigation relied on physical wreckage analysis and survivor testimony. After mandatory recorders, investigation relied on continuous data capture that recorded the aircraft's state and the crew's actions regardless of whether anyone anticipated the incident.

The AI governance evidence architecture is the black box for regulated AI operations. The E-1 through E-4 evidence tiers record the governance state, the verification results, the changes, and the attribution chain continuously, as a byproduct of normal operations, without requiring anyone to predict which records will matter.

Organizations with evidence architecture recover from incidents faster, for three measurable reasons.

Investigation timelines compress. The evidence already exists, so the investigation team queries it rather than reconstructing it. An investigation that takes six weeks without evidence architecture takes two weeks with it, because four of those six weeks were spent finding and assembling records in the first place.

Regulatory response is stronger. The organization shows not merely that it responded to the incident, but that it was governing the AI system throughout its operation. Governance posture before the incident matters as much as the response during it, and the evidence architecture captures both.

Insurance claims proceed without dispute. The carrier receives the documentation it requires, in the format it expects, on the timeline the policy demands. Coverage fights over late notification or incomplete documentation never start, because the documentation was produced automatically.

The maritime parallel closes the argument. Lloyd's Register did not prevent shipwrecks. It ensured that when a ship went down, the underwriter knew what the ship was, what it was carrying, what condition it was in at last survey, and whether the classification certificate was current. The organization that maintained its classification recovered its insurance claim. The organization that had let its survey lapse, or could not produce its certificate, absorbed the loss.

Evidence architecture does not prevent AI governance incidents. It ensures that when an incident occurs, the organization knows what happened, can prove what it knew, can demonstrate what it did, and can show that governance was operational throughout. Not the difference between safety and danger. The difference between an incident and a catastrophe.

CHAPTER SUMMARY

AI governance incident response in regulated industries requires the same architectural discipline as maritime distress protocols and ISS emergency procedures: pre-defined response sequences executed under time pressure, not improvised reactions to unanticipated failures. The AI incident taxonomy covers data exposure, fairness violations, consent boundary breaches, vendor compromise, model poisoning, and evidence integrity failures. The seven-step response sequence proceeds through detection, triage, containment, notification, investigation, remediation, and regulatory response. Regulatory notification timelines vary dramatically across frameworks: NERC CIP requires one-hour reporting, GDPR requires 72-hour notification, HIPAA allows up to 60 days, and state breach notification laws impose varying requirements based on the residence of affected individuals. Containment must balance stopping additional damage against preserving the evidence that the investigation, regulatory response, and insurance claim will require. The E-4 attribution chain is the decisive artifact in regulatory examination, connecting every governance decision and response action to an accountable party with a timestamp. Cross-sector incident contagion occurs when an incident in one jurisdiction triggers notification obligations in others because the AI system's data processing crosses jurisdictional boundaries. The harness evidence architecture functions as the black box for regulated AI operations, compressing investigation timelines, strengthening regulatory response, and preventing insurance coverage disputes by producing the documentation that every stakeholder requires from records that already exist.

Chapter 24: From Government Station to Commercial Platform

Part VI: Commercial LEO Destinations

> *"The station was never meant to last forever. It was meant to prove that permanent human presence in orbit was possible, and then to hand that capability to the organizations that would operate it commercially. The governance architecture was the thing worth inheriting."*

On January 30, 2020, NASA published the International Space Station Transition Report, formally acknowledging what the agency had been signaling for a decade: the International Space Station (ISS) would not operate indefinitely as a government facility. Structural certification extends to 2030. After that, the United States government intends to purchase services from commercial space stations rather than operating its own. The Commercial LEO Destinations (CLD) program, announced in 2021 and funded through Space Act Agreements, selected four companies to develop commercial replacements: Axiom Space, Blue Origin's Orbital Reef consortium, Nanoracks' Starlab, and Northrop Grumman.

The transition model is precise and instructive. NASA does not hand the keys to a private company and walk away. It defines the safety requirements, the crew health standards, the environmental monitoring thresholds, and the emergency response protocols. Commercial operators design, build, and operate stations that satisfy those requirements. NASA then purchases crew time, research berths, and logistics services. The government sets the governance standard; the commercial sector determines how to meet it. Operational execution transfers. Governance architecture persists.

This is the exact model this book proposes for AI governance in regulated commercial environments. The Harness Doctrine was developed in federal classified environments and tested against the most restrictive governance requirements the United States government imposes. Its harness architecture, evidence model, P-properties, and the AGENTS.md instruction surface were designed for places where a single misconfiguration produces a national security incident. The framework works. Twenty-five years of continuous ISS operation proves the governance model holds. The open question was never whether the architecture is sound. It was how the architecture transfers.

The Federal Validation Argument

An organization that has built AI governance capable of satisfying National Institute of Standards and Technology (NIST) 800-53, the FedRAMP authorization process, and the continuous monitoring requirements of the Department of Defense (DoD) Risk Management Framework has already solved the harder version of every problem commercial governance presents.

Not an abstract claim. The structural comparison is direct.

Federal authorization requires that every control be implemented, tested, and evidenced before the system receives its Authorization to Operate (ATO). Commercial System and Organization Controls (SOC) 2 Type II attestation requires the same sequence in different vocabulary: control implementation, operating effectiveness testing, evidence that covers the attestation period. The federal process is more demanding because it requires government assessor review; the commercial process is more fragmented because it must satisfy multiple assessors at once. But the evidence architecture that satisfies the federal requirement, E-1 through E-4, satisfies every commercial requirement too. Commercial requirements ask for subsets of what the federal process already demands.

Harness components H-1 through H-5 were designed for classified networks where the AI system operates on data that an adversarial nation-state is actively trying to exfiltrate. Commercial data protection requirements are real; material non-public information (MNPI) isolation, protected health information (PHI) minimum necessary, BES Cyber System Information (BCSI) boundary enforcement, and PCI cardholder data prohibition are all enforceable with serious consequences. But none carry the threat model of a classified network. An organization that can enforce data access boundaries against a nation-state adversary can enforce them against a misconfigured application programming interface (API) query.

Knowledge base integrity controls KB-1 through KB-4 were developed for environments where a poisoned knowledge base produces compromised intelligence assessments. In a commercial setting, the same poison produces a biased credit decision or an incorrect clinical recommendation instead. The consequences differ. The control architecture does not.

The federal-to-commercial validation argument is not that commercial governance is easy. It is that the hard engineering, the constraint architecture, the evidence pipeline, the continuous verification model, has already been done. Commercial deployment extends it. It does not reinvent it.

What Transfers and What Extends

Chapter 2 introduced the distinction: the inherited federal baseline of the Three-Body governance architecture, the harness components, the evidence tiers, the knowledge base controls, and the eight federal security properties (P1 through P8), transfers without modification from federal to commercial environments. This volume's Lagrange governance properties (LP1 through LP14) sit on that baseline, and the four-property commercial extension within them (LP9 through LP12) addresses requirements that federal environments do not face.

The inherited federal baseline includes the complete harness architecture (H-1 through H-5), the eight federal security properties (P1 through P8, defined in Volumes I and II), all four evidence tiers (E-1 through E-4), all four knowledge base integrity controls (KB-1 through KB-4), and the AGENTS.md instruction surface structure. Sector-independent, all of it. These components govern the relationship between an autonomous AI system and the constraints placed on it. That relationship holds whether the constraints come from Health Insurance Portability and Accountability Act (HIPAA) instead of NIST 800-53, or from Payment Card Industry Data Security Standard (PCI-DSS) instead of Security Technical Implementation Guides (STIGs). This volume defines its Lagrange governance properties (LP1 through LP14) on top of that baseline. LP1 through LP8 are the core governance-worthiness criteria, several of which restate a federal security property in governance terms; they are this volume's governance layer, not a copy of the federal eight transferred unchanged.

The four-property commercial extension consists of the four commercial properties defined in Chapter 8: LP9 (Consent-Bounded Processing), LP10 (Jurisdictional Consistency), LP11 (Fairness Verification), and LP12 (Transparency and Explainability). These properties address requirements that do not exist in federal classified environments because federal environments do not process data under individual consent regimes, do not face multi-jurisdictional regulatory conflicts, do not make consumer-facing decisions subject to disparate impact analysis, and do not owe explanation artifacts to affected individuals.

The practical migration maps the sector-independent core first, then adds the four-property commercial extension.

The first phase maps the sector-independent core by translating the vocabulary. The Authorizing Official (AO) becomes the Board Audit Committee plus Chief Risk Officer (CRO). The Information System Security Officer (ISSO) becomes the International Organization for Standardization (ISO) or HIPAA Security Officer. The ATO becomes SOC 2 Type II attestation or NERC CIP audit clearance. Classification levels become data sensitivity tiers; STIGs become CIS Benchmarks and sector-specific technical standards. The architecture stays put. Only the labels change. Chapter 2's translation table provides the complete mapping.

Next comes the four-property commercial extension. Add LP9 through LP12 to the AGENTS.md instruction surface. Connect LP9 to the consent management platform,

LP10 to the jurisdictional routing layer, LP11 to the fairness evaluation pipeline, LP12 to the explanation generation checkpoint. Each connection is a new enforcement rule in the routing layer (H-2), not a new architecture. The harness already holds the enforcement mechanism; the commercial extension adds four rules to it.

The third phase activates multi-framework rendering. Connect the evidence pipeline to the Open Security Controls Assessment Language (OSCAL) rendering layer described in Chapter 18, then configure rendering templates for every applicable regulatory framework: SOC 2, PCI-DSS, HIPAA, North American Electric Reliability Corporation (NERC) CIP, Sarbanes-Oxley Act (SOX), General Data Protection Regulation (GDPR), EU AI Act. The evidence stream itself never changes. The rendering layer produces framework-specific output from that common stream. Pack the evidence once at origin; render it at each destination port.

The entire migration, from federal-validated architecture to commercial-deployed governance, follows the ISS CLD model. NASA proved the architecture works. Commercial operators inherit it, extend it for their operating environment, and run it at competitive pace.

What Transfers Directly

The following table identifies the specific architectural components that transfer without modification, and the commercial context in which they operate.

Federal Component	Commercial Equivalent	Why It Transfers
H-1 Instruction Surface	AGENTS.md with commercial regulatory bindings	The instruction encoding mechanism is framework-independent
H-2 Routing Enforcement	H-2 with LP9-LP12 routing rules added	The enforcement layer accepts new rules without structural change
H-3 Persistent State	Session state with consent and jurisdiction attributes	State management extends to include commercial data elements
H-4 Verification Checkpoints	H-4 with fairness and explanation checkpoints	The checkpoint mechanism accepts new verification types
H-5 Accountability Records	E-4 attribution chain with commercial regulatory mappings	Attribution records gain new regulatory context fields
E-1 Foundation Evidence	Governance configuration baseline	The baseline concept is universal

Federal Component	Commercial Equivalent	Why It Transfers
E-2 Operational Evidence	Continuous verification including LP9-LP12 checks	The monitoring pipeline accepts new check types
E-3 Compliance Evidence	State transition records with framework-specific annotations	Transition records gain multi-framework rendering
E-4 Accountability Evidence	Attribution chain with commercial explanation artifacts	Attribution records gain consumer-facing explanation outputs
KB-1 Verified Provenance	Knowledge base provenance controls	Provenance verification applies to any knowledge base
KB-2 Enforced Currency	Change management with consent-aware indexing	Currency controls extend to consent-scoped content
KB-3 Controlled Ingestion	Data sensitivity tier enforcement	Tier enforcement translates directly to commercial tiers
KB-4 Monitored Retrieval	Cross-domain pollution monitoring	Retrieval monitoring is sector-independent
P1-P8 federal security properties	LP1-LP8 governance properties restate these in commercial governance terms	The federal security properties are architectural, not vocabulary-specific; this volume's governance layer is built on them

Every row in this table represents engineering that has already been done. The commercial deployment does not rebuild these components. It reconfigures them.

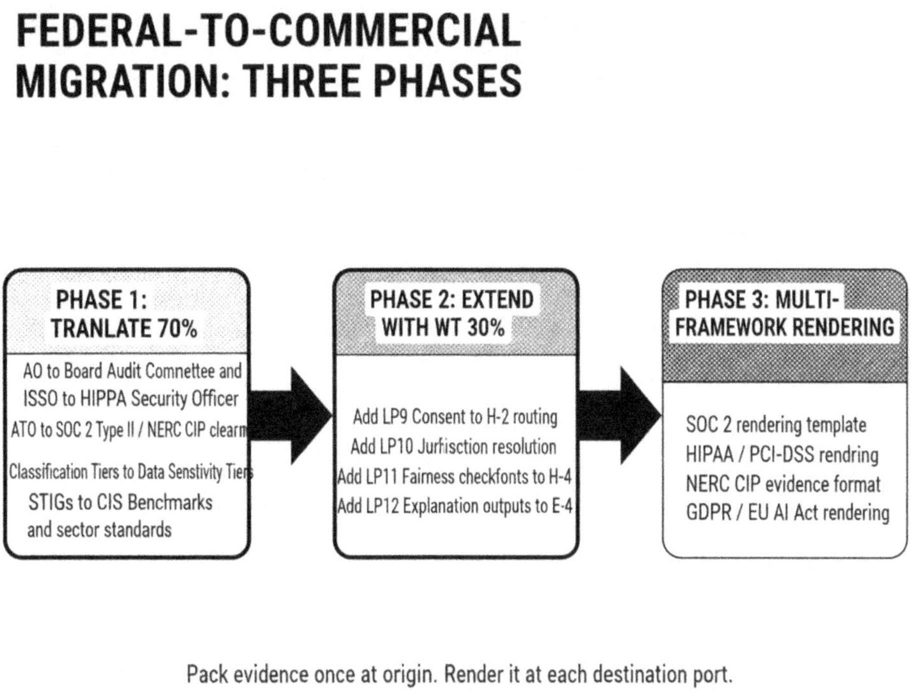

Figure 24: The three-phase federal-to-commercial migration; translate the 70 percent that transfers, extend with the commercial properties, then render evidence to each framework.

What Requires Extension

The commercial extension is not architectural. It is regulatory. The harness has the mechanisms to enforce commercial requirements. What it lacks, in its federal form, is policy content for four categories that federal environments do not encounter.

Consent management is the first gap. Federal data processing operates on institutional authorization, not individual consent, so the federal harness has no consent verification gate. The commercial extension adds one to H-2's routing enforcement: before the AI system accesses a record subject to GDPR, CCPA, HIPAA authorization, or 42 CFR Part 2, the routing layer queries the consent management platform for the current consent state. Consent absent, withdrawn, or expired means access blocked. A new gate in an existing enforcement pipeline; not a new pipeline.

Jurisdictional routing is the second. Federal systems operate under a single regulatory authority. The commercial extension adds jurisdiction determination at the data-element level: the routing layer identifies applicable regulatory frameworks for each record based on the data subject's jurisdiction, then applies the most restrictive rules in force. The mechanism is unchanged; only the rule is new.

Fairness evaluation is the third. Federal AI systems do not make consumer-facing decisions subject to fair lending or employment discrimination law. The commercial extension adds a fairness checkpoint to H-4: before an AI-assisted decision affecting an individual is finalized, the checkpoint evaluates the decision against the applicable fairness threshold (four-fifths rule, CFPB fair lending standards, or EU AI Act bias testing requirements). A new checkpoint in an existing verification pipeline.

Explanation generation is the fourth. The federal harness produces E-4 attribution records for internal accountability but not explanation artifacts for external disclosure. The commercial extension adds an explanation generation step to H-4: at the moment an AI-assisted decision is made, the harness generates a human-readable explanation artifact satisfying the applicable disclosure requirement (Fair Credit Reporting Act (FCRA) adverse action, GDPR Article 22, EU AI Act Article 13). Stored as an E-4 record; made available to the affected individual through the organization's disclosure channel.

Every extension follows the same pattern: add a rule or a checkpoint to a mechanism that already exists. None of it is new engineering. The routing enforcement, the verification pipeline, the evidence collection were all built and finished long before. Commercial deployment supplies policy content to an architecture designed to accept it.

The CLD Timeline

NASA's Commercial LEO Destinations program follows a phased timeline. Phase 1 (2021-2025): design and development, with NASA providing technical guidance and milestone-based funding. Phase 2 (2026-2028): commercial station demonstrations, with one or more modules launched and docked for testing. Phase 3 (2028-2030): transition period, with both the ISS and commercial stations operating simultaneously. Phase 4 (post-2030): full commercial operations, with NASA as a customer rather than an operator.

The commercial AI governance migration follows an analogous timeline, compressed by the urgency of regulatory enforcement dates.

Year 1, Quarters 1-2: foundation deployment. Map the sector-independent core that transfers, deploy AGENTS.md with commercial regulatory bindings, activate E-1 baseline evidence collection, verify LP1-LP4 enforcement. Functional foundation, integration verified before adding complexity.

Year 1, Quarters 3-4: commercial extension. Implement LP9-LP12, connect consent management, jurisdictional routing, fairness evaluation, and explanation generation. Activate E-2 continuous monitoring for all fourteen properties. Operational controls with commercial extensions.

Year 2, Quarters 1-2: multi-framework rendering. Deploy the OSCAL rendering pipeline. Configure templates for every applicable regulatory framework. Verify that a single evidence stream produces framework-specific output for each assessment. Evidence harmonization.

Year 2, Quarters 3-4: continuous authorization. Activate the continuous governance posture described in Phase 5 of Chapter 17; establish the board reporting cadence from Chapter 20; complete the transition from examination-driven to continuous governance. The architecture is now operating commercially.

Aggressive, but achievable for organizations starting from the federal-validated architecture. An organization starting from a clean slate, with no governance framework in place, should expect the foundation phase to run six to twelve months longer. It is building the foundation that federal organizations have already proven.

The Operator's Responsibility

In the CLD model, NASA defines what safe means. The commercial operator decides how to be safe. Not station design, manufacturing processes, or operational procedures; NASA does not dictate those. It defines crew health thresholds, environmental standards, emergency response requirements, and verification protocols. The operator's design must satisfy those requirements. How it satisfies them is the operator's decision.

This division of responsibility is the correct model for commercial AI governance. The Harness Doctrine defines governance properties (LP1 through LP14), evidence architecture (E-1 through E-4), harness components (H-1 through H-5), and knowledge base controls (KB-1 through KB-4). These are the safety requirements. They specify what governance-worthiness looks like.

Each organization decides how to implement them. A bank's implementation of LP10 (Jurisdictional Consistency) will differ from a hospital's because their regulatory sources and data flows differ. A utility's implementation of LP1 (Classification-Consistent Access) will differ from an insurer's, since data sensitivity tiers and threat models are not the same across the two. Property definitions are universal; implementations are sector-specific. Evidence that the properties hold is the proof that governance works.

This is the transition from government station to commercial platform. Government proved the architecture. The commercial sector inherits it, adapts it, and operates it. Governance requirements did not get easier; operating pace got faster; the regulatory surface got wider. The architecture that produces stability, the architecture that holds the organization at its Lagrange point, is the same architecture it always was.

The station is being decommissioned. The governance architecture is not.

Build, Rent, or Wait

The operator's responsibility frames how to satisfy the governance properties, but it does not settle who supplies the implementation. An organization facing the migration has three honest options, and each carries a defensible case. The board chapter (Chapter 20) treats this as a navigation decision with real costs on every heading; this section gives the three options the fair comparison the decision deserves.

Build means implementing the harness in-house against your own infrastructure. You own the routing layer, the evidence pipeline, and the AGENTS.md instruction surface as code your team controls. The advantage is fit. The implementation matches your data flows, your regulatory sources, and your existing controls exactly, and no external party can change the terms under you. The cost is real. You carry the engineering, the maintenance, and the on-call burden, and you reach a defensible governance posture more slowly than a vendor platform would deliver. For an organization that already holds federal-validated architecture, build is often the shortest path, because the hard engineering is done and the remaining work is the commercial extension described above. For an organization starting from a clean slate, it is the longest path and the heaviest standing commitment.

Rent means adopting a vendor platform that supplies the harness mechanisms as a product. The advantage is time and shared maintenance: the routing enforcement, the evidence collection, and the multi-framework rendering arrive built and supported, and the vendor amortizes upgrade and threat-response work across every customer. This is a legitimate engineering choice, not a near-trap. A regulated institution that buys an attested governance platform inherits a posture it would otherwise spend quarters building, and it gains a counterparty contractually answerable for the platform's behavior. The trade-offs to weigh, rather than fears to avoid, are concrete: portability of your evidence if you later change vendors, the audience scope of any delegated authorization the platform holds on your behalf, and the residual accountability that stays with you regardless of who runs the software. A rented platform supplies the mechanism; it does not transfer the institution's duty to test that the mechanism holds. The independent surveyor still tests your running evidence, and the registry is still the regulator. Renting moves the build cost, not the accountability.

Wait means deferring a full commitment until a sector standard, a regulatory safe harbor, or a more settled vendor field reduces the cost of choosing. Sometimes that is the correct call. When a binding standard is months away and would force a rework of any architecture chosen now, a deliberate hold can be cheaper than building twice. The honest cost of waiting is exposure: every quarter without an enforced harness is a quarter of AI activity governed by policy assertion rather than architecture, and regulatory enforcement dates do not pause for an institution that has not finished deciding. So waiting holds up only when it is an active decision, with a defined trigger and an interim control in place. It fails the moment it becomes a default that persists because the choice was hard.

None of the three is a trap, and none is free. Build trades time for control; rent trades control for time; wait trades present cost for the risk of accumulating exposure. The board's task, as Chapter 20 sets it out, is to choose deliberately against the institution's actual position rather than to treat any one option as the safe default. An organization with federal-validated architecture usually builds, because it is closest to done. An organization with neither the engineering capacity nor an urgent enforcement date may rent or wait with a clear conscience, provided the decision is made on the merits and revisited on a schedule.

CHAPTER SUMMARY

NASA's Commercial LEO Destinations program transitions the ISS from government-operated to commercially operated, with NASA defining safety requirements and commercial operators determining how to meet them. The same model applies to AI governance. The inherited federal baseline, proven in federal classified environments, consists of the harness components (H-1 through H-5), the evidence architecture (E-1 through E-4), the knowledge base controls (KB-1 through KB-4), and the eight federal security properties (P1 through P8, defined in Volumes I and II). That baseline transfers without modification. On top of it, this volume defines the Lagrange governance properties (LP1 through LP14): LP1 through LP8 restate the governance-worthiness criteria in commercial governance terms, and the four-property commercial extension, LP9 through LP12, extends the framework to address consent, jurisdictional consistency, fairness, and transparency requirements that federal environments do not face. Each extension adds new rules or checkpoints to existing mechanisms rather than requiring new architecture. The migration follows a phased timeline analogous to the CLD program: foundation deployment, commercial extension, multi-framework rendering, and continuous authorization. The government proved the architecture works. The commercial sector inherits it, adapts it, and operates it at competitive pace.

Chapter 25: The AGENTS.md Migration

Part VI: Commercial LEO Destinations

> *"The shipping container did not ask what was inside. It asked whether the interface was standard. The AGENTS.md does not ask which regulatory framework applies. It asks whether the governance properties are satisfied. The contents vary. The container does not."*

Most organizations deploying AI in regulated environments are not starting from a clean slate. They already hold vendor-supplied governance documentation, internally developed AI usage policies, risk committee charters, and compliance programs that predate the current generation of AI systems. Some of it is rigorous. Much of it is aspirational. Nearly all of it was written for a previous generation of technology risk, and the AI adaptation amounted to appending sections that reference AI-specific concerns without restructuring the governance architecture to enforce them.

Migration from this existing state to the AGENTS.md architecture is the practical challenge this chapter addresses. Not a theoretical exercise. A project plan with phases, deliverables, obstacles, and timeline estimates calibrated to the operational realities of regulated commercial organizations.

The containerization metaphor from Chapter 18 applies directly. Before Malcom McLean's *Ideal X* sailed from Port Newark in 1956, cargo moved from truck to ship to truck through a series of format translations at every transfer point. Each translation cost money, invited error, and left cargo open to theft. The container eliminated the translations by standardizing the interface. The same logic governs here. Replace the multiple governance documentation formats, vendor security questionnaires, internal policy documents, risk register entries, compliance checklists, with a single machine-enforceable instruction surface that the harness routing layer can execute without human interpretation at each decision point.

Step 1: Inventory the Current Governance State

Before any migration begins, the organization must know what it has. Not a compliance documentation review. An architectural assessment that answers four questions.

First: what AI systems are operating? Every AI system in the organization, whether formally approved, informally adopted, or running without governance awareness, must be catalogued. Commercial AI coding assistants, internally developed ML models, vendor-provided AI features embedded in enterprise applications, any AI tool that employees reach through personal accounts on corporate devices. The inventory has to include systems the organization never deliberately deployed, because shadow AI adoption is the norm in regulated industries, not the exception. Chapter 4 documented the progression from prohibition to shadow adoption. The inventory reveals where on that progression each business unit sits.

Second: what governance documentation exists for each system? For each AI system, collect the existing governance artifacts: vendor System and Organization Controls (SOC) 2 reports, data processing agreements, internal risk assessments, AI usage policies, model validation reports, and any compliance-specific documentation (Health Insurance Portability and Accountability Act (HIPAA) risk assessments, Payment Card Industry Data Security Standard (PCI-DSS) scope documentation, Sarbanes-Oxley Act (SOX) control descriptions). Not an evaluation of quality. A determination of which governance properties are currently addressed, which are addressed in documentation only (policy exists but enforcement does not), and which are not addressed at all.

Third: what vendor dependencies exist? For each AI system, identify whether the governance architecture is owned by the organization or supplied by the vendor. A vendor-supplied SOC 2 report describes the vendor's controls, not the organization's. A vendor's AI ethics statement describes the vendor's intentions, not the organization's enforcement architecture. The distinction matters because vendor-supplied governance creates a structural dependency. When the vendor changes its model, its terms, or its architecture, the organization's governance posture changes whether the organization intended it to or not. Chapter 10's Voight-Kampff problem applies here. A vendor who supplies both the AI system and the governance attestation for that system carries a conflict of interest, and that conflict has to be resolved architecturally, not contractually.

Fourth: what evidence does the current architecture produce? For each AI system, determine whether existing governance produces evidence that an auditor, examiner, or assessor can verify independently. If governance "evidence" consists of policy documents that describe intended behavior rather than records demonstrating actual behavior, the evidence architecture has a gap the migration must close. Description versus demonstration: that distinction is the core of the evidence model from Chapter 6.

Three to six weeks for a mid-sized regulated organization (500-5,000 employees, 20-100 AI systems). Large enterprises with thousands of AI touchpoints may require eight to twelve weeks. Sampling does not shorten the inventory. The systems excluded from the sample are precisely the ones most likely to operate outside governance boundaries.

Step 2: Classify Vendor Dependencies by Tier

Not all vendor dependencies are equal. Some are commodity. Some are existential. The National Institute of Standards and Technology (NIST) SP 800-161r1 supply chain risk framework supplies a tiering model that maps directly onto AI governance migration.

Tier 1 capabilities are those the organization cannot safely delegate to a vendor because they define the organization's regulatory posture. The AGENTS.md instruction surface is Tier 1. So is the evidence collection pipeline. So are jurisdictional routing rules that determine which regulatory framework applies to each data element. LP9 consent verification is Tier 1 because consent state is a property of the organization's relationship with the data subject, not a property of the AI vendor's system. All Tier 1 capabilities must migrate from vendor-supplied documentation to organization-owned architecture. That transfer is not negotiable.

Tier 2 capabilities are those where vendor involvement is operationally necessary but must be governed by the organization's architecture. The AI model itself is typically Tier 2: the organization may use a vendor-supplied model but must wrap it in the harness architecture that enforces organizational governance. Model performance monitoring is Tier 2: the vendor may provide monitoring tools, but the organization defines the thresholds, owns the alert pipeline, and retains the evidence. Vendor application programming interface (API) integrations are Tier 2: the vendor provides the API; the organization's H-2 routing enforcement layer governs what data flows through it.

Tier 3 capabilities are those where vendor substitution is feasible without governance disruption. Compute infrastructure is typically Tier 3. Authentication services are Tier 3 if the organization controls the identity provider; logging infrastructure is Tier 3 if the organization controls the log schema and retention policy. No migration required for Tier 3. Contractual controls that preserve the organization's ability to substitute vendors without rebuilding governance architecture.

The tiering exercise usually surfaces Tier 1 dependencies on vendors that the organization had filed under Tier 2 or Tier 3. The most common finding is that the AGENTS.md equivalent, the instruction surface defining what the AI system may and may not do, sits embedded in the vendor's system configuration rather than owned by the organization. This is the governance equivalent of letting the shipbuilder paint the Plimsoll mark wherever it chooses. Migration has to transfer Tier 1 capabilities into organizational ownership.

A clarification is owed here, because the strongest objection to this whole architecture is that the institution self-certifies. In the maritime model the shipowner does not paint his own Plimsoll line. The owner submits the vessel; an independent classification society surveys it against a published standard and assigns the mark; the registry records the class. The commercial AI model preserves all three roles. AGENTS.md is the shipowner's self-declared survey, the institution's written account of what its system is built to do. It is not a self-issued certificate. The independent surveyor is the external auditor, examiner, Qualified Security Assessor (QSA), or assessor (SOC 2 auditor, Office of the Comptroller of the Currency (OCC)/Fed/Federal Deposit Insurance Corporation (FDIC) examiner, PCI QSA, HIPAA/Office for Civil Rights (OCR) assessor, North American Electric Reliability Corporation (NERC) Critical Infrastructure Protection (CIP) auditor, state Department of Insurance (DOI) examiner) who tests the declaration against the running evidence (E-1 through E-4). The registry is the regulator. The institution authoring its own AGENTS.md is exactly the shipowner filing a survey declaration; the line is only validated when an independent party tests it. The architecture's independence does not come from who writes AGENTS.md; it comes from the fact that the evidence (E-2 through E-4) is generated as a byproduct of operation and is tested by a party the institution does not control. Tamper-evidence (LP8) is what stops the owner from repainting the line after the fact.

Step 3: Build the Evidence Architecture

Evidence architecture is the infrastructure that converts governance from documentation to demonstration. Chapter 6 established that the evidence model transfers completely from federal to commercial environments. This step implements it.

E-1 (Foundation Evidence) comes first. Deploy AGENTS.md as the canonical governance configuration document. Record initial configuration state: which P-properties are active, which regulatory frameworks are mapped, which data sensitivity tiers are defined, which access boundaries are enforced. The E-1 baseline is the reference against which all later evidence is measured. Without it, E-2 continuous monitoring has nothing to monitor against.

E-2 (Operational Evidence) instruments the harness to generate verification records as a byproduct of normal operation. Every routing decision produces an E-2 record confirming that applicable governance properties were checked. Every verification checkpoint produces an E-2 record confirming that AI output was evaluated against configured criteria. Continuous operation, not sampling schedules. An E-2 pipeline that fires quarterly is an audit preparation project wearing a monitoring architecture's clothes.

E-3 (State Transition Evidence) configures state transition recording. Every change to AGENTS.md configuration, every update to the knowledge base, every modification to routing rules, every adjustment to fairness thresholds produces an E-3 record with a timestamp, the identity of the person who authorized the change, and the reason. E-3 answers the examiner's question: what changed since the last time I looked?

E-4 (Accountability Evidence) deploys the attribution chain recording mechanism. Every AI-assisted action producing a consequential output, whether credit decision, clinical recommendation, code change affecting a regulated system, or pricing determination, generates an E-4 record linking output to input, model version, governance configuration active at processing time, data sources consulted, and the human who authorized the action. E-4 answers the regulator's question: who is accountable for this specific output?

Evidence architecture deployment usually runs concurrent with Step 4 (LP9-LP12 implementation). The E-1 baseline must be established before E-2 monitoring begins. E-3 and E-4 can deploy in parallel once AGENTS.md configuration and the harness routing layer are operational.

AI-Assisted Compliance Documentation During Migration

Step 3 (Build the Evidence Architecture) requires converting existing governance documentation into the AGENTS.md instruction surface. The conversion is not mechanical transcription. Existing documentation describes intentions; AGENTS.md encodes enforcement rules. A policy document stating "AI systems shall not access material non-public information (MNPI) without appropriate authorization" becomes a LP1 classification-consistent access rule with defined data sensitivity tiers, explicit authorization criteria, and enforcement checkpoints. The conversion from prose to machine-enforceable constraints requires regulatory expertise, technical architecture knowledge, and substantial manual effort.

AI-assisted governance-documentation conversion tooling eases this bottleneck. It helps generate structured documentation that preserves regulatory compliance while producing machine-parseable governance artifacts. This is the maritime surveyor's documentation standard applied to AI governance: every finding must be documented in the survey report format that Lloyd's Register can verify, regardless of which surveyor conducted the survey or which vessel was inspected. The format is standardized; the content varies by vessel; the format's machine-readability enables verification without manual interpretation.

A first-party scanner or conversion tool does not establish independence. An institution running its own conversion tooling is the shipowner inspecting his own hull; the independent survey is still the external assessor, not the tool. The conversion tooling accelerates the shipowner's survey declaration; it does not replace the surveyor.

Structured Governance Artifact Generation

The containerization metaphor from Chapter 18 applies to documentation format. Before containerization, cargo manifests were free-text descriptions that varied by shipping line, port authority, and customs jurisdiction. The manifest for the same cargo differed depending on who prepared it. Standardized container documentation eliminated this variability: the same goods moving in the same container produce the same documentation regardless of which carrier operates the route.

Conversion tooling applies this standardization to governance documentation generation. When Step 1 (Inventory) produces existing governance artifacts, policy documents, vendor security questionnaires, risk assessments, and compliance checklists in various formats, Step 3 must convert these into AGENTS.md format. The conversion follows defined templates: P-property definitions follow a structured schema (property name, enforcement rule, verification checkpoint, failure condition, remediation path); data sensitivity tier definitions follow another schema (tier name, regulatory basis, data elements included, access control requirements, disclosure restrictions); jurisdictional routing rules follow a third schema (jurisdiction identifier, applicable frameworks, conflict resolution rules, evidence rendering templates).

The tooling generates these structured artifacts from existing documentation while preserving the regulatory content. Take a HIPAA risk assessment documenting access controls for a clinical documentation AI system. The assessment describes who may access protected health information (PHI), under what conditions, and what safeguards apply. Converting it to AGENTS.md format means extracting the enforcement rules (who, what conditions, what safeguards), mapping them to LP1 classification-consistent access requirements, and encoding them as verification checkpoints in the H-2 routing layer. The tooling performs this extraction and mapping under human review. Compliance teams verify draft AGENTS.md sections rather than writing them from scratch.

Preserving Regulatory Lineage During Conversion

Classification societies maintain survey report lineage: every survey finding references the regulatory standard it verifies, the inspection method applied, and the surveyor's certification. A finding that a vessel's load line marking complies with International Convention for the Safety of Life at Sea (SOLAS) Chapter II-1 includes the specific regulation, the measurement method, and the surveyor's identity. This lineage enables verification: an auditor can trace the finding back to the standard, confirm the inspection method was appropriate, and verify the surveyor's authorization.

The tooling preserves equivalent lineage during governance artifact conversion. Every AGENTS.md section generated from existing documentation includes a regulatory_basis annotation referencing the source document, the applicable regulatory framework, and the conversion method. A LP9 consent-bounded processing rule generated from an existing General Data Protection Regulation (GDPR) privacy notice includes annotations referencing GDPR Article 6 (lawful basis for processing), Article 7 (conditions for consent), and the specific sections of the privacy notice that established those conditions. The annotation is not decorative; it is the verification path. An examiner reviewing the AGENTS.md configuration can trace each rule back to its regulatory basis without requiring the compliance team to reconstruct the reasoning.

Preserving this lineage heads off a common migration failure mode: governance architectures that satisfy regulatory requirements but cannot demonstrate why. An organization may implement a fairness screening flag of 0.80 for automated credit decisions. The 0.80 ratio is borrowed from the four-fifths rule as a screening heuristic for first-pass disparate-impact review; it is not a bright-line pass/fail safe harbor. The governing tests are the applicable fair lending standard and, where the decision is actuarial, the actuarial-justification standard. The regulatory_basis annotation must record those governing tests, noting that the four-fifths ratio is used as a screening flag rather than a binding threshold. During examination, when the regulator questions the threshold choice, an organization that recorded only the bare number, with no governing basis, cannot defend it. Conversion tooling prevents this by requiring regulatory_basis annotations during conversion, making the basis part of the governance artifact rather than institutional memory.

Accelerating Multi-Framework Evidence Rendering

Chapter 18's evidence rendering layer requires that a single evidence stream be renderable into multiple regulatory formats (SOC 2 Section 3, HIPAA Security Rule, NIST Cybersecurity Framework (CSF), PCI-DSS). Each regulatory framework expects evidence in its native format: SOC 2 expects trust service criteria attestations, HIPAA expects technical safeguards documentation, PCI-DSS expects compensating controls descriptions. Organizations operating under multiple frameworks traditionally maintain parallel evidence collection systems, each producing evidence in its framework's native format.

The tooling enables single-collection, multi-rendering by generating framework-specific evidence artifacts from the unified E-2 operational evidence stream. The underlying evidence is the same: E-2 verification records documenting that governance properties were checked on every AI interaction. The rendering varies by framework: SOC 2 rendering emphasizes availability (property LP2, continuous operation) and confidentiality (property LP1, classification-consistent access); HIPAA rendering emphasizes access controls (45 CFR 164.308) and transmission security (45 CFR 164.312); PCI-DSS rendering emphasizes data protection (requirement 3, protect stored cardholder data) and access logging (requirement 10, track and monitor access).

The tooling assists with rendering template generation during Step 3. Each framework requires a rendering template mapping E-2 evidence fields to framework-native documentation format. The template is defined once during migration and applied continuously thereafter. New evidence arrives in the E-2 stream; the rendering pipeline applies the template; framework-specific artifacts are generated automatically. The compliance team reviews rendered artifacts for regulatory accuracy rather than manually assembling each framework's evidence package from scratch.

Migration Timeline Compression

Organizations using conversion tooling for documentation generation during AGENTS.md migration may finish Step 3 (Evidence Architecture) faster than the estimate for manual conversion. The figures here are illustrative planning ranges, not observed from a completed deployment. The compression comes from two sources. First, structured artifact generation produces draft AGENTS.md sections that compliance teams review and approve rather than authoring from blank templates. Second, rendering template generation produces framework-specific evidence rendering logic verified once and reused continuously, which removes the recurring cost of evidence format translation at each examination cycle.

The maritime parallel is the classification society's standardized survey checklist. Lloyd's Register surveyors conducting hull surveys use defined checklists ensuring consistent coverage regardless of which surveyor performs the survey. The checklist accelerates the survey (the surveyor does not reinvent inspection procedures on each vessel) and standardizes the output (the survey report follows the same structure regardless of vessel type). Conversion tooling applies the same acceleration to governance documentation: standardized templates ensure consistent AGENTS.md structure, accelerate conversion from existing documentation, and enable verification without manual interpretation.

Code Structure as Governance Artifact

Step 1 (Inventory) identifies AI systems operating in the organization; Step 2 (Tier Classification) determines vendor dependencies; Step 3 (Evidence Architecture) establishes E-1 through E-4 evidence collection. What these steps do not verify is whether the AI systems identified during inventory are architecturally capable of enforcing the governance properties defined in AGENTS.md. An organization may define LP1 classification-consistent access rules in AGENTS.md, deploy E-2 monitoring to verify those rules, and discover during the first examination that the AI system's retrieval layer cannot enforce classification boundaries because the classification metadata is not propagated to the retrieval query layer.

Static code-structure analysis closes this verification gap. It confirms that governance constraints defined in AGENTS.md show up in the system's implementation architecture, and that governance checkpoints sit in the execution path where they belong. This is the Lloyd's Register structural survey requirement applied to software: the survey verifies that the as-built vessel matches the approved structural plans, that load-bearing members are constructed to specification, and that structural modifications have not compromised the certified design. Code-structure analysis performs the equivalent verification for AI governance: the implemented system matches the AGENTS.md specification, governance checkpoints are present in the execution path, and code changes have not bypassed enforcement logic.

An institution running its own code-scanner does not thereby become its own surveyor. Running a first-party scanner is the shipowner inspecting his own hull; the independent survey is the external assessor, not the scanner. The scanner strengthens the shipowner's self-declared survey by showing that the structure exists; the assessor still tests the declaration against the running evidence.

Structural Verification vs. Documentation Verification

The distinction between structural verification and documentation verification is the core of the classification society model. A vessel owner may produce documentation asserting that the hull structure meets SOLAS requirements; the classification society's survey verifies that the physical hull matches the documentation. Documentation is not verification; physical inspection is verification. If the survey finds discrepancies between documentation and physical structure, the certification is withheld until discrepancies are resolved.

Code-structure analysis applies the same distinction to AI governance architecture. AGENTS.md documentation may assert that LP1 classification-consistent access is enforced; the analysis verifies that the codebase includes classification boundary checks in the data retrieval path. The verification is structural: does the code architecture include the components required to enforce the documented governance property? If AGENTS.md defines four data sensitivity tiers (public, internal, confidential, restricted) but the retrieval layer's query logic does not reference sensitivity tier metadata, the structural verification fails. The documentation asserts enforcement; the code structure demonstrates enforcement is architecturally impossible without modification.

This structural verification catches the most common AGENTS.md migration failure: governance properties defined in documentation that the underlying system cannot enforce. An organization may complete Steps 1 through 3, deploy AGENTS.md with comprehensive P-property definitions, establish E-2 monitoring, and then discover during Step 4 implementation that the AI system's vendor-supplied SDK does not expose the hooks required to inject classification checks before data retrieval. The governance architecture was designed correctly. The software architecture does not support it. Code-structure analysis detects this mismatch during Step 3, before the discovery lands downstream where it costs more.

Architectural Enforcement Path Analysis

Classification societies verify not just that structural components exist but that they are connected correctly. A vessel may have adequate bulkheads, but if the bulkheads are not watertight or not connected to the hull's load-bearing frame, they do not provide the structural integrity the design requires. The survey verifies the connection, not merely the component's presence.

Code-structure analysis performs equivalent connection verification for governance enforcement paths. AGENTS.md defines a LP13 (Output Boundary Enforcement) rule at the routing layer (H-2) requiring that AI-generated content be evaluated against disclosure restrictions before transmission to external systems. The analysis verifies that the output generation code path includes a verification checkpoint, that the checkpoint invokes the disclosure restriction evaluation logic, and that the evaluation result gates the transmission decision. If any link in this enforcement path is missing (output generation bypasses the checkpoint, checkpoint invokes a no-op function, evaluation result is logged but does not gate transmission), the structural verification fails.

The enforcement path analysis follows the code's execution flow, not its organizational structure. An AI system may hold a governance module containing all required verification functions, but if the operational code path never invokes those functions, the module is decorative. Code-structure analysis traces execution paths from entry points (API endpoints, user interactions, scheduled jobs) through AI invocation, data retrieval, output generation, and external transmission, verifying that governance checkpoints appear in the paths where AGENTS.md requires them.

Detecting Governance Bypass Patterns

The International Space Station (ISS) maintains multiple redundant systems for critical functions: primary, backup, and emergency modes. A failure in the primary system triggers automatic failover to backup. What the architecture must prevent is a failure mode where the primary system fails, backup activation is attempted, but the backup system was inadvertently disabled during maintenance, leaving no operational path. The architecture includes verification that backup systems are operational before primary systems are taken offline for maintenance.

Code-structure analysis detects the governance equivalent: code paths that bypass enforcement checkpoints during failure conditions, debugging modes, or administrative overrides. Consider a LP1 classification-consistent access rule that applies during normal operation but is bypassed when the AI system operates in maintenance mode. If maintenance mode is activated manually without E-3 state transition recording, and remains active beyond its intended maintenance window, the AI system operates without classification enforcement while appearing to be under governance. The analysis detects this by identifying conditional branches (if maintenance_mode is enabled) that bypass governance checkpoints, flagging them as architectural risks requiring either removal or E-3 state transition recording before the bypass condition activates.

The maritime parallel is the emergency bilge pump bypass. Lloyd's Register requires that vessels have emergency bilge pumps capable of operating when main power is lost. The emergency pump has its own power source, typically a diesel-driven or battery-backed electric motor. What the survey verifies is that the emergency pump cannot be inadvertently disabled by the same failure that takes the main pump offline. The emergency system is structurally independent. Code-structure analysis verifies governance checkpoint independence: a failure in the operational AI system should not disable the governance verification that detects the failure.

Migration Integration: Structural Verification Before Deployment

Step 4 (LP9-LP12 Implementation) deploys commercial properties into the harness routing layer. Each property adds enforcement logic: LP9 queries the consent management platform, LP10 applies jurisdictional routing, LP11 evaluates fairness metrics, LP12 generates explanation artifacts. Before any property is declared operational, code-structure analysis verifies that the enforcement logic is structurally present in the code paths where the property applies.

The verification produces a structural compliance report documenting, for each P-property, whether enforcement checkpoints are present, whether checkpoints are invoked in operational code paths, and whether bypass conditions exist. The report is itself an E-1 governance artifact. It records the architectural state at deployment time. When Step 5 (Continuous Authorization) begins, the structural compliance report becomes the baseline against which code changes are evaluated. A code change that removes or bypasses a governance checkpoint surfaces in E-3 state transition records as a governance architecture modification requiring reauthorization.

This structural verification before operational deployment is the pre-sailing survey requirement applied to AI systems. Lloyd's Register does not issue a classification certificate based on design drawings; the certificate requires a survey of the constructed vessel verifying that the as-built structure matches the approved design. Code-structure analysis enables the same pre-deployment verification for AI governance: the implemented system matches the AGENTS.md specification, enforcement checkpoints are structurally present, and the architecture can sustain the governance properties it documents.

Step 4: Implement LP9-LP12

Four commercial properties turn the federal architecture into a commercial governance platform. Each adds a new enforcement rule to the harness routing layer.

LP9 (Consent-Bounded Processing) connects the AGENTS.md consent verification gate to the organization's consent management platform. Map each data category to its consent requirements: GDPR Article 7 for EU personal data, California Consumer Privacy Act (CCPA) opt-out for California residents, HIPAA 45 CFR 164.508 for PHI disclosures beyond treatment-payment-operations, 42 CFR Part 2 for substance use disorder records. Configure H-2 to query the consent platform before every data access event involving consent-governed records. Default behavior when consent state is unknown: block access and escalate.

LP10 (Jurisdictional Consistency) builds the jurisdiction determination layer. Each data subject record must carry a jurisdiction attribute reflecting applicable regulatory frameworks. Configure H-2 to route each processing request through jurisdiction determination logic before applying governance rules; implement "most restrictive applicable" as the default for uncertain cases. Document conflict resolution procedures for genuinely contradictory requirements (the GDPR retention versus Securities and Exchange Commission (SEC) retention conflict described in Chapter 9 is the canonical example).

LP11 (Fairness Verification) integrates the fairness evaluation pipeline with H-4 verification checkpoints. Define the applicable fairness standard for each decision category: the four-fifths ratio as a screening flag for employment-related decisions (a first-pass diagnostic that triggers deeper review, not a binding pass/fail threshold), Consumer Financial Protection Bureau (CFPB) fair lending standards for credit decisions, the National Association of Insurance Commissioners (NAIC) Model Bulletin and state unfair-discrimination law with actuarial justification for insurance underwriting, EU AI Act bias testing for high-risk AI systems. A screening flag tripped means deeper fairness analysis is triggered; a governing-standard failure means at-scale deployment is suspended. Configure E-2 to record fairness metrics continuously, not annually.

LP12 (Transparency and Explainability) adds the explanation generation checkpoint to H-4. Every AI-assisted decision reaching an external party generates a human-readable explanation artifact at decision time. Map each decision category to its applicable disclosure standard: Fair Credit Reporting Act (FCRA) adverse action notice for credit denials, GDPR Article 22 for automated decisions affecting EU data subjects, EU AI Act Article 13 for high-risk AI systems. Store explanation artifacts as E-4 records with timestamps matching the decision records.

Step 5: Establish the Continuous Authorization Cycle

The final step transitions from examination-driven governance to continuous governance. This is the distinction Chapter 17 drew between Phase 4 and Phase 5: Phase 4 achieves multi-framework evidence rendering; Phase 5 operates continuously.

Not episodic. Continuous. Four cadences compose the authorization cycle.

Daily automated verification keeps the E-2 pipeline running on every AI interaction, checking LP1 through LP14 without interruption. Alerts fire when any property verification fails, and the security operations team triages in real time. This is the ISS telemetry model. Ground controllers do not wait for a scheduled check to learn whether life support is operating correctly.

Weekly governance posture review puts the ISO, CISO, or designated governance lead in front of the week's E-2 verification results, E-3 state transitions, and E-4 attribution records. The review identifies trends: increasing verification failure rates, configuration drift from the E-1 baseline, attribution gaps in specific AI systems. Findings exceeding the organization's risk threshold escalate to the risk committee.

Quarterly board reporting has the Chief Risk Officer (CRO) or CISO presenting governance posture to the audit committee using the risk currency translation from Chapter 20. Traffic-light summary (green/yellow/red by AI system and by P-property), financial exposure estimates for governance gaps, and the set of decisions requiring board authority. The audit committee's quarterly review is the commercial equivalent of the Authorizing Official (AO)'s periodic reauthorization decision.

Annual full governance assessment validates the architecture against external standards through an independent review: SOC 2 Type II, HIPAA risk assessment, PCI-DSS QSA review, or the applicable sector-specific evaluation. The assessment draws evidence from E-1 through E-4 rather than requesting documentation assembly. "Show me the evidence that this control is operating effectively" gets answered from the evidence dashboard, not from a document repository.

Common Migration Obstacles

The migration runs into predictable obstacles. Naming them in advance is cheaper than discovering them mid-project.

The first is vendor resistance to Tier 1 transfer. AI vendors whose business models depend on governance lock-in will fight any attempt to hand instruction surface control to the customer. The response is architectural, not negotiable. Deploy your own AGENTS.md as the governance layer wrapping the vendor's system, whatever the vendor's documentation claims. The vendor's controls operate inside the organization's harness. Not the reverse.

The second is shadow AI discovery. Step 1's inventory routinely turns up AI systems that no governance program ever addressed. The temptation is to bring them under governance retroactively through documentation. Wrong move. Bring them under governance architecturally: deploy the harness around the shadow system first, then decide whether the system should keep running or be decommissioned. Documentation-first is the very approach that opened the governance gap to begin with.

The third is multi-framework evidence format conflicts. Different regulatory frameworks expect evidence in different formats, and the temptation is to run parallel evidence collection programs. Chapter 18's containerization approach answers this: one evidence stream, multiple rendering templates. Rendering cost is trivial next to collection cost.

The fourth is organizational resistance to continuous monitoring. Teams accustomed to examination-driven governance may read continuous monitoring as surveillance. The reframe is operational. Continuous monitoring catches governance drift before it produces regulatory exposure, the same way continuous telemetry catches ISS system drift before it produces a crew safety event. The alternative is finding the drift during an examination. Worse for everyone.

The fifth is timeline pressure. EU AI Act enforcement for high-risk AI systems begins August 2, 2026. Colorado AI Act requirements take effect in 2026. Organizations that have not begun the migration face a compressed timeline. The phased approach from Chapter 17 cushions that, since partial deployment delivers governance value at each phase rather than withholding every improvement until the complete architecture is operational. Start with Step 1. The inventory itself is a governance improvement.

Migration Timeline Estimates

The durations below are illustrative planning ranges, not figures observed from a completed deployment. There is no named end-to-end reference deployment behind these numbers; they are an authored estimate of phase effort for a mid-sized regulated organization, offered to support scoping rather than to assert a measured result. Treat each per-phase week count and the total as a planning starting point to be calibrated against the organization's actual governance maturity, not as a benchmark.

Phase	Duration (illustrative)	Dependencies	Deliverable
Step 1: Inventory	3-6 weeks	None	AI system catalog, governance documentation inventory, vendor dependency map
Step 2: Tier Classification	2-3 weeks	Step 1 complete	Vendor dependency tiering, Tier 1 transfer plan
Step 3: Evidence Architecture	6-10 weeks	Steps 1-2 complete	E-1 baseline, E-2/E-3/E-4 pipeline operational
Step 4: LP9-LP12 Implementation	8-12 weeks	Steps 1-3 complete	All fourteen properties enforced, consent/jurisdiction/fairness/explanation gates operational

Phase	Duration (illustrative)	Dependencies	Deliverable
Step 5: Continuous Authorization	4-6 weeks	Steps 1-4 complete	Daily/weekly/quarterly/annual cycle operational
Total	**23-37 weeks**		**Full AGENTS.md governance architecture operational**

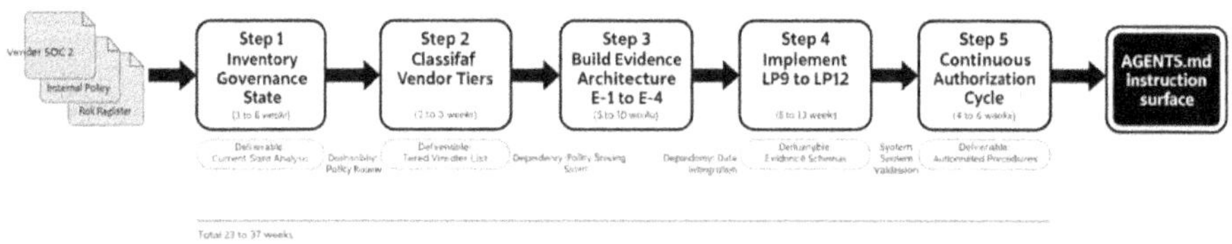

Figure 25: The five-step migration from scattered governance documentation to a single machine-enforceable AGENTS.md instruction surface, showing each step's deliverable, dependency, and illustrative duration.

Organizations with existing governance maturity, those that already hold SOC 2 Type II attestation, a functioning model risk management program, or an operational HIPAA compliance architecture, will land toward the lower end. Organizations starting from minimal governance infrastructure should plan for the upper end and add contingency for the shadow AI discovery that Step 1 will produce. It always does.

> **CHAPTER SUMMARY**
>
> The AGENTS.md migration converts existing governance documentation, whether vendor-supplied, internally developed, or absent, into a machine-enforceable instruction surface that the harness routing layer executes without human interpretation. The migration follows five steps: inventory the current governance state (including shadow AI), classify vendor dependencies by tier (Tier 1 must-own, Tier 2 acceptable-with-controls, Tier 3 commodity), build the evidence architecture (E-1 through E-4), implement LP9-LP12 commercial properties, and establish the continuous authorization cycle. The containerization metaphor from Chapter 18 applies throughout: one evidence stream, multiple rendering templates, standardized interface regardless of contents. Common obstacles include vendor resistance to Tier 1 transfer, shadow AI discovery, multi-framework format conflicts, organizational resistance to continuous monitoring, and regulatory timeline

pressure. The migration timeline ranges from 23 to 37 weeks depending on existing governance maturity, with each phase delivering governance value independently.

Chapter 26: The Stable Point

Part VI: Commercial LEO Destinations

> *"A Lagrange point is not a place where forces disappear. It is a place where forces balance. The cargo still has value. The regulators still have authority. The competitors still have speed. What changes at the Lagrange point is that these forces stop pulling the organization apart and start holding it in position."*

On a November evening in 1688, merchants crowded into Edward Lloyd's coffeehouse on Tower Street. They needed to know whether a ship was worth backing. No standardized way to answer the question existed. Information was informal, assessment was relationship-dependent, and the gap between judgment and ground truth was costing them, at intervals, everything.

Three hundred and thirty-eight years later, Lloyd's is a global institution with standardized survey requirements, actuarial models, classification society inspections, and a register that any underwriter anywhere in the world can use to verify, independently, that a specific vessel meets a specific seaworthiness standard. The merchants did not learn to trust individual ships better. They built infrastructure that made trust a function of verified compliance with a published standard, not of who knew whom.

The trajectory was never in doubt. Only the timeline.

The Five Points Exist

In any three-body gravitational system, Lagrange points are not constructed. They are discovered. They are the mathematical consequence of three forces acting on the system, and the forces determine the points. Whether anyone bothers to calculate them or not, they exist.

The five sector Lagrange points identified in this book, L1 (Financial Services), L2 (Healthcare), L3 (Energy and Critical Infrastructure), L4 (Insurance and Actuarial), and L5 (Manufacturing, Supply Chain, and Telecommunications), are similarly determined. Not inventions. Properties of the regulatory physics governing each sector: the governance configurations where data protection, regulatory compliance, and business velocity reach sustainable equilibrium for organizations operating under each sector's constraints.

L1 sits in an unstable equilibrium between five distinct data protection regimes and competitive velocity measured in milliseconds. Active station-keeping through continuous verification and continuous evidence production is the structural requirement, not an optional enhancement. Never at rest. Continuously correcting.

L2 sits at the intersection of protected health information (PHI) sensitivity tiers, Food and Drug Administration (FDA) software classification boundaries, and clinical workflow efficiency. Three forces, three corrections, one equilibrium. Minimum necessary rule corrects for data protection; FDA's predetermined change control plan corrects for regulatory compliance; clinical decision support velocity requirements correct for business speed.

L3 occupies the most restrictive regulatory environment outside classified government networks. North American Electric Reliability Corporation (NERC) Critical Infrastructure Protection (CIP)'s prescriptive controls, penalty structure, and self-reporting obligations create compliance forces that dominate the system. Think of a station in a tight orbit. The correction budget is small, drift carries immediate consequences, and the governance architecture has to keep working at the network edge where connectivity to central systems comes and goes.

L4 is the direct descendant of Lloyd's coffeehouse. Insurance faces the fairness problem in its most acute commercial form: legally permitted risk-based pricing in permanent tension with unlawful proxy discrimination. Price risk using AI while producing continuous evidence that pricing does not produce disparate impact. The four-fifths rule is the Plimsoll mark; the actuarial opinion on model reliability is the classification certificate.

L5 spans the widest operational range. Manufacturing, supply chain, and telecommunications share a common governance challenge, AI systems affecting physical infrastructure, products, and networks, but face different regulatory authorities and different consequence profiles. Product safety regulations, environmental compliance, trade restrictions, and telecommunications-specific security directives must all be accounted for simultaneously. No single framework governs the sector; each organization must determine its own equilibrium point within L5's range.

These points do not move because an organization finds them inconvenient, and they do not shift because the board would prefer a configuration that costs less. Forces determine the points. The choice is whether to find the stable configuration or to keep operating in the gravitational field between configurations, pulled toward instability by forces that cannot all be satisfied at once.

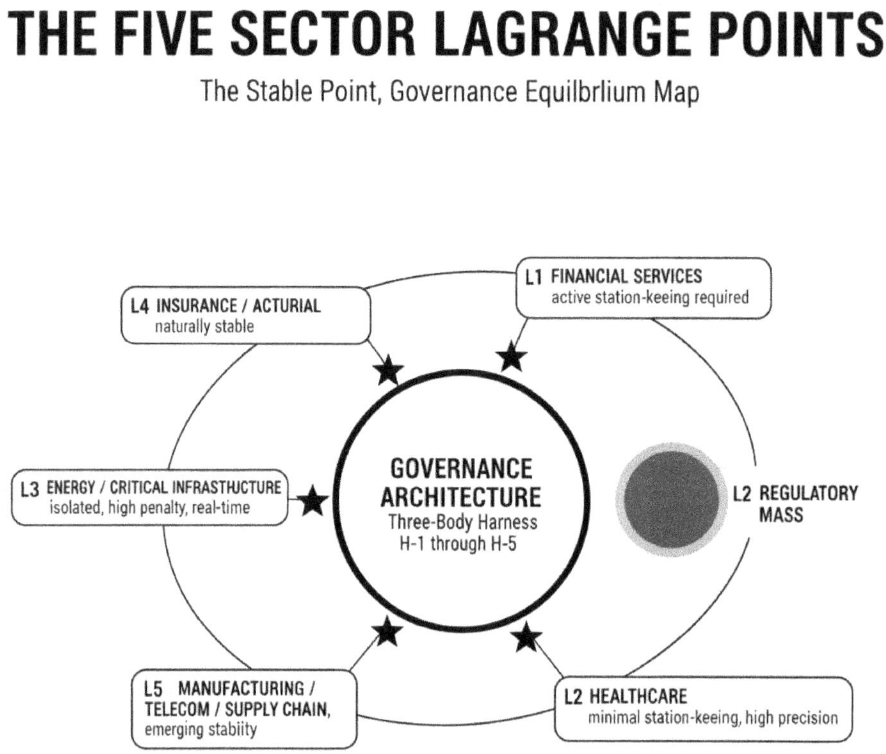

Figure 26: The five sector Lagrange points; financial services at L1, healthcare at L2, energy at L3, insurance at L4, and manufacturing, supply chain, and telecom at L5, around the governance architecture.

The Cost of Not Finding the Point

An organization that has not found its Lagrange point is not ungoverned. It is governed by drift. Every external force, a new regulatory requirement, a vendor architecture change, a competitor's deployment, a compliance examination finding, pushes governance posture in a different direction. Reactive adjustment to the most recent perturbation creates new imbalances in the other two dimensions. Correction in one axis produces instability in the other two.

The pattern is familiar. A regulatory examination finds an evidence gap. Compliance launches a documentation sprint to close it before the next examination cycle. That sprint consumes resources allocated to the AI deployment acceleration the business unit requested. The business unit, facing competitive pressure, adopts AI tools without waiting for governance to catch up. Shadow AI adoption creates new data protection risks that compliance does not know about. Next examination finds both the original evidence gap and the new shadow AI exposure. The cycle repeats.

Gravitational instability. Real budget and real staff hours expended on governance activities that do not produce stability because they are not coordinated around an equilibrium configuration. Each correction addresses one force and amplifies the others. Total governance expenditure increases; governance posture does not improve. Working harder to stay in the same unstable position.

The cost is measurable, and it falls into three categories.

Compliance cost comes first. An organization running parallel governance programs for each regulatory framework, with separate evidence collection for System and Organization Controls (SOC) 2, Health Insurance Portability and Accountability Act (HIPAA), Payment Card Industry Data Security Standard (PCI-DSS), and sector-specific requirements, pays for that redundancy on every assessment cycle. Chapter 18 established that evidence collection represents 40-60% of total compliance program cost and that the unified architecture replaces collection redundancy with rendering, which costs 5-10%. The premium for not finding the Lagrange point: the difference between collecting evidence once and collecting it four times.

Velocity cost is second. An organization whose governance architecture creates friction proportional to the number of regulatory frameworks will deploy AI more slowly than a competitor whose architecture satisfies all applicable frameworks through a single harness. Not the time spent on governance. The competitive opportunity lost while the organization navigates processes designed for a single-framework world and bolted together for a multi-framework reality.

Incident cost is third. An unstable governance posture, one continuously correcting without reaching equilibrium, has gaps between corrections. Those gaps are where incidents occur. Regulatory penalty, litigation exposure, reputational damage, remediation expense; all accumulate in the governance gaps the unstable posture cannot close.

An organization at its Lagrange point still pays for governance. The difference is that the expenditure produces stability. The continuous verification pipeline runs. Evidence renders into every applicable framework from a single stream, and the harness enforces fourteen properties without someone making a manual call at each decision point. Governance becomes an operating cost, predictable and bounded, rather than a crisis budget that spikes with every examination and every incident.

The Maritime Arc, Complete

The arc from Lloyd's coffeehouse to modern maritime governance follows a progression that AI governance is repeating on a compressed timeline.

Informal assessment came first. Lloyd's merchants assessed ships through personal inspection and reputation. AI governance in 2026 still consists largely of vendor presentations, security questionnaires, and internal policy documents. Informal, relationship-dependent, poorly standardized.

Institutional classification followed. Lloyd's Register, formalized in the 1760s, provided standardized seaworthiness criteria and independent surveys. SOC 2 Type II, PCI-DSS Qualified Security Assessor (QSA) assessments, and HIPAA risk assessments provide the AI governance equivalent: standardized criteria assessed by independent parties. The classification society model is the structural ancestor of every commercial compliance attestation.

Visible enforcement came third. The Plimsoll Mark, mandated by the Merchant Shipping Act of 1876, made a vessel's governance state visible to any observer without expert interpretation. LP9 through LP12, constructed as binary indicators enforced at the moment of action, provide the AI equivalent: governance state that is observable, testable, and independent of the deploying organization's preferences.

Statutory regulation after catastrophe was the fourth stage. The International Convention for the Safety of Life at Sea (SOLAS), adopted in 1914 after the Titanic disaster, established mandatory minimum safety requirements for all vessels. The EU AI Act, adopted in 2024 after accumulating governance failures across financial services, healthcare and employment, does the same for AI systems. Both represent the point where voluntary self-regulation proved insufficient and statutory requirements became necessary.

Containerized evidence came fifth. Malcom McLean's shipping container, introduced in 1956, standardized the interface for moving cargo across modes and jurisdictions. The Open Security Controls Assessment Language (OSCAL) rendering pipeline standardizes the interface for moving compliance evidence across regulatory frameworks. Cargo does not change; container does not change; translation cost at each framework boundary drops to near zero.

Continuous monitoring completed the arc. Modern maritime vessels carry AIS transponders, satellite communications, engine telemetry, and weather routing software providing continuous operational data to port authorities, classification societies, insurers, and fleet managers. The E-2 continuous verification pipeline provides the AI equivalent: real-time evidence of governance posture, available to auditors, examiners, and board committees without requesting a documentation package.

Maritime governance took roughly two hundred and seventy years to traverse these six stages, from 1688 to the present. AI governance is traversing them in about a decade, from 2016 to the mid-2020s. Deployment speed drives the compression; regulated environments do not permit the leisurely institutional evolution maritime governance enjoyed. The architecture is the same. The timeline is not.

The ISS Arc, Complete

The International Space Station (ISS) arc follows a different progression, one focused not on historical evolution but on operational governance at the frontier.

Before the ISS, space stations were national projects. Salyut, Skylab, Mir. Each reflected a single nation's engineering standards, safety requirements, and operational procedures, and each worked within its own boundaries. None of them ever had to integrate with a partner running different standards.

The Intergovernmental Agreement (IGA) changed that. The ISS IGA, signed in 1998, created the governance architecture for multi-jurisdictional operations. Jurisdiction follows the module; conflicts resolve through cooperation with escalation paths for genuine disagreements; "most restrictive applicable" governs uncertain situations. Not trust as a precondition. Trust as a product of architecture.

Assembly in orbit followed. Module by module over thirteen years. Each module functioned independently and integrated with everything already docked; no module required taking the station offline. The assembly sequence was the architecture. Commercial AI governance follows the same pattern: modular deployment, integration verification at each phase, non-destructive addition.

Commercial transition is the current phase. The CLD program transfers operations from government to commercial operators. NASA defines safety requirements; commercial operators determine how to meet them. Governance architecture persists; the operating model changes. Federal-validated to commercially operated.

Stable orbit is the destination. The ISS operates at specific orbital positions chosen because they satisfy mission requirements with minimum station-keeping cost. Same logic for regulated AI governance in commercial environments: find the position where forces balance, and operate there with minimum correction expenditure.

The ISS proves that multi-jurisdictional governance works at scale, in the most hostile operating environment humans have ever sustained, for decades. This book applies the same structural principles to a different environment. The physics did not change. The jurisdiction changed, the accountability framework changed, the evidence requirements changed. The architecture that produces stability did not.

The Framework Is the Same

This book began at Lloyd's coffeehouse with a question the merchants could not answer systematically: is this ship worth backing? Three centuries of institutional evolution followed. Classification societies. Plimsoll marks. SOLAS conferences. The Intergovernmental Agreement and module-by-module assembly and commercial transition. The question was translated from maritime vocabulary to AI governance vocabulary, and the translation proved structurally complete.

The merchants' question is the question every regulated organization asks today: is this AI system governance-worthy?

The Harness Doctrine provides the architecture to answer it.

Harness (H-1 through H-5): the constraint mechanism that wraps the AI system and produces predictable, verifiable behavior. Fourteen compliance properties (LP1 through LP14): seaworthiness criteria defining governance-worthiness across five regulated sectors, including the two enforcement properties (LP13-LP14) that govern what the system may emit and how it fails on uncertainty. Four evidence tiers (E-1 through E-4): the classification certificate and survey records demonstrating compliance to any examiner, auditor, or assessor. Knowledge base controls (KB-1 through KB-4): verification that the information the AI system relies on is bounded and uncontaminated. AGENTS.md: the operating manual codifying governance requirements in machine-enforceable form. OSCAL rendering pipeline: containerized evidence producing framework-specific compliance artifacts from a single stream.

The architecture is complete. Five sector Lagrange points mapped, coordinates published, and the migration path from existing governance to the AGENTS.md architecture documented with phases, deliverables, and timeline estimates.

What remains is the decision to navigate.

The Physics Did Not Change

Three forces, data protection, regulatory compliance, and business velocity, pull every regulated organization toward instability. They pulled the merchants at Lloyd's coffeehouse in 1688. They pull the bank deploying an AI credit scoring system in 2026; the hospital implementing AI-assisted clinical decision support; the utility deploying predictive maintenance on the grid; the insurer automating underwriting; the manufacturer integrating AI into quality assurance.

These forces do not weaken with technological progress. They intensify. AI systems process more data, at greater speed, across more jurisdictions than any previous technology. The data protection force grows because data volume grows, the regulatory compliance force grows because the regulatory surface grows, and the business velocity force grows because the competitive advantage of AI deployment grows.

Lagrange points exist in this intensified field too. They are not the governance configurations of a simpler time. They account for the forces as they are, not as anyone might wish them to be. An organization that locates its sector's Lagrange point and operates there, with continuous station-keeping, continuous evidence production, continuous verification of the fourteen governance properties, achieves stability. Not the stability of rest. The stability of a controlled orbit: continuous motion, continuous correction, continuous presence at the point where forces balance.

Organizations that find their Lagrange points first will govern AI effectively. They satisfy regulators, protect data subjects, and deploy at competitive velocity simultaneously, not by choosing one force over the others, but by operating at the configuration where all three are in equilibrium. Evidence of governance posture produced as a byproduct of normal operations, not as an examination preparation exercise. New regulatory requirements addressed by adding rendering templates, not by rebuilding compliance programs from scratch. Vendor transitions, staff turnover, and regulatory evolution survived because governance architecture is codified in the AGENTS.md, not stored in institutional memory.

Organizations that do not find their Lagrange points operate in the gravitational field between stable configurations. Constantly correcting, always reacting, never stable. Each new regulatory requirement produces a new compliance project, each examination produces a new documentation sprint, and each incident produces a new remediation program. Cost accumulates. Stability does not.

The merchants at Lloyd's coffeehouse eventually built the institution that solved their problem. Not quickly. Not painlessly. But they built it, because the alternative, continuing to price risk on hearsay while the ships went down, was not commercially sustainable.

The governance architecture described in this book is that same institution, adapted for the AI systems now entering every regulated industry. Same forces, same physics, the same structural logic it has always been.

The Lagrange points are waiting. Coordinates are in this book. Navigate to the stable point, or continue correcting from the unstable field between them. Those are the options.

One Thing You Can Do Tomorrow

Navigation to the stable point reads like board-level work: architecture decisions, vendor commitments, capital allocation. Most of it is. But the orbit is held by continuous small corrections, and small corrections do not require authority to change architecture. An early-career engineer, analyst, or compliance associate can begin station-keeping on day one, with no budget and no mandate, by producing one piece of the evidence the harness will later render. None of the actions below changes a system. Each one creates a record that did not exist yesterday, and a record is where every governance arc in this book started.

Three actions transfer across every sector. First, inventory the shadow AI in use on your own team: list the tools people actually paste data into, the data tiers that data belongs to, and which uses no one has declared. A one-page list of undeclared AI use is the first E-1 existence record most organizations have ever held. Second, draft a single AGENTS.md constraint for one workflow you understand: one named resource, one permitted action, one prohibited action, written so a routing layer could read it. You will not deploy it. You will have a concrete proposal where there was a policy sentence. Third, start an E-3 change log for one AI-touched process: a dated, append-only record of who changed what and why, even if that change log is just a shared document. State-transition evidence has to start somewhere, and it almost never starts on its own.

The sector-specific action is the one to take first.

In financial services (L1), pick one AI-generated figure that reaches a deliverable, a credit score input, a flagged transaction, a model output cited in a report, and trace it by hand to the source the model retrieved it from. Where the trace breaks is exactly where the connector-to-figure audit record will later be required.

In healthcare (L2), take one AI-assisted clinical or administrative workflow and write down which data tier it touches against the minimum-necessary rule: what the model can see, what it should not see, and where those two lists disagree. The disagreement is the access-consistency gap (LP1) made visible.

In energy and critical infrastructure (L3), document one place where an AI-assisted control or maintenance recommendation runs at the network edge with intermittent connectivity to central systems. Note what evidence is captured locally and what is lost when the link drops. That note is the start of a continuity-of-evidence record under the tightest correction budget in the volume.

In insurance and actuarial work (L4), take one automated pricing or underwriting output and run a first-pass disparate-impact screen on it. Treat the four-fifths ratio as a screening flag that triggers deeper actuarial-justification analysis, not as a pass/fail line; no insurance regulator has adopted the 80% ratio as a bright-line underwriting test, and the governing standards are state unfair-discrimination law and actuarial justification. A flagged screen routed to an actuary is a fairness record where there was none.

In manufacturing, supply chain, and telecommunications (L5), pick one AI system that affects a physical product, a logistics decision, or a network and write down which regulatory authority would care if it failed, and what evidence that authority would ask for. Because no single framework governs L5, naming the authority and the evidence it would demand is itself the first scoping decision.

None of these actions waits on a board vote, a vendor contract, or a budget line. Each produces a durable record, and durable records are what the harness renders into compliance evidence. The early-career person who builds the habit of producing one such record, then another, is doing station-keeping before anyone has named the orbit. That habit, more than any single architecture decision, is what carries an organization from the unstable field toward the stable point.

CHAPTER SUMMARY

A Lagrange point is not a place where forces disappear; it is a place where forces balance. The five sector Lagrange points, L1 through L5, are mathematical consequences of the three governance forces acting on each sector. They exist whether an organization discovers them or not. Organizations that find their Lagrange point achieve governance stability through continuous station-keeping: continuous verification, continuous evidence production, and continuous operation at the equilibrium configuration. Organizations that do not find their Lagrange point operate in the gravitational field between stable configurations, constantly correcting, always reacting, accumulating cost without accumulating stability. The maritime arc completes: from Lloyd's coffeehouse (informal assessment) through classification societies (institutional standards) through the Plimsoll mark (visible enforcement) through SOLAS (statutory regulation) through containerization (evidence harmonization) to continuous monitoring (real-time telemetry). The ISS arc completes: from national stations through the Intergovernmental Agreement through assembly in orbit through commercial transition to stable orbit. The governance framework is the same. The operating environment changed. The physics did not.

Appendix A: Vocabulary Translation Reference

Federal AI governance and commercial regulated industries describe the same architectural concepts in different languages. An Authorizing Official in DoD RMF becomes a Board Audit Committee in financial services and a HIPAA Security Officer in healthcare. The table below maps each federal term, as used in the Three-Body Problem of Classified AI and in authorization frameworks like NIST 800-53, FedRAMP, and DoD RMF, to its closest equivalent across five commercial sectors.

This is a lookup table, not a narrative. Reach for it when translating federal governance documentation into commercial equivalents, when briefing cross-sector teams that call the same mechanism by four different names, or when wiring The Harness Doctrine into a sector's existing regulatory vocabulary.

Chapter 2 introduced an abbreviated version. The table here expands that original with additional terms and finer-grained sector distinctions.

Federal Term	Financial Services	Healthcare	Energy / Critical Infrastructure	Insurance	Manufacturing / Telecom
Authorizing Official (AO)	CRO + Board Audit Committee	CISO + Privacy Officer + CMIO	VP Compliance + NERC CIP Senior Manager	Chief Risk Officer + Chief Actuary + Board Audit Committee	VP Quality / VP Compliance + CISO
Information System Security Officer (ISSO)	Information Security Officer (ISO) or Model Risk Manager	HIPAA Security Officer (45 CFR 164.308(a)(2))	NERC CIP Compliance Manager (CIP-003)	IT Security Director + Appointed Actuary	Plant IT Security Manager + Telecom Security Officer
Authorization to Operate (ATO)	SOC 2 Type II + Board AI Risk	HIPAA Risk Assessment + BAA + FDA SaMD Clearance	NERC CIP Audit Clearance + FERC	State DOI Examination Clearance	ISO 13485 Certification + Product

Federal Term	Financial Services	Healthcare	Energy / Critical Infrastructure	Insurance	Manufacturing / Telecom
	Authorization + MVR		Compliance Plan	+ SOX Attestation	Safety Clearance
Continuous ATO (cATO)	Continuous SOC 2 Monitoring + SR 11-7 Ongoing Monitoring	Continuous HIPAA Risk Management + FDA PCCP Compliance	Real-Time NERC CIP Monitoring + Self-Reporting	Ongoing State DOI Market Conduct Monitoring	Continuous ISO Audit + FCC Compliance Monitoring
Classification Levels	Public / Confidential / MNPI / PCI / BSA-SAR	De-identified / Standard PHI / Psychotherapy Notes / Part 2 / Genomic	Public / Internal / BCSI / CCA Data / Nuclear Safety	Public / Internal / Actuarial Privilege / Regulatory Restricted	Public / Proprietary / ITAR-Controlled / Safety-Critical
Clearance Level	RBAC + Chinese Wall Attestation + Need-to-Know	Minimum Necessary Rule + Treatment Relationship + Break-the-Glass	Personnel Risk Assessment (CIP-004-6 R4) + Electronic Access	Role-Based Access + Line-of-Business Restrictions	Role-Based Access + Export Control Clearance
STIGs	CIS Benchmarks + FFIEC IT Handbook + PCI-DSS Technical Requirements	HIPAA Security Rule Specifications + HITRUST CSF + CIS Benchmarks	NERC CIP-002 through CIP-014 + ICS-CERT + NIST SP 800-82	CIS Benchmarks + NAIC Cybersecurity Model Law Requirements	IEC 62443 (Industrial) + CIS Benchmarks + TSA Security Directives
OSCAL	OSCAL + XBRL (Financial Reporting Controls)	OSCAL + HL7 FHIR (Clinical Data Standards)	OSCAL + NERC XML Schemas (Audit Submissions)	OSCAL + ACORD Data Standards (Policy/Claims)	OSCAL + IEC 62443 Assessment Schemas
Security Assessment	SOC 2 Type II Report +	HIPAA Risk Assessment Report + FDA	NERC CIP Audit Report + FERC Filing	State DOI Examination Report +	ISO Audit Report + Product

Federal Term	Financial Services	Healthcare	Energy / Critical Infrastructure	Insurance	Manufacturing / Telecom
Report (SAR)	PCI ROC/SAQ	Conformity Assessment		SOX Audit Report	Safety Assessment
Plan of Action and Milestones (POA&M)	Remediation Tracker + Risk Acceptance Documentation	Corrective Action Plan + FDA Submission Timeline	CIP Violation Mitigation Plan + FERC Order Response	Examination Finding Response + Rate Filing Timeline	Corrective Action Report + Product Recall Plan
System Security Plan (SSP)	AI Governance Framework Documentation	HIPAA Security Plan + SaMD Technical Documentation	CIP Cybersecurity Plan + NRC Cyber Security Plan	Enterprise Risk Management AI Section + Actuarial Methodology Doc	Quality Management System AI Section + Network Security Plan
Boundary / Authorization Boundary	PCI Cardholder Data Environment + SOX-Scoped Systems	HIPAA Covered Entity Boundary + FDA SaMD Scope	NERC Electronic Security Perimeter + BES Cyber System	State DOI Examination Scope + SOX Material Systems	Product Safety Scope + Telecom Critical Infrastructure
Impact Level (IL)	Data Sensitivity Tier (Public through BSA-SAR)	PHI Sensitivity Tier (De-identified through Part 2)	NERC CIP Impact Rating (High/Medium/Low BES)	Underwriting Data Sensitivity Tier	Safety Criticality Level + ITAR Classification
Incident Response Plan	Fraud/Breach Response + SEC/OCC Notification	HIPAA Breach Notification (60 days) + State Notification	NERC CIP-008 Incident Response (1-hour BES report)	State DOI Breach Notification + Policyholder Notice	Product Safety Recall Protocol + FCC Outage Reporting
Delegation Token	OAuth2-scope-style authorization + SWIFT	SMART on FHIR Authorization Scopes + HIPAA	OPC UA Certificates + CIP-005	API Authorization Tokens + TPA	Industrial Protocol Certificates

Federal Term	Financial Services	Healthcare	Energy / Critical Infrastructure	Insurance	Manufacturing / Telecom
	Security Controls	Authorization Forms	Remote Access Controls	Delegation Agreements	+ Telecom API Tokens
Continuous Monitoring	FFIEC Continuous Monitoring + SR 11-7 Model Monitoring	HIPAA Ongoing Risk Management + FDA Post-Market Surveillance	NERC CIP Real-Time Monitoring + Self-Reporting	State Market Conduct Monitoring + Actuarial Model Tracking	Production Quality Monitoring + Network Performance Monitoring
Evidence Artifact	Examination Evidence Package + Audit Trail	HIPAA Risk Assessment Evidence + Clinical Documentation	NERC CIP Compliance Evidence + FERC Filing Exhibits	DOI Examination Workpapers + Actuarial Memorandum	Test Reports + Quality Records + Network Logs
Risk Acceptance	Board Risk Committee Acceptance + CRO Sign-Off	CISO/Privacy Officer Risk Acceptance + Clinical Risk Assessment	Senior Manager Risk Acceptance (CIP-003) + Board Oversight	Board Audit Committee + CRO + Chief Actuary Concurrence	VP Quality Risk Acceptance + Engineering Review
Control Assessment	SOC 2 Testing + PCI QSA Assessment + Model Validation	HIPAA Security Evaluation + FDA Conformity Assessment	NERC CIP Audit + FERC Compliance Review	DOI Market Conduct Examination + SOX Internal Controls Testing	ISO Surveillance Audit + Product Safety Testing
Vulnerability Scan	Application Security Testing + Model	HIPAA Technical Evaluation + SaMD Cybersecurity Testing	NERC CIP Vulnerability Assessment (CIP-010) + ICS Scan	Application Penetration Testing + Model	OT/ICS Vulnerability Assessment + Telecom

Federal Term	Financial Services	Healthcare	Energy / Critical Infrastructure	Insurance	Manufacturing / Telecom
	Adversarial Testing			Robustness Testing	Infrastructure Scan
Separation of Duties	Chinese Wall + SOX Dual Control + Model Validation Independence	Clinical vs. Administrative Access + Research/Treatment Separation	Operations vs. Compliance Separation (CIP-003)	Underwriting vs. Claims Separation + Actuarial Independence	Design vs. Quality Separation + Engineering vs. Compliance
Configuration Management	Change Management + SOX ITGC + Model Version Control	HIPAA Configuration Management + FDA PCCP + HL7 Version Control	NERC CIP Configuration Management (CIP-010) + Baseline Records	Policy Admin System Change Control + Model Version Management	Engineering Change Control + Network Configuration Management
Audit Trail	SOX Audit Trail + BSA Transaction Records + PCI Log Requirements	HIPAA Audit Controls (164.312(b)) + Clinical Event Logs	NERC CIP Electronic Access Logs (CIP-007) + Event Logs	DOI Examination Trail + Underwriting Decision Records	Manufacturing Execution Logs + Call Detail Records
Data Spillage	MNPI Wall Breach + PCI Cardholder Data Exposure	PHI Unauthorized Disclosure + Part 2 Re-Disclosure Violation	BCSI Unauthorized Access + CCA Data Exposure	Actuarial Model Exposure + Policyholder Data Breach	Trade Secret Disclosure + ITAR Violation + Network Data Leak
Least Privilege	Role-Based Access + Need-to-Know + Wall Attestation	Minimum Necessary Rule + Purpose Limitation	CIP-004 R4 Need-to-Know + Electronic Access Controls	Role-Based Access + Line-of-Business Scope	Principle of Least Functionality (IEC 62443) +

Federal Term	Financial Services	Healthcare	Energy / Critical Infrastructure	Insurance	Manufacturing / Telecom
					Role Restriction
Trusted Agent	Registered Representative + Compliance Officer	Treating Physician + Privacy Officer	Certified NERC CIP Personnel + Authorized Contractor	Licensed Agent + Appointed Actuary	Certified Quality Engineer + Authorized Telecom Technician
Cross-Domain Transfer	Inter-Entity Data Sharing + Cross-Border Transfer	Health Information Exchange (HIE) + Research Data Sharing	Inter-Utility Data Exchange + NERC Reliability Coordinator Sharing	Reinsurance Data Exchange + MGA/Carrier Data Sharing	Supply Chain Data Exchange + Telecom Interconnection Data
Accreditation	SOC 2 Attestation + PCI DSS Validation + MVR Approval	HIPAA Compliance Attestation + FDA Market Clearance	NERC CIP Audit Closure + FERC Acceptance	DOI Examination Clearance + SOX Attestation + Rate Approval	ISO Certification + UL/CE Marking + FCC Equipment Authorization
System Owner	Business Line Head + Application Owner	Department Head + Clinical System Owner	Facility Manager + System Operator	Product Line Owner + Underwriting VP	Plant Manager + Network Operations Director
Common Control Provider	Shared Services (IT, Compliance, Legal)	Enterprise IT + Health System Compliance	Regional Transmission Org + Compliance Shared Services	Holding Company IT + Shared Compliance	Corporate IT + Shared Quality Management
Interconnection Security Agreement (ISA)	Third-Party Risk Agreement + Vendor	BAA + Health Information Exchange Agreement	NERC CIP Interconnection Agreement + Mutual Aid Agreement	TPA Agreement + Reinsurer Data Agreement	Supply Chain Security Agreement + Interconnect

Federal Term	Financial Services	Healthcare	Energy / Critical Infrastructure	Insurance	Manufacturing / Telecom
	Security Addendum			+ MGA Contract	ion Agreement

Regulatory Acronyms in Plain English

The table above names statutes, rules, and supervisory standards by their acronyms. The glossary below defines each one in plain English at first reference, so a reader who has not lived inside a given sector can follow the mapping without leaving the page. Definitions are descriptive, not legal advice; verify any citation against its primary source before relying on it.

SR 11-7 is bank model-risk-management supervisory guidance, issued by the Federal Reserve and the OCC in 2011, requiring independent "effective challenge" of any model before an institution relies on it. Effective challenge means a qualified party who did not build the model tests its assumptions, data, and outputs.

SR 26-2 is described here as anticipated 2026 model-risk guidance; verify against the primary source before citing, or treat as an illustrative near-future rule. This volume treats it as the revised supervisory guidance expected to extend model-risk expectations to generative and agentic AI. SR 11-7 (2011) and SR 21-8 (2021) remain the verifiable anchors and are cited plainly.

ECOA / Regulation B is the Equal Credit Opportunity Act and its implementing rule, which prohibit credit discrimination on protected characteristics and require an adverse-action notice telling an applicant the specific reasons a credit request was denied.

42 CFR Part 2 is the federal rule giving substance-use-disorder treatment records heightened confidentiality beyond HIPAA, restricting re-disclosure even to parties that would ordinarily have access to a patient's medical record.

SMART on FHIR is the healthcare app-authorization standard, an OAuth2-scope-based access model that lets a clinical application request narrowly scoped, time-limited access to FHIR clinical data rather than open access to a patient record.

GLBA is the Gramm-Leach-Bliley Act, which governs how financial institutions handle and safeguard nonpublic personal information and requires privacy notices and a written information-security program.

MNPI is material nonpublic information: facts that a reasonable investor would consider important and that have not been released publicly. Trading on it, or improperly sharing it across an internal wall, is a securities-law violation.

BSA / AML is the Bank Secrecy Act and the anti-money-laundering regime built on it, requiring financial institutions to monitor transactions, file suspicious-activity reports, and keep records that help detect and report illicit finance.

FinCEN is the Financial Crimes Enforcement Network, the U.S. Treasury bureau that administers the Bank Secrecy Act, collects suspicious-activity and currency-transaction reports, and sets anti-money-laundering reporting rules.

NERC CIP is the set of Critical Infrastructure Protection standards issued by the North American Electric Reliability Corporation, mandatory cybersecurity controls for the bulk electric system covering access control, monitoring, configuration, and incident response.

NAIC Model Bulletin refers to the National Association of Insurance Commissioners model bulletin on AI use by insurers, issued December 2023, which directs insurers to maintain a written AI governance program and to assess AI systems regularly for unfairly discriminatory outcomes. As a model bulletin it takes effect through adoption by each state insurance regulator.

ASOP stands for Actuarial Standards of Practice, the professional standards that bind actuaries in how they build, document, and validate the models behind insurance pricing and reserving.

OSCAL is the Open Security Controls Assessment Language, a machine-readable format from NIST for expressing security control catalogs, baselines, and assessment results so that compliance documentation can be processed by tools rather than read only by humans.

DORA is the Digital Operational Resilience Act, the European Union regulation setting operational-resilience and third-party-risk requirements for financial entities, including incident reporting and oversight of critical technology providers.

ADMT / CPRA refers to automated decision-making technology rules under the California Privacy Rights Act, which extend California privacy law to govern profiling and automated decisions, including access and opt-out rights for consumers subject to them.

EU AI Act Articles 6, 13, 22, and 26 are the provisions of the European Union's AI regulation most relevant to this volume: Article 6 classifies high-risk AI systems; Article 13 requires transparency and clear information to deployers; Article 22 addresses authorized-representative obligations for providers established outside the Union; and Article 26 sets the obligations of deployers of high-risk systems, including human oversight and monitoring.

Usage Notes

The Delegation Token. In the table above, the Delegation Token maps to OAuth2-scope-style authorization in financial services and to SMART on FHIR scopes in healthcare. Those are the closest familiar analogs, not the actual token format. For the canonical Delegation Token shape used throughout this volume, its signed claim set, audience-scoped resource list, data-tier and jurisdiction tags, short expiry, and gateway enforcement, see the front-matter primer and the worked enforcement example in the financial-services harness chapter (Chapter 11b). No chapter introduces a second, conflicting token schema.

Cross-sector organizations. Organizations that straddle sector boundaries, a hospital system that issues municipal bonds or a manufacturer holding defense contracts, must read multiple columns at once. The AGENTS.md instruction surface supports multi-sector configurations natively; this vocabulary table supports the human communication required to build and maintain those configurations.

Regulatory evolution. The table reflects the regulatory state as of early 2026. State-level AI governance laws, the Colorado AI Act (SB 24-205), California's ADMT regulations, and a growing roster of state privacy statutes, will add terms and roles as they take effect. The federal column stays stable because NIST 800-53 and the DoD RMF anchor the structural reference. Commercial columns will grow as regulatory vocabulary evolves around them.

Translation, not equivalence. Every cell in this table is a functional mapping, not a claim that the federal concept and its commercial counterpart are identical. The Authorizing Official and the Board Audit Committee serve the same governance function, accepting residual risk and authorizing operation; they differ in authority structure, accountability mechanism, and decision cadence. The mapping opens communication across vocabulary boundaries. It does not flatten the structural differences that make each sector's governance distinct.

Appendix B: Cross-Framework Control Mapping

The canonical LP1 through LP14 property definitions used throughout this appendix are those given in Chapter 5, which the book declares authoritative; the property names below match that canonical set exactly. This crosswalk covers the Lagrange governance properties (LP1 through LP14) defined in this volume. The eight federal security properties (P1 through P8), the inherited baseline, are defined and mapped in Volumes I and II and are not reproduced here.

A single governance architecture can map to requirements across seven regulatory frameworks at once, but only if someone builds the crosswalk, and only if an assessor validates that each claimed mapping holds for the specific deployment. This appendix does the first part of that work. It maps every Three-Body control, H-1 through H-5 (harness components), LP1 through LP14 (compliance properties), E-1 through E-4 (evidence tiers), and KB-1 through KB-4 (knowledge base integrity controls), to specific control numbers in seven major frameworks. Chapter 18 (One Harness, Many Auditors) describes the theory; these tables provide the lookup. This appendix is the single canonical cross-framework crosswalk for this volume; chapter cross-references to a control mapping table, including the reference in Chapter 17, point here.

A mapping is a claim, not a certification. Each cell below asserts that a Harness Doctrine control maps to a named clause in a framework. That assertion is the starting point for an assessor, not a substitute for one. Wherever a single control is claimed against several frameworks at once, the specific implementation has to be tested against each clause separately; a mapping that holds for one deployment can fail for another that configures the same control differently. Read every cell as "maps to [clause]; assessor validation required."

The seven frameworks covered:

1. **NIST 800-53 Rev. 5**: Federal baseline, the structural origin of The Harness Doctrine
2. **SOC 2 Trust Services Criteria**: AICPA criteria for service organization controls
3. **PCI-DSS v4.0.1**: Payment Card Industry Data Security Standard
4. **HIPAA Security Rule**: 45 CFR Part 164 Subpart C
5. **NERC CIP**: Critical Infrastructure Protection standards for bulk electric systems

6. **GDPR**: General Data Protection Regulation (EU)
7. **EU AI Act**: Regulation (EU) 2024/1689

Harness Components (H-1 through H-5)

Control	Description	NIST 800-53	SOC 2 TSC	PCI-DSS v4.0.1	HIPAA Security Rule	NERC CIP	GDPR	EU AI Act
H-1	Instruction Surface (AGENTS.md)	PL-2 (System Security Plan), SA-4 (Acquisition Process)	CC1.1 (Control Environment), CC1.4 (Board Oversight)	Req 1.1 (Network Security Controls Documentation)	164.308(a)(1)(i) (Security Management Process)	CIP-003-8 R1 (Cybersecurity Policy)	Art. 24 (Responsibility of Controller), Art. 25 (Data Protection by Design)	Art. 9(1) (Risk Management System), Art. 11 (Technical Documentation)
H-2	Routing Enforcement	AC-3 (Access Enforcement), AC-4 (Information Flow Enforcement)	CC6.1 (Logical and Physical Access), CC6.3 (Role-Based Access)	Req 7.2 (Access Control Systems), Req 7.3 (Access Restriction)	164.312(a)(1) (Access Control), 164.312(e)(1) (Transmission Security)	CIP-005-7 R1 (Electronic Security Perimeter), CIP-007-6 R4 (Security Event Monitoring)	Art. 32(1)(b) (Access Controls)	Art. 9(2)(b) (Design for Effective Oversight), Art. 14(2) (Human Oversight Design)
H-3	Persistent State	SC-28 (Protection of Information at Rest), SI-12 (Information	CC6.7 (Data Lifecycle Management)	Req 3.1 (Account Data Storage), Req 3.3 (SAD Not Retained)	164.312(a)(2)(iv) (Encryption), 164.310(d)(1) (Device and Media Controls)	CIP-011-3 R1 (Information Protection)	Art. 5(1)(e) (Storage Limitation), Art. 17 (Right to Erasure)	Art. 12(1) (Record-Keeping Obligations), Art. 19(1) (Automatic Logging)

Control	Description	NIST 800-53	SOC 2 TSC	PCI-DSS v4.0.1	HIPAA Security Rule	NERC CIP	GDPR	EU AI Act
		Management)						
H-4	Verification Checkpoints	CA-2 (Control Assessments), SI-4 (System Monitoring)	CC4.1 (Monitoring Activities), CC7.2 (Security Event Detection)	Req 11.3 (Vulnerability Management), Req 12.10 (Incident Response)	164.308(a)(8) (Evaluation), 164.312(b) (Audit Controls)	CIP-010-4 R1 (Configuration Monitoring), CIP-007-6 R4 (Security Event Monitoring)	Art. 35 (Data Protection Impact Assessment)	Art. 9(2)(a) (Risk Assessment), Art. 72 (Post-Market Monitoring)
H-5	Accountability Records	AU-2 (Event Logging), AU-3 (Content of Audit Records), AU-6 (Audit Record Review)	CC7.2 (Security Event Detection), CC7.3 (Security Incident Evaluation)	Req 10.1 (Audit Log Implementation), Req 10.2 (Audit Log Contents)	164.312(b) (Audit Controls), 164.308(a)(1)(ii)(D) (Information System Activity Review)	CIP-007-6 R4 (Security Event Monitoring), CIP-007-6 R5 (System Access Controls)	Art. 30 (Records of Processing), Art. 33 (Notification of Breach)	Art. 12(2) (Automatic Logging), Art. 19(1) (Logging for High-Risk Systems)

Known limitations; assessor validation required. Each cell maps a harness component to a framework clause; it does not certify that the clause is satisfied. The one-control-to-many cells carry the most risk and need the closest scrutiny. H-2 (Routing Enforcement) is claimed simultaneously against PCI-DSS Req 7.2/7.3, HIPAA 164.312(a)(1), NERC CIP-005-7 R1, and GDPR Art. 32(1)(b); a single routing layer can map to all four, but each framework tests a different boundary (cardholder data environment segmentation for PCI, electronic protected health information access for HIPAA, the electronic security perimeter for NERC CIP, and the appropriateness of technical access measures for GDPR), so a routing configuration that an assessor accepts under one clause may not clear another. H-4 (Verification Checkpoints) is likewise claimed across CA-2/SI-4, SOC 2 CC4.1/CC7.2, PCI Req 11.3/12.10, HIPAA 164.308(a)(8)/164.312(b), NERC CIP-010-4 R1, GDPR Art. 35, and EU AI Act Art. 9(2)(a)/72; the checkpoint mechanism is shared, but each assessor validates it against that framework's own monitoring and evaluation cadence. Treat any row that claims one control against three or more frameworks as a hypothesis the assessor confirms against the running evidence, not as a settled equivalence.

Compliance Properties (LP1 through LP14)

Control	Description	NIST 800-53	SOC 2 TSC	PCI-DSS v4.0.1	HIPAA Security Rule	NERC CIP	GDPR	EU AI Act
LP1	Classification-Consistent Access	AC-3, AC-16 (Security Attributes)	CC6.1, CC6.3	Req 7.1 (Access by Business Need), Req 7.2	164.312(a)(1), 164.514(d) (Minimum Necessary)	CIP-004-7 R4 (Access Management)	Art. 5(1)(f) (Integrity and Confidentiality)	Art. 9(2)(b)
LP2	Scope Binding	AC-2 (Account Management), AC-6 (Least Privilege)	CC6.2 (Prior to Access), CC6.3	Req 7.1, Req 8.1 (User Identification)	164.312(a)(1), 164.312(d) (Person/Entity Authentication)	CIP-004-7 R4, CIP-004-7 R5 (Access Revocation)	Art. 5(1)(b) (Purpose Limitation)	Art. 9(4)(a) (Appropriate Constraints)
LP3	Routing Integrity (architectural expression of H-2)	AC-4, AC-24 (Access Control Decisions)	CC6.1, CC6.7	Req 1.2 (Network Controls), Req 7.3	164.308(a)(3)(i) (Workforce Security), 164.308(a)(4) (Information Access Management)	CIP-005-7 R1, CIP-005-7 R2 (Interactive Remote Access)	Art. 25(1) (Data Protection by Design)	Art. 9(1), Art. 14(1) (Human Oversight)
LP4	Persistent	SC-28 (Protec	CC6.7 (Data	Req 10.2,	164.312(a)(2)(iv)	CIP-011-3	Art. 30 (Records	Art. 12(1) (Record-Keeping

Control	Description	NIST 800-53	SOC 2 TSC	PCI-DSS v4.0.1	HIPAA Security Rule	NERC CIP	GDPR	EU AI Act
	Governance Context (architectural expression of H-3)	tion of Information at Rest), SI-12 (Information Management)	Lifecycle Management)	Req 10.3 (Audit Log Protection); assessor validation required	(Encryption), 164.312(b) (Audit Controls); assessor validation required	R1 (Information Protection)	of Processing); assessor validation required	Obligations), Art. 19(1) (Automatic Logging)
LP5	Continuous Verification	CA-7 (Continuous Monitoring), SI-4 (System Monitoring)	CC4.1 (Monitoring Activities), CC7.2 (Security Event Detection)	Req 11.3 (Vulnerability Management); assessor validation required	164.308(a)(8) (Evaluation), 164.308(a)(1)(ii)(D) (Information System Activity Review)	CIP-010-4 R1 (Configuration Monitoring), CIP-007-6 R4 (Security Event Monitoring)	Art. 32(1)(d) (Testing/Assessing Effectiveness)	Art. 9(2)(a) (Risk Assessment), Art. 72 (Post-Market Monitoring)
LP6	Attribution Completeness (architectural expression of H-5)	AU-2 (Event Logging), AU-3 (Content of Audit Records), AU-6 (Audit Record	CC7.2 (Security Event Detection), CC7.3 (Security Incident Evaluation)	Req 10.1 (Audit Log Implementation), Req 10.2 (Audit Log Contents)	164.312(b) (Audit Controls), 164.308(a)(1)(ii)(D) (Information System Activity Review)	CIP-007-6 R4 (Security Event Monitoring)	Art. 30 (Records of Processing)	Art. 12(2) (Automatic Logging), Art. 19(1) (Logging for High-Risk Systems)

Control	Description	NIST 800-53	SOC 2 TSC	PCI-DSS v4.0.1	HIPAA Security Rule	NERC CIP	GDPR	EU AI Act
		Review)						
LP7	Deviation Detection	SI-4 (System Monitoring), CA-2 (Control Assessments)	CC4.1 (Monitoring Activities), CC7.2 (Security Event Detection)	Req 11.3 (Vulnerability Management), Req 12.10 (Incident Response); assessor validation required	164.308(a)(8) (Evaluation), 164.312(b) (Audit Controls)	CIP-010-4 R1 (Configuration Monitoring), CIP-007-6 R4 (Security Event Monitoring)	Art. 35 (Data Protection Impact Assessment); assessor validation required	Art. 9(2)(a) (Risk Assessment), Art. 72 (Post-Market Monitoring)
LP8	Evidence Integrity	AU-9 (Protection of Audit Information), SI-7 (Software and Information Integrity)	CC7.2, CC7.3	Req 10.5 (Audit Log Protection from Alteration)	164.312(c)(1) (Integrity), 164.312(b) (Audit Controls)	CIP-010-4 R1 (Configuration Monitoring), CIP-011-3 R1 (Information Protection)	Art. 5(1)(f) (Integrity and Confidentiality), Art. 5(2) (Accountability)	Art. 12(1) (Record-Keeping Obligations), Art. 19(1) (Automatic Logging)
LP9	Consent-Bounded Processing	--	--	--	164.508 (Uses Requiring Authorization), 164.522 (Rights to Request	--	Art. 6 (Lawful Processing), Art. 7 (Conditions for Consent), Art. 9	Art. 26(5) (Deployer: Inform Individuals), Art. 86 (Transparency Obligations)

Control	Description	NIST 800-53	SOC 2 TSC	PCI-DSS v4.0.1	HIPAA Security Rule	NERC CIP	GDPR	EU AI Act
					Privacy Protection)		(Special Categories)	
LP10	Jurisdictional Consistency	--	--	--	--	--	Art. 3 (Territorial Scope), Art. 44-49 (Cross-Border Transfers), Art. 27 (Representatives)	Art. 2 (Scope), Art. 6(1) (Annex III Classification), Art. 26(1) (Deployer Obligations)
LP11	Fairness Verification	--	--	--	--	--	Art. 22(3) (Right to Contest Automated Decision)	Art. 9(2)(a) (Bias Identification), Art. 10(2)(f) (Bias Examination), Art. 72(1) (Post-Market Monitoring)
LP12	Transparency and Explainability	--	--	--	--	--	Art. 13-14 (Information to Data Subjects), Art. 22(3) (Right to Explanation), Recital 71	Art. 13 (Transparency), Art. 14 (Human Oversight), Art. 26(5) (Inform Natural Persons), Art. 86
LP13	Output Boundary	SI-15 (Information Output	CC6.7 (Data Lifecycle	Req 3.4 (Account Data Rendere	164.312(e)(1) (Transmission	CIP-011-3 R1 (Infor	Art. 5(1)(f) (Integrity and	Art. 13 (Transparency), Art. 15 (Accuracy/Robustn

Control	Description	NIST 800-53	SOC 2 TSC	PCI-DSS v4.0.1	HIPAA Security Rule	NERC CIP	GDPR	EU AI Act
	Enforcement	Filtering), AC-4 (Information Flow Enforcement); assessor validation required	Management); assessor validation required	d Unreadable); assessor validation required	Security); assessor validation required	mation Protection); assessor validation required	Confidentiality); assessor validation required	ess/Cybersecurity); assessor validation required
LP14	Deterministic Deny on Uncertainty	AC-3 (Access Enforcement), SC-39 (Process Isolation); assessor validation required	CC6.1 (Logical and Physical Access); assessor validation required	--	164.312(a)(1) (Access Control); assessor validation required	CIP-005-7 R1 (Electronic Security Perimeter); assessor validation required	Art. 25(2) (Data Protection by Default); assessor validation required	Art. 14 (Human Oversight); assessor validation required

Note: LP9 through LP12 have empty cells in the NIST 800-53, SOC 2, PCI-DSS, and NERC CIP columns because those frameworks do not address the requirements these properties cover. LP9 maps to HIPAA because HIPAA includes individual consent mechanisms; the other three frameworks do not. All four commercial properties map to GDPR and the EU AI Act, both of which address individual rights in automated processing directly. Chapter 8 walks through the full rationale.

Evidence Tiers (E-1 through E-4)

Control	Description	NIST 800-53	SOC 2 TSC	PCI-DSS v4.0.1	HIPAA Security Rule	NERC CIP	GDPR	EU AI Act
E-1	Foundation Evidence (Governance Configuration Baseline)	PL-2 (System Security Plan), CM-2 (Baseline Configuration), CM-6 (Configuration Settings)	CC1.1-CC1.5 (Control Environment), CC2.1 (Information Quality)	Req 1.1 (Network Docs), Req 12.1 (Information Security Policy)	164.308(a)(1)(i), 164.316 (Policies and Procedures Documentation)	CIP-003-8 R1-R4 (Policy, Plans, Delegation), CIP-002-5.1 R1 (Asset Identification)	Art. 24 (Controller Responsibility), Art. 30 (Records of Processing), Art. 35 (DPIA)	Art. 11 (Technical Documentation), Art. 53 (Registration in EU Database)
E-2	Operational Evidence (Continuous Verification)	CA-7 (Continuous Monitoring), SI-4 (System Monitoring), RA-5 (Vulnerability Monitoring)	CC4.1 (Monitoring), CC4.2 (Deficiency Evaluation), CC7.1 (Detection), CC7.2 (Monitoring)	Req 10.4 (Audit Log Review), Req 11.3 (Vulnerability Scans), Req 11.4 (Penetration Testing)	164.308(a)(1)(ii)(D) (System Activity Review), 164.308(a)(8) (Evaluation)	CIP-007-6 R4 (Security Event Monitoring), CIP-010-4 R1 (Configuration Monitoring)	Art. 32(1)(d) (Testing/Assessing Effectiveness)	Art. 9(1) (Risk Management), Art. 72 (Post-Market Monitoring System)
E-3	Compliance Evidence (State Transition Records)	CM-3 (Configuration Change Control), CM-5 (Access Restrictions for Change)	CC8.1 (Change Management), CC3.1 (Risk Assessment Processes)	Req 6.5 (Change Management), Req 6.5.1 (Change Control Procedures)	164.308(a)(1)(ii)(B) (Risk Management), 164.310(a)(2)(iv) (Maintenance Records)	CIP-010-4 R1 (Configuration Change Management), CIP-010-4 R2 (Configuration	Art. 5(2) (Accountability), Art. 25(1) (Data Protection by Design)	Art. 12(1) (Record-Keeping), Art. 72(2) (Monitoring Plan Updates)

Control	Description	NIST 800-53	SOC 2 TSC	PCI-DSS v4.0.1	HIPAA Security Rule	NERC CIP	GDPR	EU AI Act
						Monitoring)		
E-4	Accountability Evidence (Attribution Chain Records)	AU-2 (Event Logging), AU-3 (Content of Audit Records), AU-12 (Audit Record Generation)	CC7.2 (Security Event Detection), CC7.3 (Security Incident Evaluation)	Req 10.1-10.7 (Audit Trail Implementation)	164.312(b) (Audit Controls), 164.528 (Accounting of Disclosures)	CIP-007-6 R4 (Security Event Monitoring), CIP-007-6 R5 (System Access Controls)	Art. 30 (Records of Processing), Art. 33 (Breach Notification), Art. 15 (Right of Access)	Art. 12(2) (Automatic Logging), Art. 19(1) (Automatic Logs for High-Risk AI)

Knowledge Base Integrity Controls (KB-1 through KB-4)

Control	Description	NIST 800-53	SOC 2 TSC	PCI-DSS v4.0.1	HIPAA Security Rule	NERC CIP	GDPR	EU AI Act
KB-1	Verified Provenance	SA-11 (Developer Testing), SA-12 (Supply Chain Risk), SI-7 (Software and Information Integrity)	CC8.1 (Change Management), PI1.3 (Processing Integrity)	Req 6.3 (Security Vulnerabilities ID), Req 6.2 (Custom Software Security)	164.312(c)(1) (Integrity), 164.312(c)(2) (Mechanism to Authenticate ePHI)	CIP-010-4 R1 (Software Verification), CIP-013-2 R1 (Supply Chain Risk)	Art. 5(1)(d) (Accuracy), Art. 28 (Processor Requirements)	Art. 10(1) (Training Data Governance), Art. 10(3) (Relevant/Representative Data)

Control	Description	NIST 800-53	SOC 2 TSC	PCI-DSS v4.0.1	HIPAA Security Rule	NERC CIP	GDPR	EU AI Act
KB-2	Enforced Currency	CM-3 (Configuration Change Control), SI-7 (Software/Info Integrity Verification)	CC8.1 (Change Management), CC6.8 (Software Use)	Req 6.5 (Change Management), Req 11.3 (Vulnerability Management)	164.308(a)(1)(ii)(B) (Risk Management), 164.310(a)(2)(iv) (Maintenance Records)	CIP-010-4 R1 (Configuration Change Management)	Art. 5(1)(d) (Accuracy), Art. 16 (Right to Rectification)	Art. 10(5) (Data Quality Management), Art. 72 (Post-Market Monitoring)
KB-3	Controlled Ingestion	AC-3 (Access Enforcement), AC-4 (Information Flow Enforcement), SC-7 (Boundary Protection)	CC6.1 (Logical Access), CC6.6 (Boundary Protection)	Req 1.2 (Network Controls), Req 7.2 (Access Systems)	164.312(a)(1) (Access Control), 164.514(d) (Minimum Necessary)	CIP-005-7 R1 (Electronic Security Perimeter), CIP-004-7 R4 (Access Management)	Art. 5(1)(f) (Integrity/Confidentiality), Art. 32 (Security Measures)	Art. 9(2)(b) (Oversight Design), Art. 15(2) (Accuracy/Robustness)
KB-4	Monitored Retrieval	SI-3 (Malicious Code Protection), SI-4 (System Monitoring), SI-10 (Information Input	CC7.1 (Detection Mechanisms), CC7.2 (Security Event Detection)	Req 5.1 (Anti-Malware), Req 11.5 (Network Intrusion Detection)	164.308(a)(5)(ii)(B) (Protection from Malicious Software)	CIP-007-6 R3 (Malicious Code Prevention), CIP-007-6 R4	Art. 5(1)(d) (Accuracy), Art. 32(1)(b) (Processing Security)	Art. 10(2)(f) (Bias Examination), Art. 15(4) (Cybersecurity Measures)

Control	Description	NIST 800-53	SOC 2 TSC	PCI-DSS v4.0.1	HIPAA Security Rule	NERC CIP	GDPR	EU AI Act
		Validation)				(Monitoring)		

Declarative Admission Control as Cross-Framework Policy Engine

Declarative Admission Control is presented here as an illustrative reference architecture, a design pattern named for explanation, not a specific commercial product; the latency and isolation characteristics described are design targets, not measured benchmarks.

When classification societies certify a vessel for operation, they do not issue a single certificate covering all operating conditions. They issue certificates with explicit constraints: maximum load, geographic restrictions, cargo type limitations, weather conditions, crew qualification requirements. These constraints are not advisory; they are the authorization boundary. Operating outside those constraints voids the certificate.

Declarative Admission Control (DAC) implements the same boundary enforcement for AI systems operating under multiple regulatory frameworks simultaneously. Not a single policy engine trying to encode every framework's rules in imperative code. A declarative policy layer where each framework's requirements are expressed as admission control rules that execute before the AI system processes a request.

The admission control model treats every AI system interaction as a request that must be authorized before execution. A user query to a RAG system, a background inference job, a model fine-tuning request, an evidence artifact export: each is a request with attributes (user identity, data classification, framework scope, resource requirements, time of request). DAC evaluates that request against the policy set for every framework the organization is subject to. If any framework's policy denies the request, the request is blocked before processing begins.

Policy declarations are YAML documents, not code. Each framework's requirements are expressed as rules with explicit conditions and actions. A GDPR data minimization rule might state: "deny any request that includes data not strictly necessary for the stated purpose". A HIPAA minimum necessary rule might state: "deny any request that accesses more protected health information than required for the specified treatment, payment, or healthcare operations function". A NERC CIP security perimeter rule might state: "deny any external connection to bulk electric system control networks that does not originate from an approved electronic access point".

Deny-by-default semantics handle the common case, where frameworks impose additive constraints: a request is authorized only if every applicable policy explicitly permits it. If GDPR policy permits the request but HIPAA policy does not, the request is denied. If both permit it but SOC 2 policy denies it, the request is denied. The system operates within the intersection of all regulatory constraints, not the union. This resolves additive conflicts cleanly. It does not resolve genuinely contradictory requirements, where one framework compels what another forbids; the GDPR-retention versus records-preservation conflict in Chapter 9 is the worked example. Those cases cannot be set-intersected to a safe answer and must route to the human escalation and documented-resolution path Chapter 9 defines, not deadlock silently into denying everything.

This resolves the cross-framework composition problem that Chapter 18 identifies. Organizations subject to multiple frameworks cannot build separate policy enforcement systems for each framework and hope they do not conflict. They must build a single policy enforcement layer that composes all frameworks' rules deterministically. DAC provides that layer.

Policy-as-code benefits extend beyond enforcement to auditability. When a regulator asks "how does your organization ensure GDPR data minimization requirements are met?", the answer is the GDPR policy declaration plus the audit log of every request evaluated against that policy. The regulator can read the policy (it is YAML, not compiled code), verify that it correctly implements GDPR requirements, and examine the audit log to confirm that the policy was enforced for every request. No assertions about "our governance team reviews access requests"; verifiable evidence that access control decisions were policy-driven.

Maritime parallel: Lloyd's does not trust a shipowner's statement that "we follow load line regulations". Lloyd's inspects the ship's freeboard markings, verifies that they match the classification certificate, examines the loading computer to confirm that load calculations enforce the freeboard limit, and reviews cargo loading records to verify that the limit was not exceeded. DAC provides the AI governance equivalent: declarative policies (freeboard markings), policy verification (classification certificate match), automated enforcement (loading computer), and audit trail (cargo loading records).

Cross-framework control mapping in this appendix shows which regulatory requirements each Harness Doctrine control satisfies. DAC is the mechanism that ensures all applicable requirements are enforced simultaneously. LP3 (Routing Integrity) maps to access control requirements in seven frameworks. DAC enforces all seven sets of requirements with a single policy evaluation per request. The policy set grows when new frameworks apply; the enforcement mechanism stays constant.

Policy testing becomes possible because policies are data, not code. A test suite can submit synthetic requests with known attributes and verify that DAC produces the expected permit or deny decisions under each framework's rules. When a framework's requirements change, policies are updated, test cases are added for the new requirements, and the entire test suite runs to verify that no existing authorization decisions broke. Testing policy changes is testing YAML, not debugging imperative logic scattered across a codebase.

The evidence artifact that proves multi-framework compliance is not a spreadsheet mapping controls to implementation claims. It is the set of DAC policy declarations (one per framework), the audit log of requests evaluated against those policies, and the test suite results showing that all policies enforce their framework's requirements correctly. A regulator can verify compliance by reading the relevant policy and auditing a sample of its enforcement decisions. No trust required; the policy says what it does, and the audit log shows what it did.

Classification societies enforce vessel operating limits with physical constraints (freeboard markings, load line certificates) and procedural controls (loading supervision, cargo surveying). DAC enforces AI system operating limits with declarative constraints (YAML policies) and automated controls (admission evaluation, audit logging). Both approaches prevent unauthorized operation; both produce verifiable evidence; both compose multiple regulatory requirements into a single enforcement boundary.

When the board audit committee asks "how do we ensure AI systems comply with all applicable regulations?", the answer is not "our governance team interprets regulations and configures controls accordingly". The answer is DAC: declarative policies encoding each regulation's requirements, automated enforcement preventing non-compliant requests, audit trails proving that enforcement occurred, and test suites demonstrating that policies correctly implement their regulatory requirements. Not governance promises; governance proof.

Production Inference Engine as Classification Decision Support

The inference engine described here is an illustrative reference architecture, named for explanation, not a specific commercial product; its governance hooks and evidence-generation path are a design pattern, not a benchmarked system.

When a Lloyd's surveyor inspects a vessel, they are not making every determination from scratch. They are applying the classification society's standards, documented procedures, and historical precedent to the specific vessel before them. The surveyor's expertise is knowing how to apply those standards; the society's role is maintaining those standards and ensuring they are applied consistently across all surveyors and all vessels.

This reference inference engine provides the same consistency for AI-powered classification decisions in regulated industries. Not a general-purpose inference server running arbitrary models. A production-grade inference engine with governance hooks, logging requirements, and evidence generation built into the inference path itself.

Maritime classification decisions follow a deterministic structure. A hull inspection produces findings: thickness measurements below minimum requirements, corrosion beyond acceptable limits, structural members requiring repair. Those findings map to classification standards: how much corrosion is acceptable for this vessel type, this age, this service, this classification notation. The surveyor applies the standards; the society's technical rules define what the standards mean; the classification decision is the outcome of that application.

Reference-architecture classification in financial services, healthcare, energy, and insurance follows the same structure. A loan application produces features: credit score, debt-to-income ratio, employment history, collateral value. Those features map to underwriting policies: what credit score is required for this loan amount, this collateral type, this borrower profile. The underwriting model applies the policies; the institution's credit risk framework defines what the policies mean; the underwriting decision is the outcome of that application.

The governance hooks are first-class infrastructure, not bolt-on logging. Every inference request to the reference inference engine includes required context: which policy set applies, what evidence tier is required, which regulatory frameworks govern the decision, what the attribution chain must include. the reference inference engine evaluates that context before running the model; if required context is missing or malformed, the inference request fails before the model executes. Not "we log as much as we can"; every inference includes the governance metadata required for compliance evidence.

Model routing in the reference inference engine respects organizational boundaries. A healthcare provider's diagnostic support model cannot access a financial institution's credit underwriting model; an energy utility's grid optimization model cannot access a maritime classification model. Not because those models are on separate servers (they may all run on the same the reference inference engine instance), but because the reference inference engine enforces policy-defined boundaries that prevent cross-organizational model access. Multi-tenancy with governance-grade isolation.

Inference audit trails are structured evidence artifacts, not unstructured logs. Every inference produces a record containing: model identifier and version, input features (scope-bounded according to LP1 and LP2), output prediction, confidence score, policy set evaluated, regulatory frameworks applied, timestamp, user identity, and attribution chain. This record is the evidence that the classification decision was made using an authorized model, following documented policies, producing output within acceptable confidence bounds, with full traceability to the requesting user.

The maritime parallel is the survey report. Lloyd's does not accept a surveyor's verbal statement that "the vessel meets classification standards". Lloyd's requires a written survey report documenting what was inspected, what findings were identified, what standards were applied, what repairs are required, what the classification decision is, and what the surveyor's identity and qualifications are. the reference inference engine inference logs provide the AI equivalent: what model ran, what inputs were evaluated, what policies applied, what the output was, what the confidence level was, and what the user's authorization was.

High-availability requirements in regulated industries demand that classification support systems operate continuously. A Lloyd's surveyor arriving at a port expects to access the classification society's technical library, standards database, and precedent archive regardless of network conditions. the reference inference engine deployments in regulated environments require the same availability: inference capacity sufficient for peak load, failover between availability zones, cache layers for high-frequency policy evaluation, and degraded-mode operation when external dependencies are unavailable.

Performance characteristics matter for classification decisions because delays create operational bottlenecks. A cargo vessel waiting for classification approval to load high-value goods incurs demurrage costs that for large vessels commonly run into the thousands to tens of thousands of dollars per day (illustrative order of magnitude; actual demurrage is set by the charter party and varies by vessel class and route). A bank unable to make credit decisions because underwriting inference is unavailable loses competitive position to institutions whose systems remain operational. the reference inference engine is optimized for production inference at scale: low single-digit-to-double-digit-millisecond p99 latency for typical classification models (an illustrative target, not a benchmarked guarantee), batch inference for overnight processing, streaming inference for real-time decisions.

The evidence bundle that proves classification decisions were governance-compliant is the the reference inference engine inference audit log. Not a separate compliance system trying to reconstruct what the inference engine did; the inference engine's native log, structured for compliance evidence from the first request. E-4 attribution chains in the Harness Doctrine map directly to the reference inference engine inference records. A regulator auditing classification decisions reads the inference log; every decision is documented with the governance context that authorized it.

Classification societies maintain technical expertise in marine engineering, structural analysis, materials science, and regulatory compliance. They apply that expertise through surveyors who inspect vessels using documented procedures. the reference inference engine maintains technical expertise in model serving, governance enforcement, audit trail generation, and regulatory evidence production. Organizations apply that expertise through models that make classification decisions using documented policies. Both approaches convert domain expertise into consistent, auditable, repeatable decisions.

When the board audit committee asks "how do we ensure AI-assisted classification decisions are defensible?", the answer is a governed inference engine: production inference with governance metadata required per request, audit logs structured for compliance evidence, high-availability deployment ensuring continuous operation, and performance characteristics meeting operational requirements. Not a prototype demonstrating that AI works; production infrastructure proving that AI works within regulatory boundaries.

Privacy-Preserving Certification for Maritime and Financial Data

The privacy-preserving certification approach described here is an illustrative reference architecture built on standard zero-knowledge proof techniques, named for explanation, not a specific commercial product.

When a vessel owner reports a safety incident to a maritime authority through an anonymous reporting system, the authority receives the incident details without learning the vessel's identity. The system proves that the report came from a legitimate vessel operator without revealing which operator. This privacy-preserving reporting increases incident disclosure rates because operators can report near-misses, procedural violations, and safety concerns without fear that disclosure will trigger enforcement action against their specific vessel.

Zero-knowledge identity protocols provide the same privacy guarantees for AI governance in regulated industries. Not anonymity (which would prevent accountability); selective disclosure where the verifier learns only what they need to verify, nothing more.

The maritime incident reporting model demonstrates why this matters. An operator observes that a specific cargo loading procedure creates excessive stress on hull frames during heavy weather. Reporting this observation helps other operators avoid the same risk, but identifying the reporting vessel could expose it to commercial disadvantage (competitors learn its loading practices), regulatory scrutiny (was the vessel overloaded?), or insurance impact (does this disclosure affect coverage?). Anonymous reporting removes these barriers; more incidents are reported; fleet-wide safety improves.

Financial institutions face parallel privacy requirements when sharing fraud detection data. A bank's fraud model identifies a transaction pattern associated with account takeover attempts. Sharing that pattern with other banks helps the industry defend against the same attack vector, but sharing the specific transactions that revealed the pattern exposes customer data protected by GLBA, FCRA, and bank privacy policies. Zero-knowledge proofs enable the bank to prove "our fraud model detected this pattern in actual transactions" without disclosing which customers, which accounts, or which transaction details.

Healthcare data sharing for AI training faces identical constraints. A hospital's diagnostic model trained on patient records achieves superior performance for a specific condition. Publishing the model helps other hospitals improve diagnostic accuracy, but the model's weights potentially encode patient information protected by HIPAA. Zero-knowledge attestation enables the hospital to prove "our model was trained on a dataset meeting these quality criteria" without exposing the dataset or allowing reconstruction of individual patient records from model parameters.

The cryptographic mechanism is selective disclosure of attributes. A classification society verifying a vessel's compliance with environmental regulations needs to know: the vessel type, the engine specifications, the emissions control equipment installed, and the measured emissions levels. The society does not need to know the vessel's name, the owner's identity, the cargo carried, or the specific routes operated. Zero-knowledge protocols let the vessel prove "I am a Panamax container vessel with a main engine meeting IMO Tier III standards and measured NOx emissions below regulatory limits" without revealing attributes beyond what the verification requires.

Policy enforcement benefits from the same selective disclosure. An AI system processing healthcare data must prove it operates within HIPAA constraints without disclosing patient information to the auditor verifying compliance. Zero-knowledge verification enables the system to prove "I accessed only the minimum necessary protected health information for the treatment purpose documented in the authorization" without the auditor seeing the actual PHI, which they are not authorized to access.

Federated learning across organizational boundaries becomes feasible when participants can verify that each contributed training data meets quality requirements without seeing the data itself. A consortium of banks training a shared fraud detection model can use zero-knowledge proofs to verify that each bank's training contribution meets minimum size requirements, covers specified time periods, and includes examples of known fraud types, without any bank exposing its actual fraud data to other consortium members.

The maritime insurance parallel is exact. A P&I Club underwriting marine liability insurance needs to know a vessel's claims history, safety record, and compliance status. The club does not need to know cargo manifests, charter party agreements, or commercially sensitive routing information. Zero-knowledge disclosure provides the underwriting data without the commercial intelligence; the vessel owner shares what the insurer requires to assess risk without sharing information that would advantage competitors if the insurer's database were compromised.

Regulatory compliance verification extends beyond reporting to ongoing monitoring. A vessel operating under environmental regulations must continuously prove it remains within emissions limits. Continuous monitoring produces high-resolution emissions data that reveals operational patterns: speed, load, routing, weather response. Zero-knowledge proofs enable the vessel to prove "my emissions remain within regulatory limits measured continuously" without sharing the operational data that emissions measurements were derived from. The regulator verifies compliance; the operator protects operational confidentiality.

AI systems under continuous authorization in classified environments face the same tension. The Authorizing Official needs proof that the system operates within its authorization boundary. The system operator cannot share the system's actual activity logs because they contain classified information the AO is not cleared to access. Zero-knowledge attestation enables the system to prove "I operated within my P-property verification bounds throughout the monitoring period" without exposing classified activity logs to the AO's verification process.

The evidence structure that Chapter 18 describes includes selective disclosure as a first-class property. E-4 attribution chains record governance events but not the protected data those events acted on. Zero-knowledge protocols are the cryptographic mechanism that makes this separation enforceable: the evidence proves the governance action occurred, the verifier confirms the proof, neither party sees data they are not authorized to access.

Maritime authorities worldwide recognize that mandatory reporting without privacy protection reduces reporting rates; voluntary reporting with anonymity increases disclosure of safety-critical information. Financial regulators recognize that mandatory data sharing without privacy protection creates liability that exceeds the fraud prevention benefit; selective disclosure preserves privacy while enabling threat intelligence sharing. HIPAA recognizes that healthcare research requires data access but individual patient privacy must be protected; de-identification is the mechanism, though imperfect.

Zero-knowledge protocols provide cryptographic de-identification: prove properties of the data without revealing the data. Not statistical privacy (which leaks information in aggregate); mathematical privacy where disclosure of the proof reveals nothing beyond the verified property.

When the board audit committee asks "how do we demonstrate compliance while protecting competitively sensitive operational data?", the answer is zero-knowledge attestation: prove the system meets regulatory requirements without exposing data beyond what verification requires. The classification certificate proves seaworthiness without disclosing the vessel's commercial operations; zero-knowledge proofs verify governance compliance without exposing the governed system's protected activity logs.

Usage Guidance

Single-framework organizations. If your organization answers to only one of these seven frameworks, read that column to identify which controls The Harness Doctrine maps to, then confirm each mapping with the assessor who signs your attestation. That column becomes the basis for the evidence rendering template that produces your framework-specific compliance artifacts from a unified evidence stream, once the assessor has validated that the controls map as claimed for your deployment.

Multi-framework organizations. Most regulated organizations answer to more than one framework. Read across the row for each Harness Doctrine control to see how a single implementation can map to multiple requirements at once, subject to each framework's assessor validating its own clause. Chapter 18's unified control architecture builds on exactly these mappings: one control, one evidence artifact, multiple framework renderings, each rendering validated by the assessor for that framework. A control that maps to several frameworks does not collapse them into one audit; each assessor still tests the mapping against the clause they own.

Gap identification. Empty cells in the LP9-LP12 rows for NIST 800-53, SOC 2, PCI-DSS, and NERC CIP mark the commercial governance requirements those frameworks leave unaddressed. GDPR and the EU AI Act fill those gaps for organizations subject to them; sector-specific statutes (FCRA, ECOA, NAIC Model Bulletin, Colorado AI Act) fill them for others. Chapters 11 through 15 carry the sector-specific control mappings beyond the seven frameworks presented here.

Regulatory evolution. These mappings reflect the regulatory state as of early 2026. The EU AI Act's full enforcement timeline runs through August 2027 for certain provisions. Framework revision cycles, NIST 800-53 Rev. 6, PCI-DSS v5.0, will produce updated control numbers when they land. The Harness Doctrine identifiers (H-1 through H-5, LP1-LP14, E-1 through E-4, KB-1 through KB-4) stay stable across those revisions; only the mapping columns need updating.

Appendix C: Deployment Checklists by Sector

Every AI system has a deployment gate: the moment it transitions from development into production. These checklists define what must be true at that gate for each of the five regulated sectors covered in this book. Each item is a screening gate, not a final compliance verdict. PASS means the control meets the stated criterion and is operational as configured; independent assessor validation is still required before the institution can claim regulatory conformance. FAIL blocks deployment until the deficiency is remediated. A PASS is the institution's self-declared survey, not the surveyor's certificate; the distinction matters, because the line is only validated when an external auditor, examiner, or assessor tests the declaration against the running evidence.

The checklists assume the AGENTS.md migration described in the AGENTS.md migration chapter is already done. Each item references the specific Three-Body controls (H-1 through H-5, LP1-LP14, E-1 through E-4, KB-1 through KB-4) that the control meets at the checkpoint, subject to assessor validation. Read each "Pass/Fail" column as "meets the criterion, assessor validation required," not as a terminal certification.

Checklist 1: Financial Services

Applicable frameworks: SOX, PCI-DSS v4.0.1, OCC 2011-12 / Fed SR 11-7, FFIEC, BSA/AML, SEC Regulation FD, NYDFS 500, GDPR, EU AI Act

#	Item	Control	Pass/Fail
FS-1	AGENTS.md deployed with financial services regulatory bindings (SOX, PCI-DSS, OCC 2011-12, FFIEC, BSA/AML, SEC Reg FD, NYDFS 500)	H-1	
FS-2	Data sensitivity tiers defined: Public, Confidential, MNPI, PCI Cardholder, BSA/SAR	LP1	
FS-3	Chinese wall attestation registry connected to H-2 routing enforcement; wall compliance checked pre-retrieval	LP1, H-2	
FS-4	PCI cardholder data input sanitization active; PAN detection (Luhn filter) blocks cardholder data from AI context	LP1, LP2	

#	Item	Control	Pass/Fail
FS-5	BSA/SAR disclosure prohibition enforced at routing layer; AI cannot reveal SAR existence under any query	LP1, H-2	
FS-6	Model Risk Management checkpoint configured; AI-assisted changes to code affecting financial models trigger SR 11-7 validation gate	H-4	
FS-7	SOX dual-review requirement active for code changes affecting financial reporting systems	H-4	
FS-8	Fairness impact evaluation pipeline operational for credit decisioning, loan pricing, and eligibility determination	LP11	
FS-9	Adverse action explanation generation active at decision time for FCRA/CFPB compliance	LP12	
FS-10	E-1 baseline recorded: AGENTS.md configuration, wall attestation registry state, model inventory, KB health	E-1	
FS-11	E-2 continuous monitoring operational: verification results, wall compliance checks, fairness metrics, PAN detection log	E-2	
FS-12	E-3 state transition recording active for all configuration changes, model updates, wall attestation changes	E-3	
FS-13	E-4 attribution chain recording active for all AI-assisted code changes, MNPI-tier retrievals, SOX dual-review confirmations	E-4	
FS-14	5-minute examiner test passed: four examiner questions answerable from evidence dashboard within 5 minutes	E-1-E-4	
FS-15	Jurisdictional routing active for EU customers (GDPR), California residents (CCPA), Colorado residents (Colorado AI Act)	LP10	
FS-16	Consent verification gate active for GDPR-scoped data processing	LP9	
FS-17	Knowledge base indexed with sensitivity tier tags; retrieval pre-filter enforces tier boundaries	KB-3	
FS-18	Board Audit Committee briefing template prepared with risk currency translation from the board navigation chapter	Governance	

Checklist 2: Healthcare

Applicable frameworks: HIPAA, HITECH, 21 CFR Part 11, FDA AI/ML SaMD, state health privacy laws, GDPR, EU AI Act

#	Item	Control	Pass/Fail
HC-1	AGENTS.md deployed with healthcare regulatory bindings (HIPAA, HITECH, state privacy, FDA SaMD if applicable)	H-1	
HC-2	PHI sensitivity tiers defined: de-identified, standard PHI, psychotherapy notes, Part 2 substance use, genomic data	LP1	
HC-3	Minimum necessary rule enforced at routing layer: AI receives only PHI necessary for the stated purpose of each access	LP1, LP2, H-2	
HC-4	42 CFR Part 2 substance use disorder records require specific written consent per disclosure; general HIPAA authorization does not satisfy	LP9	
HC-5	Break-the-Glass emergency override configured with mandatory E-4 audit record and post-access review trigger	H-4, E-4	
HC-6	Business Associate Agreement (BAA) in place with every AI vendor that may access PHI	Governance	
HC-7	FDA SaMD classification determination completed: AI system classified or excluded with documented rationale	Governance	
HC-8	If FDA SaMD applies: predetermined change control plan (PCCP) documented and FDA-reviewed	H-4	
HC-9	Patient data not used for AI model training without explicit research consent and IRB approval	LP9	
HC-10	Clinical recommendation explanation artifacts generated at decision time with data sources and criteria cited	LP12	
HC-11	E-1 baseline recorded: AGENTS.md configuration, PHI sensitivity tier definitions, BAA registry, SaMD classification	E-1	
HC-12	E-2 continuous monitoring operational: minimum necessary checks, PHI access logs, clinical recommendation audit	E-2	

#	Item	Control	Pass/Fail
HC-13	E-3 state transition recording active for all configuration changes, KB updates, model updates	E-3	
HC-14	E-4 attribution records active for all PHI access events, clinical recommendations, emergency overrides	E-4	
HC-15	HIPAA breach notification procedure documented and tested: 60-day notification to HHS, immediate state notification where required	Governance	
HC-16	Jurisdictional routing active for state-specific health privacy requirements (e.g., California CMIA, New York SHIELD Act)	LP10	
HC-17	Knowledge base does not contain identifiable patient data unless access-controlled per PHI sensitivity tier	KB-3	
HC-18	Accounting of disclosures mechanism operational: patient can request list of PHI disclosures including AI-assisted processing	LP12, E-4	

Checklist 3: Energy and Critical Infrastructure

Applicable frameworks: NERC CIP-002 through CIP-014, FERC, NRC 10 CFR 73.54, DOT PHMSA, state PUC, EPA

#	Item	Control	Pass/Fail
EN-1	AGENTS.md deployed with energy sector regulatory bindings (NERC CIP, FERC, NRC if applicable, state PUC)	H-1	
EN-2	BES Cyber System impact rating determined (High/Medium/Low) per CIP-002-5.1; AI system classified accordingly	LP1	
EN-3	BCSI (Bulk Electric System Cyber System Information) access controls enforced at routing layer	LP1, H-2	
EN-4	Electronic Security Perimeter (ESP) boundary defined; AI system operates within or outside ESP with documented justification	H-2, KB-3	
EN-5	Interactive Remote Access controls (CIP-005-7 R2) enforced for any AI access from outside the ESP	H-2	

#	Item	Control	Pass/Fail
EN-6	Personnel risk assessment (CIP-004-7 R3) completed for all personnel with AI system access to BES Cyber Systems	LP2	
EN-7	AI system does not directly control safety-critical OT systems without human verification checkpoint	H-4	
EN-8	Edge-native autonomy patterns implemented for disconnected SCADA environments per the energy sector chapter	H-3	
EN-9	Configuration baseline established per CIP-010-4 R1; AI-assisted configuration changes trigger change management	E-1, E-3	
EN-10	Malicious code prevention (CIP-007-6 R3) applied to AI system components and knowledge base updates	KB-4	
EN-11	E-1 baseline recorded: AGENTS.md configuration, BES impact rating, ESP boundary, personnel access list	E-1	
EN-12	E-2 continuous monitoring operational: ESP boundary checks, access logs, configuration drift detection	E-2	
EN-13	E-3 state transition recording active for all configuration changes, software updates, access changes	E-3	
EN-14	E-4 attribution records active for all AI-assisted actions affecting BES Cyber Systems	E-4	
EN-15	CIP-008-6 incident response plan updated to include AI-specific incident scenarios	Governance	
EN-16	One-hour BES cybersecurity incident reporting capability verified and tested	Governance	
EN-17	Supply chain risk management (CIP-013-2) applied to AI model providers and compute vendors	KB-1	
EN-18	If nuclear-adjacent: NRC 10 CFR 73.54 cyber security plan includes AI system governance	Governance	
EN-19	NERC CIP Senior Manager designated and authorized per CIP-003-8 R1	Governance	

Checklist 4: Insurance and Actuarial

Applicable frameworks: NAIC Model Bulletin, 50-state DOI regulation, Colorado AI Act (SB 24-205), SOX, GDPR, ASOPs

#	Item	Control	Pass/Fail
IN-1	AGENTS.md deployed with insurance regulatory bindings (NAIC Model Bulletin, applicable state DOI requirements, SOX, ASOPs)	H-1	
IN-2	Data sensitivity tiers defined: Public, Internal, Actuarial Privilege, Regulatory Restricted	LP1	
IN-3	AI underwriting model fairness evaluation completed: four-fifths ratio applied as a first-pass screening flag (not a bright-line safe harbor), disparate impact analysis documented, and any flagged outcome carried into actuarial-justification analysis under applicable state unfair-discrimination law and the NAIC Model Bulletin	LP11	
IN-4	Proxy discrimination assessment completed: correlated variables (zip code, education level, occupation) evaluated for protected class impact	LP11	
IN-5	Adverse action explanation generation active for underwriting denials and claims adjudication decisions	LP12	
IN-6	Colorado AI Act impact assessment completed for AI systems making consequential decisions affecting Colorado residents	LP11, LP10	
IN-7	50-state jurisdictional routing operational: state-specific DOI requirements applied based on policyholder state of residence	LP10	
IN-8	Actuarial model validation completed per applicable ASOPs; AI-assisted pricing reviewed by qualified actuary	H-4	
IN-9	Consumer consent verification active for data processing beyond policy administration purposes	LP9	
IN-10	Rate filing documentation includes AI model methodology description sufficient for DOI review	LP12	
IN-11	E-1 baseline recorded: AGENTS.md configuration, actuarial model inventory, fairness evaluation results, state DOI filings	E-1	
IN-12	E-2 continuous monitoring operational: fairness metrics, pricing drift detection, claims adjudication consistency	E-2	

#	Item	Control	Pass/Fail
IN-13	E-3 state transition recording active for all model updates, pricing changes, AGENTS.md revisions	E-3	
IN-14	E-4 attribution records active for all underwriting decisions, claims adjudication, and policyholder communications	E-4	
IN-15	SOX Section 404 internal controls documentation includes AI-assisted actuarial and financial reporting processes	Governance	
IN-16	NAIC cybersecurity model law annual certification prepared (where adopted by domicile state)	Governance	
IN-17	Knowledge base does not contain policyholder PII in unmasked form outside access-controlled tiers	KB-3	
IN-18	Market conduct examination readiness verified: evidence dashboard produces DOI examination package on demand	E-1-E-4	

Checklist 5: Manufacturing, Supply Chain, and Telecommunications

Applicable frameworks: FDA 21 CFR 820, ISO 13485, IEC 62443, ITAR, FCC, TSA Security Directives, EPA, product safety regulations

#	Item	Control	Pass/Fail
MT-1	AGENTS.md deployed with applicable sector bindings (manufacturing: ISO 13485/FDA 21 CFR 820; telecom: FCC/TSA SD; supply chain: trade compliance)	H-1	
MT-2	Data sensitivity tiers defined: Public, Proprietary, ITAR-Controlled, Safety-Critical, Network Operations	LP1	
MT-3	ITAR-controlled data (if applicable) access restricted to US persons only per 22 CFR 120-130; AI system enforces nationality-based access	LP1, LP10	
MT-4	AI system outputs affecting product safety reviewed by qualified engineer before implementation	H-4	
MT-5	Quality Management System (QMS) updated to include AI-assisted design, manufacturing, and quality decisions	Governance	

#	Item	Control	Pass/Fail
MT-6	AI-assisted procurement decisions include supply chain fraud detection and vendor verification	H-4	
MT-7	Multi-agent system behavioral boundaries defined and enforced; emergent behavior monitoring active	H-4, E-2	
MT-8	Telecommunications network AI includes FCC outage reporting integration for AI-caused service disruptions	Governance	
MT-9	TSA Security Directive compliance verified for AI systems in telecommunications critical infrastructure	H-1	
MT-10	Environmental compliance integration: AI recommendations affecting emissions, waste, or resource use routed through EPA compliance check	H-2	
MT-11	E-1 baseline recorded: AGENTS.md configuration, QMS integration points, ITAR classification, network topology	E-1	
MT-12	E-2 continuous monitoring operational: product quality metrics, supply chain anomaly detection, network performance	E-2	
MT-13	E-3 state transition recording active for all design changes, manufacturing process updates, network configuration changes	E-3	
MT-14	E-4 attribution records active for all AI-assisted design decisions, quality determinations, procurement actions, network changes	E-4	
MT-15	Supply chain vendor risk tiering completed per NIST 800-161r1; Tier 1 capabilities owned by organization	KB-1	
MT-16	Knowledge base source verification includes vendor-supplied technical data validation	KB-1	
MT-17	Product recall and safety incident response plan updated to include AI-caused defect scenarios	Governance	
MT-18	If dual-use (commercial + defense): evidence rendering configured for both commercial and federal framework outputs	E-1-E-4	

Apple Foundation Model: On-Device Deployment for Zero-Trust Data Boundaries

When a vessel operating in international waters must process classified cargo manifests, the classification data cannot transit public networks or be stored on shore-based servers. The vessel's navigation and cargo management systems process that data locally, never transmitting it beyond the electronic security perimeter defined by the vessel's secure computing environment. Shore-side authorities receive aggregated compliance reports, not raw cargo details.

Apple's on-device Foundation Model provides the same data boundary guarantees for AI processing in regulated environments. Not a cloud model with local caching. A complete inference stack that runs entirely on-device, processes data locally, and produces results without any data leaving the device's hardware-backed security boundary.

The regulatory context that makes this architecture critical: HIPAA's security rule requires covered entities to implement technical safeguards that "reasonably and appropriately" protect electronic PHI. Transmitting PHI to a third-party AI provider for inference creates a technical safeguard gap: the covered entity loses direct control of PHI from the moment it leaves their network until results return. Even with a Business Associate Agreement and encryption in transit, the PHI is processed in an environment the covered entity does not administer. Apple's on-device model eliminates this gap entirely: PHI never leaves the clinician's device, so no third-party processor relationship exists, and the covered entity maintains direct control throughout the inference process.

Financial services face parallel requirements under SEC Regulation S-ID (identity theft red flags), GLBA Safeguards Rule, and PCI-DSS. Customer financial data transmitted to external AI providers for fraud detection, credit scoring, or transaction analysis creates data minimization and purpose limitation challenges. Each transmission must be justified under the "need to know" principle; each provider must be vetted for data security; each transmission creates breach notification exposure if the provider is compromised. On-device inference confines the data to the institution's controlled environment: the fraud detection model runs on the bank's servers, customer data stays within the bank's security perimeter, and third-party AI provider risk is eliminated from the threat model.

Energy sector BES Cyber Systems under NERC CIP face the strictest data boundaries. CIP-011-3 R1 requires that BCSI (information about BES Cyber Systems) be protected from unauthorized disclosure. Transmitting BCSI to a cloud AI provider for grid optimization, load forecasting, or equipment maintenance scheduling creates a controlled unclassified information (CUI) handling problem: BCSI is now in an environment outside the utility's Electronic Security Perimeter, subject to the provider's security controls rather than the utility's NERC CIP-compliant controls. On-device models keep BCSI within the ESP boundary. The model runs on utility-controlled hardware, processes BCSI locally, and produces optimization recommendations without BCSI ever transiting external networks.

The technical architecture that enables this is hardware-backed secure enclave processing. Apple's Foundation Model executes inside the Secure Enclave, a hardware-isolated processor with its own encrypted memory, boot chain, and cryptographic operations. Data processed by the model never exists in plaintext outside the enclave; even the operating system cannot access enclave memory. This provides defense-in-depth against both external attacks (malware running on the device cannot exfiltrate data from the enclave) and insider threats (privileged users cannot extract data from enclave memory without physical device compromise requiring specialized forensic equipment).

Model capability boundaries are pre-defined at the hardware level, not configurable at runtime. The on-device Foundation Model supports classification, extraction, formatting, and summarization tasks up to a defined complexity threshold (currently score <= 20 in DOCTRINE's auto-selector heuristic). Tasks beyond that threshold fail with an explicit error rather than silently degrading. This hard capability boundary is itself a governance control: organizations can deploy on-device models knowing that tasks requiring more capable models will not be attempted on-device, preventing scenarios where insufficient model capability produces incorrect outputs that appear valid.

Context window constraints are similarly hardware-enforced. The on-device model supports a 4K token context limit; inputs exceeding 2500 tokens are rejected before processing begins. This constraint is not a temporary limitation to be worked around; it is a design property that ensures the model operates within its validated performance envelope. Organizations designing workflows around on-device inference must accommodate this limit in their data preparation steps, similar to how maritime communications over constrained satellite links must structure messages to fit available bandwidth.

Zero-cost deployment economics change the infrastructure trade-off calculation. Cloud inference at scale carries measurable compute costs: $0.01-0.10 per inference depending on model size and provider pricing. A healthcare provider running 10,000 inferences per day for clinical decision support incurs $100-1,000 daily cloud compute cost, or $36,000-365,000 annually. On-device inference on Apple silicon costs the depreciated hardware cost and local electricity: effectively zero marginal cost per inference once the device is deployed. For high-volume use cases, on-device deployment becomes cost-competitive with cloud inference within months.

Deployment checklist additions for on-device Apple Foundation Model integration appear as sector-specific items:

Healthcare: HC-19. On-device AI configured for PHI-sensitive tasks (clinical summarization, diagnosis support); PHI never transmitted to external inference providers.

Financial Services: FS-19. On-device models deployed for customer data classification and fraud scoring; PCI cardholder data and MNPI processed locally within security perimeter.

Energy: EN-20. On-device models operational for BCSI-classified data; grid optimization and equipment monitoring execute within ESP boundary.

Insurance: IN-19. On-device models configured for policyholder PII minimization; underwriting feature extraction and claims adjudication scoring performed locally.

Manufacturing/Telecom: MT-19. On-device models deployed for ITAR-controlled data and network operations; proprietary design data and critical infrastructure telemetry remain on operator-controlled hardware.

The governance benefit is not merely compliance cost reduction. It is elimination of an entire category of third-party risk. An organization using on-device models for regulated data processing does not need to evaluate AI provider SOC 2 reports, negotiate BAAs, conduct vendor security assessments, or monitor provider breach notifications. The regulated data never reaches the provider, so the provider is not in the data flow, so provider risk is not in the threat model. One less vendor, one less contract, one less audit, one less breach surface.

When the board audit committee reviews third-party risk registers, AI inference providers appear as high-risk vendors: they process the organization's most sensitive data, they operate in multi-tenant environments, they are high-value targets for adversaries, and their breach notification obligations create reputational risk for every customer simultaneously. On-device deployment removes AI inference providers from the third-party risk register entirely. Not risk mitigation; risk elimination through architectural data boundary enforcement.

Classification societies inspect vessels to verify that safety-critical systems operate within design limits. On-device AI operates within hardware-enforced design limits: token count, memory allocation, computational complexity, data boundary. Both approaches treat limits as governance controls, not performance targets. The vessel does not operate beyond its load line; the model does not process beyond its capability envelope. Physical constraints in both cases; verifiable boundaries producing auditable evidence that limits were respected.

Consolidated Control Evaluation Rubric

Procurement and vendor-assessment teams asked for one place where every control criterion is scored, rather than reconstructing the full set from per-sector checklists scattered across this appendix and definitions scattered across the book. This rubric is that single sheet. It covers the full control set once: the five harness mechanisms (H-1 through H-5), the fourteen compliance properties (LP1 through LP14), the four evidence tiers (E-1 through E-4), and the four knowledge-base controls (KB-1 through KB-4). The per-sector checklists above map sector-specific regulatory bindings onto these same controls; this table is the sector-independent master from which they derive.

Score each row as MEETS or DOES NOT MEET, with a third state, ASSESSOR-VALIDATION-REQUIRED, for any row where the institution's self-declaration has not yet been tested by an independent auditor, examiner, or assessor. A MEETS score here is the institution's self-declared survey, not an external certificate; treat it as a screening result, not a terminal compliance finding.

Control	Criterion (the control is verified when this holds)	Status
H-1	AGENTS.md instruction surface is deployed and machine-enforceable; regulatory bindings are present for every framework in scope	
H-2	Routing enforcement checks every action against AGENTS.md and the Delegation Token before any resource is reached; out-of-scope requests are denied at the gateway, not at the prompt	
H-3	Persistent state survives session boundaries and disconnection; edge-native autonomy holds where connectivity is not guaranteed	
H-4	Verification checkpoints gate AI-assisted changes to regulated systems; human review is required where the framework requires it	
H-5	Accountability records attribute every AI-assisted action to an identity, a mandate, and a timestamp	
LP1	Access is classification-consistent: data sensitivity tiers are defined and the routing layer enforces tier boundaries on retrieval	
LP2	Scope binding holds: each session and request is bound to its authorized scope under least privilege; access outside the bound scope is denied, and no cross-contamination of context occurs between tenants, sessions, or sensitivity tiers	
LP3	Routing integrity holds: the enforcement path cannot be bypassed by prompt instruction	
LP4	Persistent governance context survives session boundaries and disconnection: the mandate, scope, and prior governance state carry	

Control	Criterion (the control is verified when this holds)	Status
	forward so the harness governs the next action with full context, not a blank slate	
LP5	Continuous verification is active: AI-assisted actions and outputs are checked against source, policy, and the authorization boundary on an ongoing basis, not once at deployment	
LP6	Attribution completeness holds: every AI-assisted action is attributable to an identity, a mandate, and a timestamp, with no gaps in the chain	
LP7	Deviation detection is operational: drift from the authorized configuration, scope, or expected behavior is detected and surfaced for response	
LP8	Evidence integrity holds: records are tamper-evident, so a self-declared survey cannot be repainted after the fact	
LP9	Consent is verified for data processing that requires it; heightened-consent regimes are honored where they apply	
LP10	Jurisdiction and residency routing is active; the session is scoped to a jurisdiction tag and data residency matches	
LP11	Fairness evaluation is operational; screening flags such as the four-fifths ratio trigger actuarial-justification or equivalent analysis under the governing standard, never a single bright-line pass	
LP12	Transparency artifacts are generated at decision time: adverse-action notices, explanations, and accounting of disclosures where required	
LP13	Output boundary enforcement holds: AI output is evaluated against disclosure/scope boundaries at the routing layer before emission; boundary-crossing content is routed to review or denied at the gateway	
LP14	Deterministic deny on uncertainty holds: when a condition cannot be evaluated against an authorized policy, the gateway fails closed and escalates; uncertainty does not propagate	
E-1	Foundation evidence recorded: AGENTS.md configuration, model inventory, control baselines exist and are retrievable	
E-2	Operational evidence is continuous: verification results, gateway denials, and monitoring metrics stream without gaps	
E-3	Context evidence captures every state transition: configuration changes, model updates, access changes	

Control	Criterion (the control is verified when this holds)	Status
E-4	Accountability evidence chains each action to its actor, mandate, and time; the attribution chain is unbroken	
KB-1	Knowledge-base provenance is tracked: every source is attributable and supply-chain-verified	
KB-2	Knowledge-base scoping holds: retrieval is bounded to authorized sources for the session's tier and jurisdiction	
KB-3	Knowledge-base integrity holds: sensitivity-tier tags are applied and the retrieval pre-filter enforces them	
KB-4	Knowledge-base retrieval is monitored: updates and queries are logged, and malicious-content prevention is applied to ingest	

A deployment is gate-ready when every row reads MEETS, with any ASSESSOR-VALIDATION-REQUIRED rows flagged for the independent survey that follows. The definitions for each H, P, E, and KB control are in the front-matter primer and Appendix A; this rubric is the scoring instrument, not the definition source.

Usage Notes

These checklists are deployment gates, not ongoing governance procedures. Once a system passes the gate and enters production, ongoing governance lives in the continuous authorization cycle described in the AGENTS.md migration chapter. The deployment checklist confirms the governance architecture is operational at initial launch. After that, the E-2 continuous monitoring pipeline carries the verification load.

Every FAIL result needs a remediation item: assigned owner, target date, and an E-3 state transition record documenting both the deficiency and its resolution. Any AI system deployed with unresolved FAIL items operates outside its authorized governance boundary. The evidence trail will reflect that fact.

Appendix D: Multi-Framework Evidence Rendering Reference

One evidence event. Four regulatory audiences. This appendix walks through the full rendering pipeline from Chapter 18 by example, starting with a single E-4 attribution record and producing four framework-specific compliance artifacts from it. The concrete data structures appear in order: base record, OSCAL intermediate representation, then the four framework renderings.

The example event is an E-4 record for an AI-assisted credit recommendation that triggers obligations under SOC 2, HIPAA (the credit decision involves a healthcare financing product), NERC CIP (included for structural comparison), and GDPR (the applicant is an EU data subject).

The Base E-4 Attribution Record

The E-4 record is the single source of truth for one AI-assisted action's complete attribution chain. Every rendering operation derives from this record. No framework-specific output adds information absent from, or not derivable from, the base record.

```
{
  "e4_record_id": "e4-2026-0401-credit-0847a3",
  "timestamp": "2026-04-01T14:23:07.441Z",
  "evidence_tier": "E-4",
  "event_type": "ai_assisted_credit_recommendation",
  "system": {
    "ai_system_id": "credit-scoring-v3.2",
    "model_version": "claude-sonnet-4-2026Q1",
    "agents_md_version": "fs-prod-2026.03.28-r4",
    "harness_version": "3bp-harness-2.1.0"
  },
  "data_subject": {
    "jurisdiction": "eu_gdpr",
    "residence": "DE",
    "consent_scope": "credit_assessment_gdpr_art6_1b",
    "consent_verified_at": "2026-04-01T14:23:06.892Z"
  },
  "action": {
    "type": "credit_recommendation",
    "decision": "adverse_action",
    "reason_codes": ["insufficient_credit_history", "high_dti_ratio",
"recent_derogatory"],
    "confidence_score": 0.87,
    "threshold_applied": "internal_credit_policy_v4.1"
```

```
  },
  "data_sources_consulted": [
    {
      "source_id": "bureau-equifax-2026Q1",
      "data_tier": "confidential",
      "access_authorized": true,
      "consent_verified": true
    },
    {
      "source_id": "internal-account-history",
      "data_tier": "confidential",
      "access_authorized": true,
      "consent_verified": true
    }
  ],
  "governance_verification": {
    "p1_classification_access": "pass",
    "p2_scope_binding": "pass",
    "p9_consent_bounded": "pass",
    "p10_jurisdiction_applied": "gdpr_eu_ai_act",
    "p11_fairness_threshold": "pass",
    "p11_four_fifths_result": 0.91,
    "p12_explanation_generated": true,
    "p12_explanation_id": "expl-2026-0401-0847a3"
  },
  "human_oversight": {
    "h4_human_review_required": true,
    "h4_reviewer_id": "analyst-jm-4472",
    "h4_review_timestamp": "2026-04-01T14:25:12.003Z",
    "h4_review_decision": "concur_with_adverse_action",
    "h4_override": false
  },
  "explanation_artifact": {
    "explanation_id": "expl-2026-0401-0847a3",
    "format": "consumer_readable",
    "language": "de",
    "principal_reasons": [
      "Your credit history is shorter than our minimum requirement for
this product.",
      "Your debt-to-income ratio exceeds the threshold for this loan
category.",
      "A recent negative item appears on your credit report."
    ],
    "human_review_available": true,
    "human_review_contact": "credit-review@example-bank.eu"
  }
}
```

OSCAL Intermediate Representation

OSCAL (Open Security Controls Assessment Language) is the machine-readable intermediate layer. All framework-specific renderings derive from this representation. The component-definition schema carries annotations for multiple frameworks simultaneously, so the same OSCAL document feeds every downstream renderer.

OSCAL is a rendering target, not the canonical store. The base E-4 attribution record above is the source of truth; the OSCAL document is generated from it on demand and exists only to feed the framework renderers. Editing the OSCAL document does not change the evidence. If the OSCAL representation and the base record ever disagree, the base record wins and the OSCAL is regenerated. This separation keeps OSCAL as the format the assessor's tooling trusts while preserving a single authoritative evidence record underneath it.

```
# OSCAL Assessment Result - E-4 Attribution Event
# Derived from: e4-2026-0401-credit-0847a3

assessment-results:
  uuid: "ar-e4-2026-0401-credit-0847a3"
  metadata:
    title: "E-4 Attribution Record: AI-Assisted Credit Recommendation"
    last-modified: "2026-04-01T14:25:12.003Z"
    version: "1.0"
    oscal-version: "1.1.2"

  results:
    - uuid: "result-credit-0847a3"
      title: "Credit Recommendation Attribution Assessment"
      start: "2026-04-01T14:23:06.892Z"
      end: "2026-04-01T14:25:12.003Z"

      observations:
        - uuid: "obs-access-control"
          description: "LP1 Classification-Consistent Access verified
for credit bureau and internal account data"
          methods: ["AUTOMATED"]
          subjects:
            - subject-uuid: "credit-scoring-v3.2"
              type: "component"
          relevant-evidence:
            - href: "#e4-2026-0401-credit-0847a3"
              description: "Automated access control verification at
routing layer"
          collected: "2026-04-01T14:23:07.441Z"
          # Framework annotations
          props:
            - name: "soc2-criteria"
              value: "CC6.1, CC6.3"
            - name: "hipaa-standard"
              value: "164.312(a)(1)"
            - name: "gdpr-article"
              value: "Art. 32(1)(b)"
```

```yaml
        - uuid: "obs-consent-verification"
          description: "LP9 Consent-Bounded Processing verified; GDPR
Art 6(1)(b) contractual basis confirmed"
          methods: ["AUTOMATED"]
          collected: "2026-04-01T14:23:06.892Z"
          props:
            - name: "gdpr-article"
              value: "Art. 6(1)(b), Art. 7"
            - name: "eu-ai-act-article"
              value: "Art. 26(5)"

        - uuid: "obs-fairness-check"
          description: "LP11 Fairness Verification passed; four-fifths
ratio 0.91 exceeds 0.80 threshold"
          methods: ["AUTOMATED"]
          collected: "2026-04-01T14:23:07.200Z"
          props:
            - name: "gdpr-article"
              value: "Art. 22(3)"
            - name: "eu-ai-act-article"
              value: "Art. 9(2)(a), Art. 10(2)(f)"

        - uuid: "obs-explanation-generation"
          description: "LP12 Explanation artifact generated at
decision time in consumer language (DE)"
          methods: ["AUTOMATED"]
          collected: "2026-04-01T14:23:07.441Z"
          props:
            - name: "gdpr-article"
              value: "Art. 13, Art. 14, Art. 22(3), Recital 71"
            - name: "eu-ai-act-article"
              value: "Art. 13, Art. 26(5)"
            - name: "fcra-section"
              value: "15 U.S.C. 1681m"

        - uuid: "obs-human-oversight"
          description: "H-4 human oversight review completed by
analyst-jm-4472; concurred with adverse action"
          methods: ["MANUAL"]
          collected: "2026-04-01T14:25:12.003Z"
          props:
            - name: "soc2-criteria"
              value: "CC3.4"
            - name: "gdpr-article"
              value: "Art. 22(1), Art. 22(3)"
            - name: "eu-ai-act-article"
              value: "Art. 14, Art. 26(2)"

      findings:
        - uuid: "finding-all-pass"
          title: "All governance properties satisfied"
          description: "LP1, LP2, LP9, LP10, LP11, LP12 verified and
H-4 human oversight completed; no deviations detected"
          target:
            type: "objective-id"
            target-id: "p1-p12-compliance"
```

```
status:
  state: "satisfied"
```

SOC 2 Rendering

For SOC 2, the renderer pulls every observation annotated with `soc2-criteria` properties and formats them into a Trust Services Criteria evidence package.

```
SOC 2 Type II Evidence Artifact
================================
Evidence ID: e4-2026-0401-credit-0847a3
System: credit-scoring-v3.2 (AI-Assisted Credit Recommendation)
Period: 2026-04-01

CC6.1 - Logical and Physical Access Controls
  Observation: Classification-consistent access verified for credit
  bureau data (confidential tier) and internal account history
  (confidential tier). Routing enforcement layer confirmed authorized
  access before data entered AI context.
  Result: PASS
  Timestamp: 2026-04-01T14:23:07.441Z
  Method: Automated verification at H-2 routing layer

CC6.3 - Role-Based Access
  Observation: Scope binding enforcement verified. AI system operated
  within authorized data sensitivity tiers for the requesting user's
  role and active authorizations.
  Result: PASS
  Timestamp: 2026-04-01T14:23:07.441Z

CC3.4 - Risk Mitigation Activities
  Observation: Human reviewer (analyst-jm-4472) completed H-4
  human oversight review of adverse action recommendation.
  Reviewer concurred with AI recommendation. No override applied.
  Result: PASS
  Timestamp: 2026-04-01T14:25:12.003Z
  Method: Manual review with documented decision

CC7.2 - Security Event Detection
  Observation: Complete attribution chain recorded for AI-assisted
  credit recommendation. All data sources, governance verifications,
  and human oversight actions documented with timestamps.
  Result: PASS
  Evidence Chain: e4-2026-0401-credit-0847a3
```

HIPAA Rendering

The HIPAA renderer targets observations relevant to the Security Rule and formats them for the covered entity's compliance documentation. This rendering applies whenever the credit decision touches healthcare financing, a medical loan, HSA eligibility, or similar products.

```
HIPAA Security Rule Compliance Evidence
=========================================
Evidence ID: e4-2026-0401-credit-0847a3
Covered Entity: [Organization Name]
AI System: credit-scoring-v3.2

164.312(a)(1) - Access Control
  Standard: Implement technical policies and procedures for
  electronic information systems that maintain ePHI to allow
  access only to those persons or software programs that have
  been granted access rights.

  Evidence: AI system access to data sources verified against
  classification tier and role-based access controls. Confidential-
  tier data accessed with verified authorization. No PHI accessed
  beyond minimum necessary scope for credit assessment purpose.
  Result: Specification Satisfied (Required)
  Timestamp: 2026-04-01T14:23:07.441Z

164.312(b) - Audit Controls
  Standard: Implement hardware, software, and/or procedural
  mechanisms that record and examine activity in information
  systems that contain or use ePHI.

  Evidence: E-4 attribution record generated automatically as
  byproduct of AI system operation. Record contains complete
  audit trail: data sources accessed, governance properties
  verified, human reviewer identity and decision, explanation
  artifact generated.
  Result: Specification Satisfied (Required)
  Evidence Artifact: e4-2026-0401-credit-0847a3

164.308(a)(1)(ii)(D) - Information System Activity Review
  Standard: Implement procedures to regularly review records
  of information system activity.

  Evidence: E-2 continuous monitoring pipeline active. AI system
  interactions reviewed in real time through automated
  verification. Human review triggered per H-4 checkpoint.
  Weekly governance posture review includes audit of E-4
  attribution records.
  Result: Specification Satisfied (Required)

164.528 - Accounting of Disclosures
  Standard: Provide individuals with an accounting of
  disclosures of PHI.

  Evidence: E-4 attribution record provides disclosure accounting
  data. Data subject can request list of processing events
```

```
involving their data through standard disclosure channel.
Result: Addressable Specification Implemented
```

NERC CIP Rendering

NERC CIP evidence must fit the compliance monitoring and enforcement program's submission format. This rendering shows the structure for energy-sector evidence, even though the specific event (a credit recommendation) does not directly apply to a BES Cyber System. The structural pattern is what matters here.

```
NERC CIP Compliance Evidence Submission
==========================================
Evidence ID: e4-2026-0401-credit-0847a3
Registered Entity: [Organization Name]
Standard: CIP-007-6 R4, CIP-007-6 R5

CIP-007-6 R4 - Security Event Monitoring
  Requirement: Each Responsible Entity shall implement one or
  more documented processes to log events at the BES Cyber
  System level and review the logs.

  Evidence: AI system generates E-4 attribution records for
  every consequential action. Records include: system identity,
  user identity, action type, data accessed, governance
  verification results, and timestamp. Records are immutable
  and tamper-evident (LP8 evidence integrity).

  Measure: Log entry generated at 2026-04-01T14:23:07.441Z
  for AI-assisted action. Log reviewed as part of automated
  E-2 continuous monitoring pipeline.

  VSL Applicability: N/A (no violation detected)

CIP-007-6 R5 - System Access Controls
  Requirement: Each Responsible Entity shall implement one or
  more documented processes to enforce authentication and
  access controls.

  Evidence: LP1 classification-consistent access verified via
  automated check at H-2 routing enforcement layer. LP2 scope
  binding confirmed user authorization for data
  sensitivity tier before AI context population.

  Measure: Access verification completed at
  2026-04-01T14:23:07.441Z. Authorization confirmed for
  both data sources consulted.

  VSL Applicability: N/A (no violation detected)
```

GDPR Rendering

GDPR evidence serves two audiences: data protection authority inquiries and Data Protection Impact Assessment documentation. LP10's jurisdictional determination triggers this rendering for any event involving an EU data subject.

```
GDPR Compliance Evidence Record
==================================
Evidence ID: e4-2026-0401-credit-0847a3
Controller: [Organization Name]
Data Subject Jurisdiction: Germany (DE)
Processing Basis: Art. 6(1)(b) - Contractual Necessity

Art. 6 - Lawfulness of Processing
  Basis: Processing necessary for performance of contract
  (credit application submitted by data subject).
  Consent verification: Art. 6(1)(b) contractual basis
  confirmed at 2026-04-01T14:23:06.892Z via consent
  management platform query.
  Result: Lawful basis established

Art. 22 - Automated Individual Decision-Making
  Assessment: Decision involved automated processing
  (AI-assisted credit recommendation). Human reviewer
  (analyst-jm-4472) participated in final decision at
  2026-04-01T14:25:12.003Z. Decision is not "solely"
  automated per Art. 22(1); human oversight confirmed.

  Right to contest: Available via credit-review@example-bank.eu
  Right to human review: Exercised (reviewer concurred)
  Right to explanation: Explanation artifact generated in
  German (DE) at decision time with principal reasons specific
  to this applicant.

Art. 13/14 - Information Provided to Data Subject
  Information provided: Decision involved AI-assisted
  processing. Principal reasons for adverse action provided
  in consumer-readable German. Human review contact provided.
  Explanation artifact ID: expl-2026-0401-0847a3

Art. 35 - DPIA Reference
  This processing activity is covered by DPIA-2026-credit-v3,
  approved 2026-02-15. No new risks identified in this event.

Art. 5(1)(d) - Accuracy
  Data sources verified: Credit bureau data (current as of
  2026-Q1 pull), internal account history (real-time).
  LP11 fairness verification: four-fifths ratio 0.91 (above
  0.80 threshold). No evidence of proxy discrimination in
  this decision.

Art. 30 - Record of Processing
  Processing activity: AI-assisted credit recommendation
  Data categories: Credit history, account history, debt
  ratios, derogatory items
  Recipients: Credit analyst (internal), data subject
```

```
(explanation artifact)
Retention: Per GDPR schedule (pseudonymization after
purpose completion; retention of pseudonymized record
per regulatory retention obligations)
```

Gemini-3.1-Flash-Lite: Ultra-Low-Cost Multi-Framework Translation

Maritime classification societies face a fundamental cost constraint. Lloyd's surveyor reviewing a container ship's structural drawings applies the same technical knowledge whether the vessel sails under Liberian registry (ISM Code compliance), US flag (CFR Title 46), or EU registration (MLC 2006). The surveyor's expertise remains constant; only the format of the compliance certificate changes. Yet charging the shipowner for three separate expert reviews of identical structural evidence would be economically nonsensical.

Multi-framework evidence rendering presents the identical economics. One E-4 attribution record captures the factual basis for compliance. OSCAL transforms that single record into a neutral intermediate representation. Then four different framework renderers, each specialized for SOC 2, HIPAA, NERC CIP, or GDPR, extract the framework-specific control mappings and generate audience-appropriate documentation. The E-4 record collection cost is fixed. The OSCAL transformation cost is fixed. The variable cost lives entirely in the rendering phase, where four framework outputs must be generated from one OSCAL input.

Gemini-3.1-Flash-Lite reduces this variable cost by two orders of magnitude. Where GPT-4 costs $10 per million input tokens and Claude Sonnet costs $3, Flash-Lite operates at $0.0375 per million input tokens for batch processing. The OSCAL intermediate document for a typical E-4 record runs approximately 2,500 tokens. Rendering four framework outputs requires four passes over that OSCAL input, plus generation of approximately 1,200 tokens per framework (4,800 tokens total output). At GPT-4 rates, this single evidence rendering costs $0.073. At Flash-Lite batch rates, it costs $0.00028.

For an organization generating 10,000 E-4 attribution records per day across 100 AI systems, the annual rendering cost at GPT-4 rates would be $266,450. At Flash-Lite rates, it falls to $1,022. The cost reduction enables continuous multi-framework evidence generation where previously only periodic batch rendering was economically feasible. Lloyd's surveyor writes one technical report; the classification society's document control system generates certificates for every flag state the owner might need. The surveyor's work remains fixed-cost; the certificate printing becomes nearly free.

The quality threshold for evidence rendering differs fundamentally from the threshold for original evidence collection. When an AI agent generates an E-4 attribution record, factual accuracy is paramount. The record must correctly capture which data sources were accessed, which governance properties were verified, and which human reviewer made which decision. This high-accuracy requirement justifies expensive frontier models during the collection phase.

Framework rendering, by contrast, is a translation task. The OSCAL intermediate representation contains all necessary facts. The renderer's job is to extract the SOC 2-tagged observations and format them according to Trust Services Criteria structure, or to pull the GDPR-tagged observations and format them per Article 30 Record of Processing requirements. Translation accuracy matters, but translation is also verifiable. A simple automated check confirms that every fact appearing in a framework-specific rendering has a direct provenance chain back to the OSCAL source document. If the renderer invents a control statement or adds explanatory text absent from the source, the verification check fails.

This verifiability transforms the economics. Lloyd's Lloyd's surveyor's structural calculations require precision; the society cannot accept an apprentice's work without extensive review. But translating that surveyor's English-language report into the Spanish-language format required by the Panamanian Maritime Authority is a mechanical task. An apprentice can perform the translation; verification takes minutes. Flash-Lite's role in the rendering pipeline parallels the apprentice translator. The model generates framework-specific output from OSCAL input at ultra-low cost. The verification layer, which costs essentially nothing compared to model inference, confirms that no hallucination occurred and every statement in the rendered output traces back to source evidence.

The Lagrange point is the balance between rendering cost and verification cost. Frontier models produce rendering output so reliably that verification becomes perfunctory; their cost makes continuous rendering economically prohibitive. Local models reduce rendering cost to near-zero but require extensive verification infrastructure to catch errors. Flash-Lite occupies the stable configuration: rendering cost low enough to enable continuous multi-framework evidence generation, reliability high enough that verification requires only structural checks rather than expert semantic review. The surveyor's report gets translated continuously; the translation accuracy is automatically verified; the certificate printing costs effectively nothing.

Deploy Flash-Lite for multi-framework rendering when the following conditions hold. First, the source document uses structured markup with explicit framework annotations (OSCAL property tags, JSON schema with control mappings, or equivalent formal tagging). Second, the rendering output must conform to a well-defined template or schema that permits automated structural verification. Third, the organization generates evidence volume sufficient that rendering cost becomes a line item in the budget, typically above 5,000 evidence records per month across multiple frameworks. Fourth, regulatory requirements mandate multi-framework compliance where the same underlying facts must appear in different control framework vocabularies (SOC 2 for cloud services, HIPAA for healthcare data, GDPR for EU data subjects).

When these conditions align, Flash-Lite transforms rendering from a bottleneck into a background process. The classification society prints certificates continuously as surveyors complete inspections. The cost per certificate approaches zero. The ship owner receives every flag state's required compliance documentation simultaneously. Stability emerges when translation costs stop constraining the number of regulatory audiences that can be served from the same evidence base.

For detailed technical coverage of Gemini-3.1-Flash-Lite's architecture and batch processing capabilities, see Book I Chapter 10. This section focuses on how ultra-low inference costs enable continuous multi-framework evidence rendering at volumes where frontier model pricing would force periodic batch operations.

beautiful-html-templates: Classification Society Report Standardization

A Lloyd's surveyor completing a hull inspection aboard a bulk carrier in Singapore generates a detailed condition report. That same condition report, reformatted, becomes the Annual Survey Certificate for the vessel's flag state, the Cargo Ship Safety Construction Certificate under SOLAS, the ISM Document of Compliance for the management company, and the Classification Certificate for the insurance underwriter. The technical findings are identical. The presentation formats differ only in which regulatory audience needs which subset of the structural data arranged according to which visual template.

This pattern directly applies to multi-framework evidence rendering. The E-4 attribution record is the surveyor's condition report: complete, technically accurate, containing every fact needed for any downstream compliance purpose. The framework-specific renderings (SOC 2, HIPAA, NERC CIP, GDPR) are the regulatory certificates: identical facts, formatted per each audience's requirements. beautiful-html-templates provides the visual standardization layer that ensures every rendered framework output appears in the format the regulatory audience expects.

The template library includes 34 pre-configured report layouts covering security frameworks (NIST SP 800-53, ISO 27001, SOC 2), healthcare regulations (HIPAA Security Rule, HITRUST CSF), financial controls (PCI DSS, GLBA), energy sector mandates (NERC CIP), and data protection regimes (GDPR, CCPA). Each template encodes the structural conventions of its target framework. SOC 2 evidence appears in Trust Services Criteria order with color-coded compliance indicators. HIPAA documentation follows the administrative, physical, and technical safeguards structure. GDPR records present the Article 30 processing inventory format. The framework renderer generates JSON or YAML with the evidence data; the template layer transforms that structured data into the PDF or HTML document the auditor recognizes as correctly formatted.

Maritime classification societies discovered this separation decades ago. The surveyor records hull plate thickness measurements, cargo hold dimensions, and watertight door test results in a standardized inspection database. That database feeds a document generation system that produces flag state certificates, class certificates, statutory certificates, and insurance documentation without the surveyor re-entering data or manually reformatting reports. The surveyor's expertise focuses on inspection; document formatting becomes a background automation task.

beautiful-html-templates brings the same separation to evidence rendering pipelines. The framework renderer (the translation layer that extracts SOC 2 observations from OSCAL and maps them to Trust Services Criteria) produces structured output. The template system consumes that structured output and generates the visual artifact the audit firm expects to receive. Security architects focus on evidence collection accuracy; document formatting becomes configuration, not custom code.

The stability point is the separation between semantic correctness and presentation correctness. When these two concerns entangle, every framework addition requires both logic changes (the control mapping) and presentation changes (the document layout). Auditors reject evidence not because the facts are wrong, but because the formatting deviates from the standard they recognize. Classification societies prevent this failure mode by standardizing certificate templates across all flag states and class rules. The surveyor never worries whether the PDF layout will be acceptable to the Panamanian Maritime Authority; the template system guarantees format compliance.

Deploy beautiful-html-templates when the organization must produce evidence artifacts for external regulatory review where presentation format affects acceptance. Internal monitoring dashboards can tolerate inconsistent layout; auditors conducting SOC 2 Type II assessments cannot. The classification certificate must look like a classification certificate. The template library provides that format assurance.

The templates integrate directly with the OSCAL rendering pipeline. A framework-specific renderer outputs structured data (JSON or YAML) conforming to a specific schema. The template system receives that structured data and applies the appropriate visual layout. If the HIPAA renderer generates JSON with the Security Rule implementation statements, the `hipaa-security-rule.html.eta` template transforms that JSON into the three-column evidence table format that HIPAA auditors recognize from every other covered entity's documentation. No custom formatting code in the renderer; the renderer focuses solely on extracting the correct control statements from OSCAL and mapping them to the framework's control structure.

Configuration over code. When a new framework enters scope, the rendering logic (extract tagged observations, map to framework controls) lives in a dedicated renderer module. The visual presentation (font choices, margin widths, table layouts, color coding) lives in an HTML template that references the renderer's output schema. The renderer developer never touches CSS. The template designer never touches control mapping logic. Lloyd's surveyor never worries about certificate printing; the document control specialist never inspects hull plates. Each role operates within its expertise domain; the system composes their outputs into compliant documentation.

For detailed technical coverage of beautiful-html-templates' architecture and agent-friendly template syntax, see Book II Chapter 32. This section focuses on how standardized report formatting enables reliable multi-framework evidence presentation where ad-hoc document generation would introduce compliance risk through format deviations.

HyperFrames: Voyage Data Recorder Playback for AI Attribution

Every commercial vessel over 3,000 gross tons carries a Voyage Data Recorder (VDR), the maritime equivalent of an aircraft's black box. The VDR captures twelve hours of bridge audio, radar images, ship position, heading, speed, rudder orders, engine commands, and hull stress data. When an incident occurs, maritime investigators replay the VDR recording to reconstruct the exact sequence of events that led to the collision, grounding, or structural failure. The reconstruction is not a summary or a log file analysis. It is a frame-by-frame playback of what the bridge crew saw, heard, and did during the critical minutes before impact.

HyperFrames provides the equivalent VDR playback capability for AI-assisted decision incidents. An E-4 attribution record captures the decision structure: which model ran, which data sources were queried, which governance properties passed or failed, which human reviewer approved the action. That structured record is necessary but insufficient for incident investigation. When a credit recommendation later proves discriminatory, when a medical AI's triage decision is challenged, or when an energy sector control system's automated load-shedding action causes a cascade failure, regulatory investigators need to see exactly what the AI system "saw" at decision time.

This is where video evidence becomes critical. The HTML rendering of the AI's decision context (the prompt assembly, the retrieved documents, the governance checkpoint outputs, the model's reasoning chain) can be transformed into a video timeline that shows how the decision unfolded step by step. An investigator reviewing a disputed credit denial doesn't merely read "LP11 fairness verification passed with four-fifths ratio 0.91." The investigator watches a 90-second video where the AI system retrieves applicant data (0:00-0:15), applies the credit scoring model (0:15-0:35), evaluates the four-fifths fairness threshold against the distribution of prior decisions (0:35-0:60), and generates the explanation artifact (0:60-0:90). Each frame corresponds to a logged step in the E-4 attribution record. The timestamps align exactly. The video is not a simulation; it is a deterministic playback of the actual decision process.

The Plimsoll Line established that a ship's seaworthiness cannot be evaluated by inspecting the final manifest or reviewing the cargo plan. An investigator must see the loading sequence, the ballast adjustments, the stability calculations performed at each stage. The final freeboard mark painted on the hull proves nothing if the load line was submerged during critical phases of the journey. Similarly, an AI system's final recommendation (approve/deny, classify as benign/malignant, shed load/maintain grid) proves nothing about whether proper governance constraints were active during reasoning.

HyperFrames converts the E-4 attribution record's JSON structure into an HTML document that visualizes the decision timeline, then renders that HTML document into an MP4 video at 30 frames per second. The visualization includes the prompt context window (which documents were retrieved from RAG, which portions were highlighted for relevance), the governance checkpoints (LP1 access control verification, LP2 scope binding, LP11 fairness validation), the model's internal reasoning steps (if available via chain-of-thought or tool-calling traces), and the human oversight interaction (the specific query the reviewer saw, their approval or rejection, the timestamp of their decision).

The video format serves three functions. First, it provides a sequential narrative that matches how incident investigators think. An NTSB investigator reconstructing an aviation accident works chronologically: what the crew saw, what alarms fired, what actions were taken, in order. A JSON attribution record presents the same information as a key-value dictionary, which is accurate but cognitively expensive to sequence. The video transforms spatial data (JSON keys) into temporal data (sequential frames), matching the investigator's need for "what happened next."

Second, video evidence is portable across regulatory frameworks in ways that log files are not. A VDR recording plays on any standard video player; maritime authorities in Panama, Liberia, Singapore, and Norway can all review the same MP4 file without specialized forensic tools or proprietary log parsing software. Similarly, a HyperFrames-generated video of an AI decision timeline can be reviewed by a GDPR data protection authority, a HIPAA covered entity's compliance officer, a SOC 2 audit firm, and an insurance underwriter without requiring any of them to install attribution record parsing libraries or learn the E-4 JSON schema. The video is the universal evidence format.

Third, video evidence provides a definitive timestamp chain that resists post-hoc manipulation. Maritime VDRs record directly to write-once media in a protected enclosure. Tampering with the VDR after an incident is immediately evident. HyperFrames implements equivalent protection by generating the video deterministically from the E-4 attribution record's hash, the OSCAL intermediate representation, and the rendering timestamp. If the attribution record is modified post-incident, the video regenerates with different content, and the hash mismatch proves the record was altered. The video serves as a cryptographic commitment to the original decision timeline.

Deploy HyperFrames when the organization must defend AI-assisted decisions to external investigators who do not have access to proprietary logging tools. Internal monitoring teams can review JSON attribution records directly; external auditors, data protection authorities, and legal discovery processes cannot. The classification society's surveyor reviews the inspection database directly; the flag state inspector receives the printed certificate and the VDR playback. The AI operations team reviews E-4 records in their native format; the regulatory authority receives the HyperFrames video.

The video generation pipeline integrates directly with the multi-framework evidence rendering system. After the OSCAL intermediate representation is generated and the four framework-specific renderings (SOC 2, HIPAA, NERC CIP, GDPR) are produced, the HyperFrames renderer receives the same OSCAL document and generates an HTML timeline visualization. That HTML document includes embedded markers for each observation, finding, and control verification in the OSCAL structure. The HTML-to-video conversion preserves those markers as frame boundaries in the output MP4.

A maritime investigator reviewing a VDR recording can see exactly when the helmsman received the rudder order, exactly when the engine room received the speed change command, and exactly when the bridge team recognized the collision risk. The frame-level precision matters because causation depends on sequence. HyperFrames provides the same frame-level precision for AI decision timelines. An investigator reviewing a disputed medical AI triage decision can see exactly when the patient's allergy information was retrieved from the EHR, exactly when the LP1 access control verification occurred, and exactly when the human physician approved the recommended treatment. If the allergy data was retrieved after the treatment recommendation was generated, the video proves the temporal sequencing error that the JSON log obscures.

This temporal precision distinguishes HyperFrames from static evidence artifacts. A PDF rendering of a GDPR compliance record presents the facts in a logical order (Article 6 lawfulness basis, Article 22 automated decision-making assessment, Article 13 information provided to data subject). That logical order may not match the chronological order in which those compliance steps occurred during the actual decision process. If the Article 13 information notification was generated after the automated decision completed, but the PDF presents information notification first, the evidence artifact misrepresents the temporal reality. The video cannot misrepresent sequence; playback is inherently chronological.

For detailed technical coverage of HyperFrames' HTML-to-MP4 rendering pipeline and evidence artifact generation capabilities, see Book II Chapter 32 and Book I Chapter 14. This section focuses on how video evidence enables external regulatory investigation of AI attribution chains where static document formats would require investigators to reconstruct temporal sequences from spatial data structures.

Implementation Notes

Rendering is stateless. Each renderer operates on the OSCAL intermediate representation with no access to external state. Every piece of information a framework-specific rendering needs lives in the OSCAL document's property annotations. Same OSCAL input, same framework-specific output, regardless of when the rendering executes.

Renderers do not modify evidence. Rendering contextualizes evidence for a specific audience. It never adds, removes, or alters the underlying facts. If a framework rendering appears to need information absent from the E-4 base record, the correct fix is to capture that information at collection time, not to patch it in during rendering.

New framework support. Adding a regulatory framework to the pipeline takes two steps: annotate the OSCAL observations with the new framework's control references (property tags), then create a rendering template that extracts and formats those annotations. Evidence collection does not change. The OSCAL schema does not change. Only the rendering layer extends.

How the Evidence Records Are Emitted and Stored

The following describes an illustrative reference implementation, not a verified production deployment. The four evidence tiers (E-1 Foundation/Existence, E-2 Operational/Continuous monitoring, E-3 Context/State-transition, E-4 Accountability/Attribution chain) are not authored by a separate logging process; they are emitted as a byproduct of operation by the H-2 routing enforcement layer at the moment each action is checked. When the routing layer evaluates an AI request against AGENTS.md and the Delegation Token, the verdict (permitted action or gateway denial) is the event, and the record of that verdict is the evidence. A denied request to a resource outside the token's audience field emits an E-2 operational evidence record proving the boundary held; a completed attribution chain like the base record above is emitted as an E-4 record. Nothing downstream invents evidence; the renderers in this appendix only reformat what the routing layer already wrote.

Records are written in a structured, machine-readable format (the base E-4 record above is illustrative JSON) to an append-only store. LP8 evidence integrity is the tamper-evidence property that makes the records trustworthy as audit material, and the reference pattern is to make the store append-only and to chain each record to its predecessor with a content hash, so any edit, deletion, or reordering breaks the chain and is detectable after the fact. Two equivalent reference approaches satisfy this: a write-once / WORM-backed store where records cannot be overwritten in place, or a hash-chained append-only log where each entry carries the hash of the prior entry and is signed with the institution's signing key. Either way, the institution that generated a record cannot quietly repaint it later; tamper-evidence is what stops the owner from rewriting the survey after the line is drawn. Because the evidence is generated as a byproduct of operation and its integrity is independently checkable, an external assessor can test the records without trusting the institution's narrative about them.

The same append-only evidence stream is what the ch11b finance-harness reference implementation queries. The audit-event query in that worked example, the one proving that every cited figure traces to a connector-access record, runs against this stream: it joins each figure to its connector-access audit event (connector name, retrieval timestamp, Delegation Token audience entry) and flags any figure with no backing connector event. The pre-merge pipeline policy in ch11b reads the same records when it checks that every_figure_traces_to_connector_record and that no_connector_outside_delegation_token_audience_was_called. Those checks consume the E-1 through E-4 records described here; the renderings in this appendix are the assessor-facing view of the very stream those checks gate against. A prompt instruction is interpreted; a gateway denial is enforced, and the enforced denial is what these records preserve.

Appendix E: Regulatory Timeline

Ten years of AI governance regulation, compressed into a single chronological reference. This timeline covers major regulatory events from 2017 through 2027 across all five sectors in this book, plus federal and international authorities. Each entry names the regulatory source, the sectors it hits, and the governance implication for organizations deploying AI in regulated environments.

One pattern will become obvious before you reach the halfway mark: the pace is accelerating. Organizations that design their governance architecture for a static compliance target will find themselves rebuilding before the architecture is finished.

2017

June 2017: New York Department of Financial Services Cybersecurity Regulation (23 NYCRR 500) takes effect.
Sector: Financial Services. The first state-level cybersecurity regulation for financial institutions, and a template that every subsequent state AI governance law borrows from. It requires CISO designation, risk assessment, penetration testing, audit trail, and 72-hour incident notification.

2018

May 25, 2018: General Data Protection Regulation (GDPR) enforcement begins.
Sectors: All (extraterritorial). Data subject rights for EU residents, regardless of where the processing happens. Article 22 (automated decision-making), Article 7 (consent conditions), and Articles 13-14 (transparency obligations) ground the requirements for LP9, LP10, and LP12 in any organization that touches EU data. Extraterritorial scope makes this unavoidable: if you have EU customers, you are subject.

May 2018: NIST publishes Cybersecurity Framework v1.1.
Sectors: All (voluntary, becoming de facto standard). Adds a supply chain risk management category to the existing risk management framework. Not AI-specific, but it establishes the risk-based governance model that every subsequent AI framework adopts.

2020

January 1, 2020: California Consumer Privacy Act (CCPA) enforcement begins.
Sectors: All (California consumers). Grants consumers the right to know, delete, and opt out of the sale of personal information. Other states copy the model. Any AI system processing California consumer data falls within scope.

February 2020: NIST publishes AI Risk Management Framework concept paper.
Sectors: All. Kicks off the multi-year development process that produces the NIST AI RMF 1.0 in January 2023. The signal is clear: the federal government intends to establish voluntary AI governance guidance, and the framework is already in motion.

2021

April 2021: European Commission publishes proposed AI Act.
Sectors: All (EU operations). The first comprehensive AI-specific regulatory proposal anywhere in the world. Introduces risk-based classification (unacceptable, high-risk, limited-risk, minimal-risk) and sector-specific requirements. Three years of legislative process begin here and conclude in 2024.

December 2021: Colorado SB 21-169 (Life Insurance Algorithm Governance) signed.
Sector: Insurance. First US state law requiring algorithmic bias testing for insurance AI. Requires life insurers to demonstrate that AI-assisted underwriting does not produce unfairly discriminatory outcomes. Precursor to the broader Colorado AI Act (SB 24-205).

2022

October 2022: White House publishes Blueprint for an AI Bill of Rights.
Sectors: All (non-binding). Five principles: safe and effective systems, algorithmic discrimination protections, data privacy, notice and explanation, and human alternatives. None of it is binding, but it establishes the policy direction that shapes every subsequent executive order and legislative proposal.

December 2022: NAIC adopts Model Bulletin on the Use of Artificial Intelligence Systems by Insurers.
Sector: Insurance. Sets expectations for insurer AI governance: bias testing, documentation, human oversight, and regulatory examination readiness. It is a model bulletin, not a model law, so it carries no binding force until individual states

adopt it. State DOIs are already weaving its expectations into their examination procedures.

2023

January 26, 2023: NIST publishes AI Risk Management Framework (AI RMF 1.0).
Sectors: All. Voluntary AI governance framework organized around four functions: Govern, Map, Measure, and Manage. MEASURE 2.11 (bias evaluation) is the reference standard that LP11 fairness verification builds on. Within months this becomes the de facto US AI governance reference.

June 2023: CFPB publishes Circular 2023-03 on adverse action and AI.
Sector: Financial Services. The CFPB is plain: creditors using AI for credit decisions must provide specific reasons in adverse action notices. "Complex algorithm" and "model output" are not sufficient explanations. This circular directly shapes the LP12 explainability requirements for financial services.

October 30, 2023: Executive Order 14110 on the Safe, Secure, and Trustworthy Development and Use of Artificial Intelligence.
Sectors: All (federal agencies, with downstream effects on government contractors and regulated industries). AI safety and security requirements for federal agencies, mandatory red-teaming for dual-use foundation models, and a directive to NIST to develop AI safety guidelines. Developers of large AI models gain new reporting obligations overnight.

December 2023: SEC finalizes Regulation S-K Item 106 cybersecurity governance disclosure rules.
Sector: Financial Services (public companies across all sectors). Public companies must now disclose their cybersecurity governance structure, including board oversight, in 10-K filings. This is the disclosure obligation that Chapter 20's board reporting framework satisfies.

2024

January 17, 2024: Digital Operational Resilience Act (DORA) enters into force.
Sector: Financial Services (EU). ICT risk management requirements land on EU financial entities: third-party AI vendor governance, operational resilience testing, and incident reporting. Full application begins January 17, 2025.

March 13, 2024: European Parliament adopts the AI Act.
Sectors: All (EU operations). The world's first comprehensive AI regulation clears its final legislative hurdle. Risk-based classification, specific requirements for high-risk

AI systems in financial services, healthcare, employment, education, and essential services. Enforcement phases in through 2027.

May 17, 2024: Colorado SB 24-205 (Concerning Consumer Protections for Artificial Intelligence) signed.
Sectors: All (Colorado consumers). Developers and deployers of high-risk AI systems making consequential decisions now face impact assessments, consumer notification, and algorithmic discrimination prevention requirements. First US state statute with AI-specific deployer obligations comparable to EU AI Act conformity requirements. Effective June 30, 2026.

August 1, 2024: EU AI Act enters into force (published in Official Journal).
Sectors: All (EU operations). The clock starts. Prohibited AI practices are banned from February 2, 2025. GPAI model obligations kick in August 2, 2025. High-risk AI system requirements land August 2, 2026.

September 2024: California AB 2013 (AI Transparency) signed.
Sectors: All (California operations). Generative AI developers must publish summaries of training data on their websites. Effective January 1, 2026.

October 2024: NIST publishes AI RMF Generative AI Profile (NIST AI 600-1).
Sectors: All. Extends the AI RMF into generative AI territory: confabulation, data privacy, information integrity, and value chain risks all get explicit treatment. This profile becomes the governance reference for any generative AI deployment in the US.

2025

January 17, 2025: DORA full application begins.
Sector: Financial Services (EU). EU financial entities must have ICT risk management frameworks, third-party risk governance, and incident reporting procedures fully operational. Every AI vendor contract must satisfy DORA provisions or be renegotiated.

February 2, 2025: EU AI Act prohibited practices enforcement begins.
Sectors: All (EU operations). Subliminal manipulation, vulnerability exploitation, public-authority social scoring, and real-time remote biometric identification in public spaces (with narrow exceptions) are now prohibited. This is the AI Act's first enforcement milestone.

March 2025: HIPAA Security Rule update proposed rule published.
Sector: Healthcare. HHS proposes updates to the HIPAA Security Rule that bring cloud computing, AI-assisted processing, and modern cybersecurity into scope. Mandatory encryption, network segmentation, and AI-specific risk assessment requirements are all on the table.

August 2, 2025: EU AI Act general-purpose AI model obligations take effect.
Sectors: All (EU operations). GPAI model providers must now meet transparency obligations: technical documentation, copyright compliance, and content provenance. Providers whose models carry systemic risk face additional obligations, adversarial testing and incident reporting among them.

2026

January 1, 2026: California AB 2013 (AI Transparency) effective.
Sectors: All (California operations). Generative AI training data summaries must be published.

June 30, 2026: Colorado AI Act (SB 24-205) effective; Attorney General enforcement authority activates.
Sectors: All (Colorado consumers). Deployers of high-risk AI must complete impact assessments, notify consumers when AI is involved in consequential decisions, and run algorithmic discrimination prevention programs. The obligations are live. The effective date and the start of enforcement now coincide on this date; this is the first US state enforcement of AI-specific deployer obligations. The original February 1, 2026 effective date was amended during Colorado's 2025 legislative session to June 30, 2026; dates current as of publication, subject to revision.

August 2, 2026: EU AI Act high-risk AI system requirements take effect.
Sectors: All (EU operations). Conformity assessment procedures, post-market monitoring, human oversight, and technical documentation become mandatory for deployers of high-risk AI systems. Article 26 deployer obligations are enforceable. For organizations deploying AI in regulated sectors with EU exposure, this is the milestone that matters most. Financial services credit scoring, healthcare clinical decision support, insurance underwriting, and employment AI all sit within Annex III high-risk categories.

2026 (anticipated): SR 26-2 model-risk guidance.
Sector: Financial Services. SR 26-2 (described here as anticipated 2026 model-risk guidance; verify against the primary source before citing, or treat as an illustrative near-future rule). This volume treats SR 26-2 as guidance the agencies, the Federal Reserve, the OCC, and the FDIC, are expected to issue to revise model-risk management for generative and agentic AI, building on the verifiable SR 11-7 (2011) and SR 21-8 (2021) anchors. The issuance, date, and any supersession are not settled fact; the entry is subject to verification.

2027

January 2027: California Automated Decision-Making Technology (ADMT) regulations effective (projected; as of publication, subject to revision).
Sectors: All (California consumers). Pre-use notice, opt-out rights, and access rights for AI profiling and significant decisions affecting California residents. In practice, this extends CCPA/CPRA to cover automated decision-making specifically.

August 2, 2027: EU AI Act full application, including obligations for high-risk AI systems embedded in products regulated under Union harmonization legislation.
Sectors: Manufacturing, Healthcare (medical devices), Energy (machinery). AI systems embedded in products subject to existing EU product safety directives must comply with all AI Act requirements. Conformity assessment procedures under the applicable product legislation now absorb AI-specific obligations. The regulatory circle closes.

Reading This Timeline

Three patterns emerge.

Acceleration. The interval between major regulatory events is shrinking. GDPR (2018) to CCPA (2020): two years. EO 14110 (2023) to the EU AI Act (2024) to Colorado SB 24-205 (2024): less than one year. From 2025 forward, multiple enforcement deadlines arrive in the same calendar year. The pace is not slowing.

Convergence. Early regulations addressed privacy (GDPR, CCPA) or cybersecurity (NYDFS 500) without mentioning AI. Current regulations name AI explicitly: the EU AI Act, Colorado AI Act, NAIC Model Bulletin. The next generation will fold AI governance into sector-specific frameworks as an integrated requirement, not a separate compliance domain tacked on afterward.

Extraterritoriality. GDPR set the precedent in 2018: regulation follows the data subject, not the organization. The EU AI Act adopted the same model. Colorado adopted it for Colorado residents. LP10, jurisdictional consistency, is not an edge case. It is the default operating condition for any organization with cross-border or multi-state operations.

Organizations that design their governance architecture for one regulatory framework, or for the regulatory state as of a fixed date, will rebuild continuously as new frameworks arrive. The Harness Doctrine absorbs new regulatory requirements through rendering template additions rather than architectural reconstruction. That is the difference between a governance program that ages and one that compounds.

Jurisdiction-Aware Model Routing Under Accelerating Compliance

When Lloyd's Register certifies a tanker for international trade, the vessel must simultaneously satisfy Liberian flag state requirements, US Coast Guard inspections, EU port state control, and IMO conventions. No single certificate suffices. The classification society maintains relationships with every major maritime authority and knows which structural evidence satisfies which jurisdiction's particular interpretation of common standards. The surveyor records one set of structural facts; the document control system routes those facts through jurisdiction-specific certification pipelines.

The regulatory timeline above documents an identical pattern emerging for AI governance. GDPR enforcement began in 2018, targeting EU data subjects regardless of where processing occurs. Colorado SB 24-205 takes effect in 2026, imposing deployer obligations for AI systems affecting Colorado consumers. The EU AI Act begins enforcing high-risk AI system requirements in August 2026. California AB 2013 mandates generative AI training data transparency from January 2026. Each jurisdiction defines its own risk categories, documentation requirements, and conformity assessment procedures, but they all draw from the same underlying evidence base: the E-4 attribution records, LP1-LP14 governance properties, and human oversight verifications that the Harness generates during operation.

Jurisdiction-aware model-routing layers enable jurisdictional compliance without multiplying infrastructure cost linearly with the number of jurisdictions. When an AI system processes a request from an EU data subject, the governance checkpoint that generates GDPR-compliant explanation artifacts (Article 13, Article 22) must use a model that satisfies the EU AI Act's transparency requirements for general-purpose AI models. When the same system processes a request from a Colorado consumer subject to SB 24-205, the explanation generation must satisfy Colorado's algorithmic discrimination prevention program requirements. The technical requirements differ; the underlying explanation generation task is identical.

Rather than deploying separate model infrastructure for each jurisdiction, a routing layer brokers access to multiple AI providers through a unified interface. The governance engine specifies the required model characteristics (GDPR-compliant training data documentation, Colorado-compliant bias mitigation measures, specific model version with published adversarial testing results) as part of the routing request. The routing layer selects the appropriate model from its provider network based on those characteristics, routes the inference request, and returns the result. The E-4 attribution record logs which model was selected, which version, and which provider, establishing the provenance chain for regulatory review.

The maritime parallel is direct. A classification society does not maintain separate inspection teams for every flag state. One surveyor performs the hull inspection. The document control system routes the inspection report through flag-specific certification workflows based on the vessel's registry and intended trading routes. If the owner later changes the vessel's flag, the underlying structural evidence remains valid; only the certificate formatting changes. A jurisdiction-aware routing layer provides the equivalent routing intelligence for model selection under multi-jurisdictional constraints.

The Lagrange point is the balance between model capability and compliance overhead. Deploying the most capable frontier model for every task regardless of jurisdiction satisfies all compliance requirements but burns budget on over-provisioned capability. Deploying the cheapest local model for every task minimizes cost but fails jurisdictional requirements where specific model transparency, bias testing, or adversarial-resilience documentation is mandatory. A routing layer enables selective routing: use frontier models when regulatory requirements mandate them (GDPR Article 22 explanation generation for EU subjects, Colorado SB 24-205 impact assessment documentation), use economy models when general capability suffices (internal monitoring dashboards, development environment testing).

Deploy a jurisdiction-aware routing layer when the organization operates under multiple jurisdictional regimes with differing model transparency or documentation requirements. Internal AI systems serving a single market can use direct provider APIs. Systems processing cross-border requests, multi-state consumer data, or operating in rapidly shifting regulatory environments gain routing flexibility that direct provider relationships cannot match. The classification society's document control system does not just print certificates; it routes structural evidence through the appropriate jurisdiction-specific workflows based on vessel characteristics and trading routes. A routing layer does not just call models; it routes inference requests through jurisdiction-appropriate model selections based on governance characteristics and data subject context.

Configuration is claim-based. Each routing request carries governance claims that specify the regulatory context (GDPR data subject, Colorado consumer, HIPAA covered transaction, NERC CIP BES Cyber System). The model selection logic evaluates those claims against each provider's compliance documentation. If the request includes a `gdpr-data-subject: true` claim, only models with published EU AI Act conformity documentation remain in the selection pool. If the request includes a `colorado-consumer: true` claim, only models with algorithmic discrimination testing documentation survive filtering. The routing decision is deterministic and auditable; the E-4 attribution record proves which model was selected based on which claims.

The acceleration pattern from the timeline above compounds routing complexity. In 2018, GDPR was the only extraterritorial AI governance requirement. In 2024, the EU AI Act, Colorado SB 24-205, and California AB 2013 all entered the regulatory stack. In 2026, two major enforcement deadlines arrive within roughly two months (Colorado effective and enforcement June 30, 2026, EU high-risk AI systems August 2, 2026), with California ADMT projected for January 2027. By 2027, organizations operating across these jurisdictions face continuous compliance monitoring rather than periodic audits. A system that processes 10,000 consumer requests per day cannot manually route each request through jurisdiction-specific model selection. A routing layer automates that routing based on governance claims extracted at request time.

The alternative is geographic model deployment: separate infrastructure stacks per jurisdiction, each with locally compliant models. This works for organizations with discrete geographic markets (EU subsidiary with EU-only operations, US division with US-only customers). It fails for organizations with cross-border operations where a single customer interaction might trigger multiple jurisdictional requirements simultaneously (EU data subject financing healthcare equipment through a US lender covered by HIPAA, triggering GDPR plus EU AI Act plus HIPAA obligations in one transaction). The classification society cannot maintain separate survey ships for Panamanian, Liberian, and Singaporean flags. One surveyor inspects the vessel; document routing handles jurisdictional certification variations.

Multi-Framework Evidence-Latency Optimization Under Accelerating Compliance

Maritime classification societies measure survey completion time as a critical operational metric. When new IMO regulations require additional hull inspections, the classification society must demonstrate that expanded inspection scope does not proportionally expand survey duration. A five-day port visit for annual survey cannot balloon to ten days simply because the inspection checklist doubled. Port time costs the ship owner thousands of dollars per hour in lost revenue. If expanded regulatory scope makes classification surveys economically prohibitive, owners will seek alternative compliance pathways or register under flags with less stringent enforcement.

This economic constraint forces classification societies to continuously optimize survey procedures. New ultrasonic testing equipment reduces hull thickness measurement time from four hours to ninety minutes. Digital documentation replaces paper forms, eliminating re-keying delays. The surveyor's technical work remains rigorous, but procedural efficiency improvements prevent regulatory expansion from making compliance economically untenable.

The regulatory timeline in this appendix documents an identical acceleration pattern for AI governance. Organizations deploying AI systems in 2018 faced one extraterritorial requirement: GDPR. By 2024, that grew to GDPR, the EU AI Act, Colorado SB 24-205, and California AB 2013. By 2026, enforcement deadlines for high-risk AI systems (EU AI Act Article 26) and deployer impact assessments (Colorado SB 24-205) arrive within roughly two months, with automated decision-making transparency (California AB 2013, ADMT) following in projected sequence. Each new regulation adds documentation requirements, conformity assessment procedures, and human oversight checkpoints to the E-4 attribution pipeline.

If evidence generation latency scales linearly with regulatory scope, compliance becomes economically prohibitive. As an illustrative model, suppose generating one GDPR-compliant E-4 attribution record takes 200 milliseconds of model inference time (explanation generation, fairness verification, consent validation). Adding SOC 2 rendering adds 150 milliseconds (access control logging, audit trail formatting). HIPAA rendering adds another 120 milliseconds (minimum necessary verification, accounting of disclosures). NERC CIP rendering adds 180 milliseconds (security event logging, access control verification). GDPR Article 30 processing inventory rendering adds 140 milliseconds. A decision that required 200 milliseconds of compliance overhead now requires 790 milliseconds, a 395 percent latency increase for regulatory scope expansion alone. These figures are illustrative order-of-magnitude estimates, not measured benchmarks.

Multi-framework evidence-latency optimization benchmarks this latency scaling to identify where procedural optimization can prevent regulatory expansion from making real-time AI governance economically infeasible. The approach measures end-to-end decision latency across three phases: evidence collection (E-4 record generation), intermediate processing (OSCAL transformation), and multi-framework rendering (SOC 2, HIPAA, NERC CIP, GDPR outputs). Baseline measurements establish how long each phase takes with one framework. Scaled measurements reveal whether latency grows linearly, sub-linearly, or super-linearly as additional frameworks enter scope.

Maritime surveyors discovered that inspection time does not scale linearly with checklist length when proper optimization applies. Ultrasonic hull thickness testing equipment can measure multiple points simultaneously rather than sequentially. Digital forms eliminate duplicate data entry. Procedural improvements break the linear scaling assumption. Latency benchmarking reveals where similar optimization opportunities exist in multi-framework evidence rendering pipelines.

The most significant optimization target is parallel rendering. The SOC 2, HIPAA, NERC CIP, and GDPR renderers all operate on the same OSCAL intermediate representation. If rendering executes sequentially (finish SOC 2, then start HIPAA, then start NERC CIP, then start GDPR), total latency is the sum of individual renderer times. If rendering executes in parallel (start all four renderers simultaneously), total latency is the maximum of individual renderer times. For the illustrative latency profile above, sequential rendering takes 590 milliseconds; parallel rendering takes 180 milliseconds (NERC CIP is the slowest renderer). That is roughly a 3.3 times latency reduction for identical evidence output, achieved purely through procedural restructuring.

A benchmark harness quantifies this difference. It measures evidence generation latency under both sequential and parallel rendering modes. Organizations experiencing compliance latency bottlenecks gain empirical data for justifying infrastructure changes. If baseline measurements show 800ms end-to-end latency with sequential rendering and 240ms latency with parallel rendering, the business case for parallel rendering infrastructure becomes quantifiable. Port time costs the ship owner thousands of dollars per hour; if survey optimization saves two days per vessel inspection, the surveyor's new ultrasonic equipment pays for itself in three surveys.

The second optimization target is model selection for rendering tasks. OSCAL-to-framework rendering is a translation task with verifiable output. The renderer extracts observations tagged with framework-specific control references and formats them according to framework templates. Output correctness can be verified structurally: every fact in the rendered output must trace back to the OSCAL source. This verifiability means rendering does not require frontier model capability. A fast, inexpensive model with reliable formatting output suffices if verification catches errors.

Latency-and-cost benchmarking compares rendering across model tiers. If a frontier model produces SOC 2 rendering output in 180ms at a higher per-request cost, and an economy model produces equivalent output in 95ms at a fraction of that cost, the performance and cost differences are both quantifiable; these are illustrative figures, not measured results. The verification step (which runs regardless of model choice) proves whether the cheaper model's output quality matches the frontier model's reliability. If verification passes at identical rates for both models, the economy model becomes the rational choice. The surveyor's structural calculations require expertise; translating the English-language report into the Spanish-language format for Panama does not.

The Lagrange point is the balance between regulatory scope expansion and procedural optimization. If optimization cannot keep pace with regulatory growth, compliance latency eventually makes real-time AI-assisted decisions economically infeasible. A credit approval system that took 200ms to generate compliant E-4 records cannot tolerate 2,000ms latency when many jurisdictional requirements stack. The customer's application times out. The loan origination throughput collapses. The organization abandons AI-assisted credit decisioning not because the model failed, but because compliance overhead made the model too slow to use in production.

Deploy multi-framework evidence-latency optimization when the organization operates under multiple regulatory frameworks with overlapping evidence requirements, or when new regulatory frameworks enter scope faster than infrastructure scales naturally. Organizations facing single-framework compliance (SOC 2 only, or HIPAA only) gain less value from multi-framework latency optimization. Organizations facing the acceleration pattern documented in this timeline (GDPR in 2018, plus EU AI Act in 2024, plus Colorado in 2026, plus California ADMT projected in 2027) face compounding latency risk that benchmarking quantifies before it becomes a production incident.

The benchmark suite runs in production environments, not isolated test harnesses. Synthetic benchmarks that generate mock E-4 records and synthetic OSCAL documents miss the real-world latency contributions from network round trips, database writes, and cross-service authentication. The benchmark instruments the actual evidence rendering pipeline, measuring latency at every integration boundary (E-4 record serialization, OSCAL transformation, framework renderer invocation, verification execution, artifact storage). Optimization targets become empirically obvious rather than theoretically plausible. If 60 percent of end-to-end latency lives in OSCAL transformation but only 15 percent lives in framework rendering, optimizing rendering yields marginal gains while OSCAL optimization produces step-function improvements.

The maritime surveyor's rule: measure the constraint before you optimize the wrong thing. If port time cost drives economics, optimize survey duration. If dry dock availability is the bottleneck, optimize survey scheduling flexibility. If flag state paperwork delays vessel certification, optimize document processing. Latency benchmarking makes the constraint visible for multi-framework AI governance compliance. The timeline above documents that regulatory scope is expanding faster than most organizations anticipated. Latency benchmarking proves whether current infrastructure can absorb that expansion or whether procedural optimization must intervene.

Appendix F: The Maritime-to-AI Governance Genealogy

Maritime governance and AI governance share more than metaphor. They share architecture. This appendix traces the structural lineage, pairing each maritime milestone with its AI governance equivalent in chronological order. The problems are identical. How do you deploy capable systems in regulated environments, produce verifiable evidence of compliance at commercial velocity, and operate without requiring every participant to trust every other participant's self-assessment?

The maritime industry took roughly 270 years to build this governance architecture. The AI governance community is compressing that arc into a decade. Compression does not change the structure. It changes the consequences of delay.

1688: Lloyd's Coffee House

Maritime. Edward Lloyd opens a coffeehouse on Tower Street in the City of London. Merchants, ship captains, and underwriters gather to share intelligence on ships, cargoes, and voyages. Risk assessment is informal, relationship-based, and unverifiable. Seaworthiness depends on personal reputation and word of mouth. No standardized criteria exist. Ships sail with insurance written against descriptions nobody has independently confirmed.

AI Governance Equivalent (2020-2024). Organizations deploy AI systems on the strength of vendor presentations, security questionnaires, and internal risk assessments. Governance-worthiness is a judgment call, filtered through vendor reputation, analyst reports, and procurement team experience. No standardized criteria define what "governance-worthy" means for an AI system. Risk acceptance decisions get written against descriptions of intended behavior, not verified evidence of actual behavior. The coffeehouse, in other words, is still open.

1760: Lloyd's Register Formalized

Maritime. The Society of Lloyd's publishes the first edition of the Lloyd's Register of Shipping. Ships are classified by independent surveyors against published criteria. The classification ("A1" for hull condition, "100" for equipment) provides a standardized assessment that any underwriter can rely on without personally inspecting the vessel. The classification society model separates the assessor from the assessed; the surveyor's income does not depend on the ship receiving a favorable classification.

AI Governance Equivalent (2023-2025). SOC 2 Type II attestation, PCI-DSS Qualified Security Assessor reviews, HIPAA risk assessments, and NERC CIP audits now provide standardized, independent assessments of organizational controls. Assessor and assessed are structurally separated. The evidence architecture (E-1 through E-4) is the classification certificate for AI governance: a structured evidence package that any examiner, auditor, or assessor can verify independently, without trusting the organization's self-report. No personal inspection required. The classification speaks for itself.

1771: Lloyd's Committee Rules

Maritime. Lloyd's underwriters establish the first Committee rules governing who may participate in the market and under what conditions. The rules standardize the terms on which insurance is written and create accountability for underwriters who fail to honor their commitments. An informal gathering becomes a governed institution with published rules and enforcement mechanisms.

AI Governance Equivalent (2022-2023). The NAIC Model Bulletin on AI systems (2022), NIST AI RMF 1.0 (2023), and sector-specific guidance from the OCC and FFIEC publish the rules for AI governance in commercial environments. These are not statutory requirements yet. They are institutional standards that define, in writing, the terms on which AI deployment is acceptable in regulated industries. The Committee rules are on the wall. Enforcement follows.

1807: Lloyd's Act of Parliament

Maritime. Parliament grants Lloyd's a formal charter, establishing it as a corporation with legal standing. A voluntary association becomes a legally recognized institution. The charter gives Lloyd's authority to regulate its own members and standing to enforce its rules in court.

AI Governance Equivalent (2023-2024). Executive Order 14110 (October 2023) sets federal AI governance policy. The EU AI Act (March 2024) provides the legislative charter for AI governance across Europe. Colorado SB 24-205 (May 2024) delivers the first US state-level legislative framework for AI deployer obligations. In eighteen months, AI governance crosses from voluntary guidance into statutory territory. The charter exists. The court cases follow.

1834: Lloyd's Register Becomes Independent

Maritime. Lloyd's Register of Shipping separates from the Lloyd's insurance market and becomes an independent classification society. The separation eliminates the conflict of interest inherent in having the insurance market control the classification of ships it insures. Independent classification establishes a structural principle: the assessor and the economic beneficiary of the assessment cannot be the same party.

AI Governance Equivalent. Chapter 10 calls this the Voight-Kampff problem. The vendor who supplies both the AI system and the governance attestation for that system holds a structural conflict of interest. Resolution follows the same logic that separated Lloyd's Register from the insurance market. Transfer governance ownership to the deploying organization through the AGENTS.md migration (Chapter 25) and the conflict dissolves. Not mitigated. Dissolved.

1876: The Plimsoll Mark (Merchant Shipping Act)

Maritime. Samuel Plimsoll campaigns against "coffin ships," vessels overloaded beyond safe limits by unscrupulous owners who profit from insurance payouts when the ships sink. The Merchant Shipping Act of 1876 mandates a load line, the Plimsoll Mark, painted on every vessel's hull at the maximum safe draft. The mark is visible, binary, enforced at the moment of departure, and independent of the shipowner's commercial preferences. Any observer at the dock can determine whether the ship is within its authorized operating boundary.

AI Governance Equivalent (2024-2026). LP9 through LP12, the four commercial governance properties from Chapter 8, follow the Plimsoll principle exactly: visible, binary, enforced at the moment of action, independent of the deploying organization's commercial preferences. Consent is verified or it is not. The correct jurisdiction's rules are applied or they are not. The fairness threshold is met or it is not. The explanation artifact is generated at decision time or it is not. Anyone looking at the evidence dashboard can read the governance state of an AI system without expert interpretation. Same structural function. Different hull.

1914: International Convention for the Safety of Life at Sea (SOLAS)

Maritime. After the Titanic sinks in April 1912, killing 1,517 passengers and crew, the international community convenes to adopt mandatory safety requirements. The first SOLAS Convention (1914) requires lifeboats for all passengers, continuous radio watch, and ice patrol. These are not voluntary industry guidelines. They are statutory obligations with enforcement. Maritime's failure to self-regulate adequately, specifically the commercial incentive to allocate lifeboat space to revenue-generating deck areas rather than safety requirements, produces mandatory external regulation.

AI Governance Equivalent (2024-2026). The EU AI Act is the SOLAS of AI governance. Algorithmic bias in credit decisions, AI-generated claims denials overturned on appeal, chatbot liability incidents, AI drive-thru shutdowns: failures accumulated until mandatory requirements followed. Risk classification, technical documentation, human oversight, post-market monitoring, and deployer obligations under Article 26 are not voluntary. They are enforceable. Chapter 22 maps this parallel in detail.

1948: International Maritime Organization (IMO) Established

Maritime. The United Nations establishes the Inter-Governmental Maritime Consultative Organization (later renamed the IMO) as the specialized agency responsible for maritime safety, security, and environmental protection. The IMO provides the international institutional framework for developing and enforcing maritime governance standards. National maritime authorities implement IMO conventions through domestic law, creating a multi-jurisdictional governance architecture with a common set of standards.

AI Governance Equivalent (2023-present). OECD AI Principles (2019), the G7 Hiroshima AI Process (2023), and emerging coordination through the UN, ISO/IEC, and bilateral agreements are assembling the institutional framework for international AI governance. ISO/IEC 42001 (AI Management System) provides the international standard. National authorities, NIST in the US, BSI in the UK, CNIL in France, implement through domestic frameworks. LP10 (Jurisdictional Consistency) addresses the same multi-jurisdictional governance problem the IMO's convention system resolved for shipping. Different domain. Same architecture.

1956: Containerization (SS Ideal X)

Maritime. Malcom McLean's *Ideal X* sails from Port Newark to Houston carrying 58 aluminum truck bodies. The standardized shipping container eliminates the cargo repacking that occurred at every port transfer. Intermodal transport becomes economically viable. The container standardizes the interface, not the contents. Cranes at every port handle the same container regardless of what is inside.

AI Governance Equivalent. Chapter 18's OSCAL rendering pipeline and evidence harmonization architecture. The unified evidence stream is the cargo inside the container. The OSCAL intermediate format is the container specification. Framework-specific rendering templates are the cranes at each port. Collect the evidence once, render it for SOC 2, PCI-DSS, HIPAA, NERC CIP, GDPR, and EU AI Act assessments, and never repack it at a framework boundary. One collection run. Six outputs.

1974: SOLAS Convention Updated

Maritime. The 1974 SOLAS Convention, which remains in force today, establishes the "tacit acceptance" procedure: amendments take effect automatically unless a specified number of Contracting Governments object within a specified period. This mechanism allows safety standards to evolve without requiring the years-long ratification process that earlier amendments demanded. The maritime governance framework becomes self-updating.

AI Governance Equivalent. Rendering layer extensibility in the evidence harmonization architecture. When a new regulatory framework takes effect, the governance architecture does not need rebuilding. Add a rendering template to the pipeline; the evidence collection infrastructure stays stable. Extension, not reconstruction. The same logic as tacit acceptance: absence of objection is consent to move.

1993: International Safety Management (ISM) Code

Maritime. The IMO adopts the ISM Code following a series of ferry disasters, including the *Herald of Free Enterprise* (1987) and the *Scandinavian Star* (1990). The ISM Code requires ship operators to implement a Safety Management System with documented procedures, defined responsibilities, regular internal audits, management review, and continuous improvement. Maritime governance shifts from prescriptive technical standards (how the ship must be built) to management system standards (how the ship must be operated). Compliance is assessed through audits of the management system, not just inspection of the vessel.

AI Governance Equivalent. AGENTS.md combined with the continuous authorization cycle from Chapter 25. AGENTS.md is the Safety Management System for AI: documented governance procedures, defined responsibilities, continuous monitoring, management review. The shift from prescriptive controls ("what the AI system must not do") to management system governance ("how the AI system must be governed") mirrors the ISM Code's move from hull inspection to management system audit. The vessel is not the point. The operation is.

1998: ISS Intergovernmental Agreement

Maritime/Space. The IGA, signed by the United States, Russia, Japan, Canada, and eleven European nations, establishes the governance architecture for the most complex multi-jurisdictional operating environment humans have ever sustained. Jurisdiction follows the module. Cross-module activities follow defined protocols. Conflicts resolve through cooperation with escalation paths. The architecture has sustained 25+ years of continuous operation.

AI Governance Equivalent. LP10 (Jurisdictional Consistency) and the multi-jurisdictional governance architecture from Chapter 9. Jurisdiction follows the data subject, not the system. Cross-jurisdiction processing follows defined routing rules. Conflicts resolve through the "most restrictive applicable" default, with escalation and documented resolution when that default is insufficient. Continuous operation across regulatory boundaries the deploying organization neither created nor controls. The IGA proved this is solvable. The question is whether AI governance will reach for the same tools.

2006: Maritime Labour Convention (MLC)

Maritime. The ILO adopts the MLC, establishing minimum standards for working conditions of seafarers, covering employment agreements, hours of work, accommodation, health protection, and social security. The Convention creates a "no more favorable treatment" clause: ships flying flags of non-ratifying states receive no benefit from non-ratification when visiting ports of ratifying states. Regulatory arbitrage through flag selection is eliminated.

AI Governance Equivalent. The EU AI Act's extraterritorial scope and the "most restrictive applicable" default for jurisdictional uncertainty. An AI system processing EU data subjects' data does not escape AI Act obligations because the server sits outside Europe, just as a ship flying a non-ratifying flag does not escape MLC requirements when it docks in a ratifying state's port. LP10's default, applying the most restrictive applicable framework when jurisdiction is uncertain, eliminates the governance arbitrage incentive entirely. The port determines the rules. Not the flag.

2024: EU AI Act

Maritime Parallel. The EU AI Act occupies the same structural position in AI governance that SOLAS occupies in maritime governance: the first comprehensive, mandatory, internationally significant safety regulation adopted after accumulated governance failures demonstrated that voluntary self-regulation was insufficient. Both establish risk-based classification, mandatory documentation, independent assessment, continuous monitoring, and penalties for non-compliance.

AI Governance Equivalent. The EU AI Act itself. Risk classification (Annex III high-risk categories) maps to SOLAS vessel classification. Technical documentation (Article 11) maps to the ship's safety certificate. Human oversight obligations (Article 14) map to the requirement for qualified officers. Post-market monitoring (Article 72) maps to the continuous survey system. Deployer obligations (Article 26) map to the shipowner's obligations under the ISM Code. The structural parallels are not approximate. They are precise.

Academic-Research-Skills: Citation Integrity in Regulatory Knowledge Graphs

Maritime governance rests on citation integrity. When a classification society writes a survey report referencing IMO Convention MARPOL Annex I Regulation 13, that citation must be precise, current, and traceable to the authoritative source. A surveyor cannot cite "the pollution prevention requirements" without specifying which regulation, which amendment, and which publication date applies. Regulatory citations work because every participant can verify them against the same authoritative corpus. An inspector in Singapore reading "SOLAS Chapter II-2 Regulation 10" retrieves identical regulatory text to an inspector in Rotterdam.

AI governance knowledge graphs face the same citation integrity requirement. When an E-4 attribution record asserts that a credit decision satisfied "LP11 fairness verification per NIST AI RMF MEASURE 2.11," that citation must reference the specific NIST publication, the specific measurement function, and the specific threshold applied. When a GDPR compliance artifact claims "Article 22(3) human oversight requirement met," the citation must trace to the exact GDPR Article text, not a paraphrased summary or vendor interpretation.

Academic-Research-Skills applies academic research patterns to governance knowledge graph construction. Every node representing a regulatory requirement carries provenance: source document URI, publication date, amendment version, relevant recitals or preambles, and superseded-by relationships when regulations update. Every edge connecting a governance property to a regulatory requirement carries evidence. Which E-4 attribution records instantiated that property, which framework renderers referenced it, and which conformity assessments relied on it.

The pattern comes from academic citation networks. A research paper citing "Einstein (1905)" provides sufficient information for any reader to retrieve the original source. A governance knowledge graph citing "EU AI Act Article 26(5)" must provide equivalent precision. Academic-Research-Skills enforces citation completeness: no dangling references, no circular citations, no claims to regulatory compliance without traceable provenance back to the authoritative source text.

Maritime classification societies learned this lesson when different flag states interpreted identical IMO conventions differently. Bureau Veritas cites "SOLAS Chapter V Regulation 15" in a bridge equipment survey report. The shipowner later faces a port state inspection in a jurisdiction that adopted SOLAS Chapter V with a national variance. Without precise citation to the IMO convention's exact text and the flag state's specific variance, the survey report's regulatory claim becomes ambiguous. Academic citation standards prevent that ambiguity. The citation must be precise enough that any examiner can retrieve the exact regulatory text the surveyor relied on.

Deploy Academic-Research-Skills when the organization's governance knowledge graph references multiple regulatory frameworks with overlapping control vocabularies. SOC 2 Trust Services Criteria CC6.1 and NIST Cyber security Framework PR.AC-1 both address access controls; the regulatory texts differ, but governance artifacts claiming compliance with both must cite each distinctly. GDPR Article 32 and HIPAA Security Rule 164.312(a)(1) both mandate access controls for personal data. An E-4 attribution record claiming dual compliance must trace provenance to both regulatory sources independently.

The tool performs citation verification through graph traversal. If an E-4 record claims LP1 classification-consistent access was verified, the knowledge graph must contain an edge from LP1 to the applicable regulatory control (NIST AI RMF GOVERN 1.3, GDPR Article 32, HIPAA 164.312(a)(1), or equivalent). That edge must carry metadata proving which framework renderer generated it, which evidence artifacts support it, and which assessments validated it. Missing edges trigger citation warnings. Circular references trigger structural errors. Every governance claim becomes traceable to authoritative source text.

For detailed technical coverage of Academic-Research-Skills' citation verification algorithms and graph traversal patterns, see Book II Chapter 9. This section focuses on how academic citation integrity standards prevent the regulatory ambiguity that emerges when governance knowledge graphs reference multiple overlapping frameworks without precise provenance attribution.

MCTS-Search: Multi-Step Regulatory Reasoning with Rollback

A Lloyd's surveyor inspecting a container ship's cargo hold must determine whether the vessel satisfies multiple overlapping requirements at once. SOLAS Chapter VI mandates cargo securing standards. The vessel's flag state (Liberia) requires specific lashing configurations. The cargo type (hazardous materials) triggers IMO Code requirements, and the destination port (Rotterdam) enforces EU port state control standards. Each requirement is binary, satisfied or not, but the combination is multi-step: evaluate SOLAS compliance, then flag state requirements, then IMO codes, then port state control, backtracking whenever a constraint fails and an alternative configuration exists.

This multi-step reasoning pattern, with backtracking when dead ends appear, directly applies to multi-jurisdictional AI governance routing. An AI system processing a transaction involving an EU data subject (GDPR), healthcare data (HIPAA), financial services (GLBA), and Colorado consumer status (SB 24-205) must determine which combination of governance properties, model configurations, and evidence artifacts satisfies all four frameworks simultaneously. The search space is exponential: four frameworks, each with distinct requirements, generate 2^4 possible compliance paths before considering the interaction effects where one framework's requirement constrains another's implementation options.

Monte Carlo Tree Search provides the backtracking infrastructure that prevents exhaustive search while guaranteeing a compliant path exists if one is available. MCTS builds a decision tree where each node represents a governance configuration (which model, which P-properties active, which evidence renderers enabled) and each edge represents adding one framework's requirements to the active constraint set. The search expands promising branches (configurations that satisfied similar multi-framework combinations in prior transactions) and backtracks from branches where constraint conflicts emerge (HIPAA's minimum necessary principle conflicts with GDPR's data minimization when applied to the same retrieval query).

The maritime parallel is the surveyor's inspection sequence. The surveyor does not exhaustively evaluate every possible cargo configuration against every applicable regulation. The surveyor starts with the most constraining requirement (hazardous materials stowage rules), verifies compliance, then adds the next constraint (flag state lashing standards), verifies again, and backtracks only when a constraint conflict appears (the lashing configuration that satisfies flag state rules violates port state control requirements for this specific cargo type). MCTS automates this prioritized search with backtracking: evaluate the most constraining framework first (typically GDPR or EU AI Act for organizations with EU exposure), add the next framework, backtrack when conflicts emerge, and record successful paths for reuse in future transactions with similar multi-framework profiles.

The Lagrange point is the balance between search completeness and decision latency. Exhaustive search guarantees finding a compliant configuration if one exists but takes exponential time as framework count grows. Greedy search (apply frameworks sequentially without backtracking) runs in linear time but fails when early framework choices constrain later frameworks into impossible configurations. MCTS occupies the stable region: prioritized search with bounded backtracking, completing in logarithmic time relative to the full search space while maintaining high probability of finding compliant paths when they exist.

Deploy MCTS-Search when the organization processes transactions that simultaneously trigger three or more regulatory frameworks with potentially conflicting technical requirements. Single-framework compliance (GDPR only, or HIPAA only) requires no search; the governance configuration is deterministic. Dual-framework compliance (GDPR plus HIPAA) is usually solvable through sequential constraint application without backtracking. The exponential search space opens at triple-framework and beyond, GDPR plus HIPAA plus Colorado SB 24-205 plus sector-specific regulations, where MCTS prevents exhaustive enumeration while guaranteeing compliant paths are discovered.

The search tree persists across transactions. When MCTS discovers that transactions involving [EU data subject + HIPAA covered entity + financial services] route successfully through governance configuration [Model: Claude Sonnet with GDPR training data docs, P-properties: LP1+LP5+LP6+LP9+LP10+LP11+LP12, Renderers: GDPR+HIPAA+GLBA], that path becomes a cached template. The next transaction matching that framework profile skips the search phase entirely, routing directly through the proven-compliant configuration. The classification society's inspection database works identically: once a cargo configuration satisfying [SOLAS + Liberian flag + IMO hazmat + Rotterdam PSC] is verified, future inspections for that cargo type and trading route reference the established compliant configuration.

For detailed technical coverage of MCTS-Search's tree expansion heuristics and backtracking algorithms, see Book II Chapter 11. This section focuses on how Monte Carlo tree search enables multi-jurisdictional governance routing with bounded complexity where exhaustive search would produce unacceptable decision latency.

OpenKB: Two-Hop Citation Networks for Regulatory Cross-References

Maritime regulations reference other regulations. SOLAS Chapter II-2 (Construction - Fire Protection) references the Fire Safety Systems Code (FSS Code) for technical standards. The FSS Code, in turn, references IMO Resolution MSC.98(73) for testing procedures, and that resolution references ISO standards for equipment certification. A surveyor verifying fire suppression system compliance cannot stop at

SOLAS Chapter II-2. The full regulatory chain requires following three citation hops before reaching the testable technical specification.

This multi-hop citation pattern is universal in regulatory governance. The EU AI Act Article 26 requires deployers to "ensure that high-risk AI systems are used in accordance with the instructions of use." Article 13 defines what those instructions must contain. Article 9 specifies the risk management system that generates those instructions. Recital 71 provides interpretive guidance on how Article 13's transparency obligations interact with GDPR Article 22's automated decision-making restrictions. A compliance officer implementing Article 26 deployer obligations cannot read Article 26 in isolation; the complete obligation requires traversing the citation graph through Articles 13, 9, and GDPR Article 22, plus relevant recitals.

OpenKB implements two-hop graph traversal for regulatory knowledge retrieval. A single-hop query returns the direct text of the cited regulation. A two-hop query returns the cited regulation plus all regulations it references, plus all regulations that reference it. When an E-4 attribution record claims compliance with "GDPR Article 22," a two-hop retrieval returns Article 22's text, the six recitals that provide interpretive context (Recitals 71, 72, 73), the Articles referenced within Article 22 (Articles 6, 13, 14), and the Articles that reference Article 22 from elsewhere in GDPR (Article 35 DPIA requirements when automated decision-making is deployed).

The maritime parallel is the classification society's regulatory reference system. When a surveyor cites "MARPOL Annex I Regulation 13" in a tanker inspection report, the document control system automatically retrieves Regulation 13's text, the referenced IMO resolutions, the flag state's specific implementation guidance, and any port state control memoranda that interpreted Regulation 13 differently in prior inspections. The surveyor does not manually trace citations through four regulatory documents. The retrieval system performs the graph traversal automatically.

Two-hop traversal prevents the citation incompleteness failure mode. When a governance artifact claims compliance with a regulatory requirement, but that requirement's interpretation depends on cross-references the artifact did not retrieve, the compliance claim becomes incomplete. A credit decision claiming GDPR Article 22 compliance without retrieving Article 13's transparency requirements and Recital 71's interpretive guidance is structurally incomplete; the full obligation was not evaluated. OpenKB's two-hop retrieval ensures citation completeness by default: retrieve the cited regulation plus its immediate reference neighborhood.

Deploy OpenKB when the organization's governance knowledge graph references regulations with dense cross-citation structure. Standalone regulations, such as Colorado SB 24-205's deployer obligation for impact assessments, require single-hop retrieval. Regulations embedded in complex frameworks (GDPR, EU AI Act, HIPAA, NERC CIP) require two-hop traversal to capture cross-references that affect interpretation. The classification society's fire safety compliance verification references SOLAS, FSS Code, IMO resolutions, and ISO standards in a four-hop chain. OpenKB's two-hop limit balances completeness, capturing immediate context, against retrieval cost, which is to say preventing exponential graph expansion.

The retrieval system integrates with the OSCAL rendering pipeline. When a framework renderer extracts observations tagged with "GDPR Article 22" from an E-4 attribution record, OpenKB retrieves the two-hop citation neighborhood and includes it in the rendered compliance artifact. The auditor reviewing the GDPR rendering sees not only the Article 22 compliance claim, but also the Articles 13, 14, and 6 cross-references that establish the full obligation context. The maritime surveyor's report cites SOLAS Chapter II-2; the attached technical supplement includes the FSS Code references and ISO standards the surveyor verified against.

For detailed technical coverage of OpenKB's graph traversal algorithms and citation network construction, see Book I Chapter 7. This section focuses on how two-hop regulatory retrieval prevents the citation incompleteness that emerges when governance artifacts claim compliance with regulations whose full obligations depend on cross-references not explicitly retrieved.

The Structural Lineage

What this genealogy reveals is structural lineage, not metaphor.

Every maritime governance innovation, from Lloyd's coffeehouse through SOLAS through containerization to the ISM Code, solved a governance problem that reappears in AI governance with different vocabulary but identical architecture. The coffee house is the vendor presentation. The classification society is the independent audit. The Plimsoll mark is the binary governance property. SOLAS is mandatory regulation after catastrophe. Containerization is evidence harmonization. The ISM Code is the management system approach. The IGA is multi-jurisdictional governance. Each pair carries the same load.

This book's governance architecture is not new. It is a translation. The merchants at Lloyd's coffeehouse in 1688 began building the institutional infrastructure that this book applies to AI systems in 2026. The trajectory has not changed. The urgency has. The architecture, at last, is written down.

Appendix G: Acronyms and Abbreviations

This glossary defines the regulatory, standards, and technical acronyms used across the book. It does not repeat the framework's own identifiers, which are defined where they are introduced: the harness components (H-1 through H-5), the evidence tiers (E-1 through E-4), the knowledge-base controls (KB-1 through KB-4), the federal security properties (P1 through P8, defined in Volumes I and II), and this volume's Lagrange governance properties (LP1 through LP14) are set out in Chapter 5 and collected in Appendix A. Control identifiers of the form AC-3, AU-2, SA-9, SI-4, SC-7, CM-3, and the like are NIST SP 800-53 controls; CIP-002 through CIP-014 are NERC Critical Infrastructure Protection standards.

ADMT: Automated Decision-Making Technology.

AML: Anti-Money Laundering.

API: Application Programming Interface.

ASME: American Society of Mechanical Engineers.

ASOP: Actuarial Standard of Practice (issued by the Actuarial Standards Board).

ATO: Authorization to Operate (the federal accreditation decision); see also cATO.

BAA: Business Associate Agreement (under HIPAA).

BCS: BES Cyber System (NERC CIP).

BCSI: BES Cyber System Information.

BES: Bulk Electric System.

BSA: Bank Secrecy Act.

CALEA: Communications Assistance for Law Enforcement Act.

cATO: continuous Authorization to Operate (ongoing authorization sustained by continuous monitoring rather than periodic re-accreditation).

CCPA: California Consumer Privacy Act.

CDA: Critical Digital Asset (nuclear cyber security, under NRC 10 CFR 73.54).

CDPA: (Virginia) Consumer Data Protection Act.

CFPB: Consumer Financial Protection Bureau.

CFR: Code of Federal Regulations.

CIP: Critical Infrastructure Protection (the NERC CIP reliability standards).

CISA: Cybersecurity and Infrastructure Security Agency.

CMIA: Confidentiality of Medical Information Act (California).

CMS: Centers for Medicare and Medicaid Services.

COSO: Committee of Sponsoring Organizations of the Treadway Commission.

CPNI: Customer Proprietary Network Information (telecommunications).

CPRA: California Privacy Rights Act.

CRO: Chief Risk Officer.

CSF: Cybersecurity Framework (NIST).

DAP: (GitLab) Duo Agent Platform.

DCF: Discounted Cash Flow.

DOE: Department of Energy.

DOI: Department of Insurance (a state insurance regulator).

DORA: Digital Operational Resilience Act (European Union).

DOT: Department of Transportation.

DPIA: Data Protection Impact Assessment (under GDPR).

DUA: Data Use Agreement.

EAR: Export Administration Regulations.

ECOA: Equal Credit Opportunity Act.

EHR: Electronic Health Record.

EPA: Environmental Protection Agency.

ESP: Electronic Security Perimeter (NERC CIP).

FAA: Federal Aviation Administration.

FCC: Federal Communications Commission.

FCRA: Fair Credit Reporting Act.

FDA: Food and Drug Administration.

FD&C Act: Federal Food, Drug, and Cosmetic Act.

FDIC: Federal Deposit Insurance Corporation.

FERC: Federal Energy Regulatory Commission.

FFIEC: Federal Financial Institutions Examination Council.

FHIR: Fast Healthcare Interoperability Resources (an HL7 standard).

FINRA: Financial Industry Regulatory Authority.

FMCSA: Federal Motor Carrier Safety Administration.

GDPR: General Data Protection Regulation (European Union).

GINA: Genetic Information Nondiscrimination Act.

GLBA: Gramm-Leach-Bliley Act.

GMDSS: Global Maritime Distress and Safety System.

GPAI: General-Purpose AI (a classification under the EU AI Act).

HHS: (Department of) Health and Human Services.

HIPAA: Health Insurance Portability and Accountability Act.

HITECH: Health Information Technology for Economic and Clinical Health Act.

HITRUST: Health Information Trust Alliance.

HL7: Health Level Seven (healthcare data interoperability standards).

ICS: Industrial Control System.

ICT: Information and Communications Technology.

IEC: International Electrotechnical Commission.

IGA: Intergovernmental Agreement (the International Space Station governance instrument).

IMO: International Maritime Organization.

ISM Code: International Safety Management Code (under SOLAS).

ISO: International Organization for Standardization.

ISS: International Space Station.

ITAR: International Traffic in Arms Regulations.

KYC: Know Your Customer.

LBO: Leveraged Buyout.

LiDAR: Light Detection and Ranging.

MARPOL: International Convention for the Prevention of Pollution from Ships.

MCP: Model Context Protocol.

MGA: Managing General Agent (insurance).

MLC: Maritime Labour Convention.

MNPI: Material Non-Public Information.

MoE: Mixture of Experts (a model architecture).

NAIC: National Association of Insurance Commissioners.

NERC: North American Electric Reliability Corporation.

NHTSA: National Highway Traffic Safety Administration.

NIST: National Institute of Standards and Technology.

NRC: Nuclear Regulatory Commission.

NTSB: National Transportation Safety Board.

NYDFS: New York Department of Financial Services.

OCC: Office of the Comptroller of the Currency.

OCR: Office for Civil Rights (within HHS, the HIPAA enforcement office).

OPA: Open Policy Agent (a policy engine; its policy language is Rego).

OSCAL: Open Security Controls Assessment Language.

OSHA: Occupational Safety and Health Administration.

OT: Operational Technology.

PAN: Primary Account Number (payment cards).

PCCP: Predetermined Change Control Plan (FDA, for AI/ML medical devices).

PCI-DSS: Payment Card Industry Data Security Standard.

PHI: Protected Health Information.

PHMSA: Pipeline and Hazardous Materials Safety Administration.

PII: Personally Identifiable Information.

QSA: Qualified Security Assessor (PCI-DSS).

RAG: Retrieval-Augmented Generation.

RCRA: Resource Conservation and Recovery Act.

RMF: Risk Management Framework (NIST).

SaMD: Software as a Medical Device.

SAR: Suspicious Activity Report.

SBOM: Software Bill of Materials.

SCADA: Supervisory Control and Data Acquisition.

SEC: Securities and Exchange Commission.

SLA: Service Level Agreement.

SOC: System and Organization Controls (for example, a SOC 2 report); separately, in an operations context, a security operations center.

SOLAS: International Convention for the Safety of Life at Sea.

SOX: Sarbanes-Oxley Act.

SR: Supervision and Regulation letter (Federal Reserve guidance; for example, SR 11-7).

STIG: Security Technical Implementation Guide.

TSA: Transportation Security Administration.

UL: Underwriters Laboratories (for example, the UL 4600 autonomous-vehicle safety standard).

VDR: Voyage Data Recorder.

Download The Harness Doctrine Codex

This series runs on a shared vocabulary of coded elements: the security properties P1 through P8, the harness components H-1 through H-5, the evidence tiers E-1 through E-4, the knowledge-base controls KB-1 through KB-4, and each volume's own governance properties, the Lagrange (LP), Replicant Fleet (RF), and Shatterdome (SD) sets, along with the control families that support them.

The Harness Doctrine Codex collects every one of those elements in a single reference. Each entry is short by design: the identifier, its name, and one or two paragraphs on what it is and what it requires. It is the fastest way to look up what a property means, see how the federal core carries across the series, and understand how each volume builds its own governance layer on that baseline.

The Codex is free. Go to www.harnesstheory.com, register for a free account, and download the complete Codex and other supplemental material as a PDFs.

www.harnesstheory.com

www.ingramcontent.com/pod-product-compliance
Lightning Source LLC
LaVergne TN
LVHW060018210926
848481LV00013B/3737
9798950627156